# Particle Physics: An Introduction

# Particle Physics: An Introduction

*M. Leon*
MP DIVISION
LOS ALAMOS SCIENTIFIC LABORATORY
LOS ALAMOS, NEW MEXICO

*ACADEMIC PRESS NEW YORK and LONDON*

ACADEMIC PRESS, INC.
111 Fifth Avenue, New York, New York 10003

*United Kingdom Edition published by*
ACADEMIC PRESS, INC. (LONDON) LTD.
24/28 Oval Road, London NW1

LIBRARY OF CONGRESS CATALOG CARD NUMBER: 72-77329

PRINTED IN THE UNITED STATES OF AMERICA

*To*

***A.J.L.***

# Contents

## 8. *Pion Resonances*



## 9. *Strange Particles*



## 10. *SU(3)*



## 11. *Weak Interactions*



## 12. *Hadron Dynamics*

# Preface

The purpose of this book is to introduce seniors and graduate students to some of the very interesting and beautiful phenomena of particle physics and, no less important, to some of the very elegant and powerful theoretical ideas which are used in trying to cope with these phenomena. This volume is intended to be much less formidable than the standard particle physics texts and thus provides a much easier entry into the subject. A nontrivial acquaintance with quantum mechanics is assumed.

The book is *not* comprehensive; it is a selective introduction to the topics. Both theory and phenomena are included, but the emphasis is very much on the organizing principles; there is no detailed discussion of experiments, accelerators, etc. The first three chapters develop the use of relativity and quantum mechanics to describe scattering (and decay), wave equations, the $S$-matrix, and the connection between symmetries and conservation laws. Chapter 4 introduces field theory; the student learns what a field operator is and where the Feynman rules come from. These are applied in Chapter 5 to quantum electrodynamics, with a (descriptive) discussion of the divergencies of higher order processes and renormalization. Chapters 6–9 mainly concern the phenomenology of pions and nucleons, but continue to bring in important theoretical ideas. Chapter 10 concerns SU(3) and the mysterious quarks, while Chapter 11 surveys the weak interactions, ending with the beautiful phenomena connected with $K^0$ decay. Finally, Chapter 12, which is somewhat more demanding than the others, discusses topics in strong interaction dynamics, introducing dispersion relations and Regge poles.

This work is based largely on a one-semester course for seniors and graduate students given at Rensselaer Polytechnic Institute (for several years), with junior modern physics and one semester of senior quantum mechanics as prerequisities. The student is strongly urged to attempt the problems (many are easy) and to work through the mathematical manipulations in the text [more difficult problems are indicated by asterisks (*)].

I am indebted most of all to my students who in their innocence and enthusiasm made this book possible. I am also grateful to Miss Phyllis Kallenberg for her patient typing of several versions of the manuscript and to Professor Hans Ohanian for a critical reading of the manuscript.

# Introduction

Particle physics, like Gaul, can be divided into three parts, but there are two ways of doing this: according to the kind of *interaction* (*strong*, *weak*, *electromagnetic*) or according to the kind of *particle* (*hadron*, *lepton*, *photon*).[1] Hadrons are all those particles subject to the strong interactions, for example, *pions* and *nucleons*. The strong interactions are characterized by short range and great magnitude within that range and are the forces that bind the nucleons in nuclei. The electromagnetic interaction acts between any charged particles (or particles with magnetic moments) and is, of course, responsible for the structure of atoms and larger aggregates of matter. The weak interaction affects both hadrons and leptons, and brings about beta decay of nuclei. The lepton family includes *electrons*, *muons*, and *neutrinos*.

Like Caesar, the high energy physicists have been exploring and expanding the frontiers of a vast region, revealing many strange tribes and peoples. In this book we will not pay much attention to the weapons of conquest (impressive though they are) or the details of the campaigns, but will concentrate on what is presently understood of the language and laws of the inhabitants.

---

[1] As usual, we are ignoring *gravitation* and the *graviton*, which have yet to be integrated into the rest of particle physics.

## Bibliography

*Elementary Introductions to Particle Physics*

K. W. FORD. *The World of Elementary Particles.* Ginn (Blaisdell), Boston, Massachusetts, 1963.

D. H. FRISCH and A. H. THORNDIKE. *Elementary Particles.* Van Nostrand-Reinhold, Princeton, New Jersey, 1963.

H. H. HECKMAN and P. W. STARRING. *Nuclear Physics and the Fundamental Particles.* Holt, New York, 1963.

R. D. HILL. *Tracking Down Particles.* Benjamin, New York, 1964.

C. E. SWARTZ. *The Fundamental Particles.* Addison-Wesley, Reading, Massachusetts, 1965.

C. N. YANG. *Elementary Particles.* Princeton Univ. Press, Princeton, New Jersey, 1962.

*More Advanced General Particle Physics Texts*

W. R. FRAZER. *Elementary Particles.* Prentice-Hall, Englewood Cliffs, New Jersey, 1966.

H. MUIRHEAD. *Physics of Elementary Particles.* Pergamon, Oxford, 1965.

D. H. PERKINS. *Introduction to High Energy Physics.* Addison-Wesley, Reading, Massachusetts, 1972.

W. S. C. WILLIAMS. *An Introduction to Elementary Particles.* (2nd ed.) Academic Press, New York, 1971.

CHAPTER ONE

# Relativity and Kinematics

The special theory of relativity is a part of everyday life for particle physics. Therefore, we need to develop a certain amount of its formalism.

## 1.1. Formalism of Special Relativity

### *a. Lorentz Transformation. Scalar Invariants*

The basic ingredient of relativity is the *Lorentz transformation*, which relates events as seen in two inertial coordinate systems:

$$\begin{aligned} x' &= \gamma(x - vt), & y' &= y, \\ z' &= z, & t' &= \gamma[t - (v/c^2)x]. \end{aligned} \tag{1.1}$$

As usual, $v$ is the relative velocity and

$$\gamma \equiv [1 - (v/c)^2]^{-1/2}.$$

This simple form holds because we have chosen the coordinate systems so that (1) the origins coincide, i.e.,

$$x', y', z', t' = 0,$$

when

$$x, y, z, t = 0;$$

(2) the corresponding axes are (and therefore remain) parallel; and (3) the relative velocity is in the $x$ direction (Fig. 1.1). The Lorentz transformation

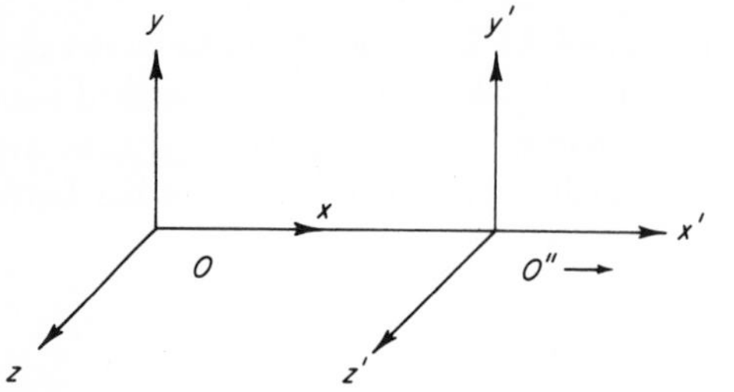


**Fig. 1.1.** Inertial systems related by the Lorentz transformation, Eq. (1.1).

follows from (1) the equivalence of inertial frames of reference and (2) the constancy of the velocity of light $c$.

We know that, in ordinary three-dimensional space, *length* is a property independent of translations or rotations of the coordinate system, while individual coordinate differences are not; that is,

$$r^2 \equiv \mathbf{r}^2 \equiv (x_a - x_b)^2 + (y_a - y_b)^2 + (z_a - z_b)^2$$

is *invariant* under such coordinate transformations, while $(x_a - x_b)$, etc., are not. In four-dimensional space–time, we define the *interval* by

$$\begin{aligned} S^2 &\equiv (x_a - x_b)^2 + (y_a - y_b)^2 + (z_a - z_b)^2 - c^2(t_a - t_b)^2 \\ &= \mathbf{r}^2 - c^2(t_a - t_b)^2. \end{aligned} \tag{1.2}$$

Then by direct computation, we find that

$$(S')^2 \equiv (x_a' - x_b')^2 + (y_a' - y_b')^2 + (z_a' - z_b')^2 - c(t_a' - t_b')^2 = S^2,$$

when the primed coordinates are related to the unprimed ones by the Lorentz transformation, Eqs. (1.1). Since Eq. (1.2) is also independent of space–time translations and ordinary spatial rotations and translations, $(S^2)^{1/2}$ constitutes the relativistic generalization of length. A quantity like $S^2$, which is the same in all inertial frames, is said to be a *scalar invariant*, a *world-scalar*, or simply a *scalar*.

### *b. Covariant Notation. Four Vectors. Tensors*

For simplicity, we can let our two points be the origin and $x, y, z, t$. It is highly advantageous to go over to *covariant notation*; to do this, let

$$x_1 \equiv x, \qquad x_2 \equiv y, \qquad x_3 \equiv z, \qquad x_4 \equiv ict. \tag{1.3}$$

Then

$$S^2 \equiv \sum_{\mu=1}^{4} x_\mu x_\mu. \tag{1.2'}$$

The quantities $x_\mu$ together form the *four-vector*

$$x \equiv (x_1, x_2, x_3, x_4) = (\mathbf{r}, x_4).$$

The $i$ is included in the definition of $x_4$ *only* to introduce the extra minus sign in front of the time component in Eq. (1.2).[1] We will use the standard conventions of understanding that summation is implied when indices are repeated, and of letting Greek letter indices run from 1 to 4, while Latin indices run from 1 to 3. Thus,

$$\mathbf{r}^2 = x_i x_i,$$

while

$$S^2 = x_\mu x_\mu. \tag{1.4}$$

A further convenience which we use henceforth is to choose our units so that $c = 1$.[2]

By introducing the matrix

$$a \equiv (a_{\mu\nu}) \equiv \begin{pmatrix} \gamma & 0 & 0 & iv\gamma \\ 0 & 1 & 0 & 0 \\ 0 & 0 & 1 & 0 \\ -iv\gamma & 0 & 0 & \gamma \end{pmatrix} \tag{1.5}$$

we can express the Lorentz transformation equations (1.1) in the simple and elegant form

$$x_\mu' = a_{\mu\nu} x_\nu, \tag{1.6a}$$

or, using the symbol $x$ for the whole column vector $(x_\mu)$,

$$x' = ax, \tag{1.6b}$$

---

[1] There are other ways of doing this. For example, introduce the *matrix*

$$g \equiv \begin{pmatrix} 1 & 0 & 0 & 0 \\ 0 & 1 & 0 & 0 \\ 0 & 0 & 1 & 0 \\ 0 & 0 & 0 & -1 \end{pmatrix}$$

and use

$$S^2 = \pm \sum g^{\mu\nu} x_\mu x_\nu,$$

where $x_4 = ct$; sometimes $x_4$ is called $x_0$. All possible conventions are in use, so one must be wary when reading the literature.

[2] For example, we take the second and the light-second as time and length units, respectively.

which is shorthand for the matrix product

$$\begin{pmatrix} x_1' \\ x_2' \\ x_3' \\ x_4' \end{pmatrix} = \begin{pmatrix} a_{11} & a_{12} & a_{13} & a_{14} \\ a_{21} & a_{22} & a_{23} & a_{24} \\ a_{31} & a_{32} & a_{33} & a_{34} \\ a_{41} & a_{42} & a_{43} & a_{44} \end{pmatrix} \begin{pmatrix} x_1 \\ x_2 \\ x_3 \\ x_4 \end{pmatrix}. \tag{1.6c}$$

Compounding transformations is then simply a matter of matrix multiplication:

$$x' = ax, \qquad x'' = bx'$$

imply that $x'' = cx$, where $c = ba$.

It is important to realize that Eq. (1.6a–c) holds not only for the very simple Lorentz transformation of Eq. (1.1) but for an *arbitrary* (homogeneous) Lorentz transformation. The general $a$ is then a much more complicated $4 \times 4$ matrix than the simple $a$ of Eq. (1.5).

Now, since

$$S^2 = x_\mu x_\mu = x_\mu' x_\mu' = (a_{\mu\nu} x_\nu)(a_{\mu\alpha} x_\alpha) = x_\nu (a_{\mu\nu} a_{\mu\alpha}) x_\alpha$$

holds for *arbitrary* $x_\mu$ (the parentheses indicating which summations are to be done first), we must have

$$a_{\mu\nu} a_{\mu\alpha} = \delta_{\nu\alpha}. \tag{1.7a}$$

Here,

$$\delta_{\nu\alpha} \equiv \begin{cases} 0, & \nu \neq \alpha, \\ 1, & \nu = \alpha, \end{cases}$$

is the *Kronecker delta*. Denoting by $\tilde{a}$ the *transpose* of $a$,

$$\tilde{a}_{\alpha\beta} \equiv a_{\beta\alpha},$$

we can express Eq. (1.7a) in the usual matrix product form

$$\tilde{a}_{\nu\mu} a_{\mu\alpha} = \delta_{\nu\alpha} \tag{1.7b}$$

or

$$\tilde{a}a = \mathbf{1},$$

where **1** is the unit matrix

$$\mathbf{1} \equiv \begin{pmatrix} 1 & 0 & 0 & 0 \\ 0 & 1 & 0 & 0 \\ 0 & 0 & 1 & 0 \\ 0 & 0 & 0 & 1 \end{pmatrix}.$$

A matrix which satisfies Eq. (1.7a) or (1.7b) is said to be *orthogonal*.[3] Note that only the invariance of $S^2$ is used in deriving Eqs. (1.7a,b).

Besides the intervals $x_\mu$, there exist other sets of four quantities which are defined in all (inertial) reference frames; if these satisfy the same transformation equations as the $x_\mu$, i.e., if

$$A_\mu{}' = a_{\mu\nu} A_\nu, \tag{1.8}$$

then the quantities $A_\mu$ are also said to form a four-vector, and the same notation as in Eqs. (1.6a–c) is used. Furthermore, we can now define the *scalar product* between any two four-vectors:

$$AB \equiv A \cdot B \equiv A_\mu B_\mu. \tag{1.9}$$

Because

$$A' \cdot B' \equiv A_\mu{}' B_\mu{}' = (a_{\mu\nu} A_\alpha)(a_{\mu\alpha} B_\alpha) = A_\nu(a_{\mu\nu} a_{\mu\alpha})B_\alpha = A_\nu B_\nu \equiv A \cdot B,$$

this scalar product is invariant.

We can also consider quantities that transform like products of four-vectors; these are known as *tensors*, and the number of indices is called the *rank* of the tensor. For example, a second-rank tensor transforms according to the prescription

$$T'_{\mu\nu} = a_{\mu\alpha} a_{\nu\beta} T_{\alpha\beta}, \tag{1.10}$$

just like the product $x_\mu y_\nu$. Scalar invariants and four-vectors can be considered to be tensors of rank 0 and 1, respectively.

Tensors play a central role when we construct physical theories, because the *principle of relativity* requires that physical laws be expressible in a way that is independent of the reference frame, and this is most conveniently done in terms of tensors. For example, the equation

$$T_{\mu\nu} A_\nu = B_\mu$$

transformed to a new reference frame has exactly the same form,

$$T'_{\mu\nu} A_\nu{}' = B_\mu{}',$$

and is therefore a candidate for being a law of physics. Expressions of this sort are called *covariant*.

---

[3] Ordinary rotations in three-dimensional space define orthogonal matrices; there $r^2 = x_i x_i$ is invariant under the rotations.

### c. *Infinitesimal Transformations. Improper Transformations*

The Lorentz transformations discussed so far can be considered to result from a sequence of many *infinitesimal transformations*, each of which has the relative velocity $v \ll 1$; correspondingly, the matrix of the transformation can be expressed as the product of many matrices, each having $v \ll 1$. Thus denoting any of the $v \ll 1$ matrices by $a_{\text{inf}}$, we write

$$a = \prod a_{\text{inf}}. \tag{1.11}$$

(The symbol $\prod$ means product.) Such Lorentz transformations are called *proper* or *continuous*. From Eq. (1.5), we see that, for an infinitesimal Lorentz transformation in the $x$ direction a power series expansion in $v$ gives

$$a_{\text{inf}} = \begin{pmatrix} 1 & 0 & 0 & iv \\ 0 & 1 & 0 & 0 \\ 0 & 0 & 1 & 0 \\ -iv & 0 & 0 & 1 \end{pmatrix} + O(v^2) \simeq 1 + v \begin{pmatrix} 0 & 0 & 0 & i \\ 0 & 0 & 0 & 0 \\ 0 & 0 & 0 & 0 \\ -i & 0 & 0 & 0 \end{pmatrix}.$$

Then, dropping $O(v^2)$ terms, we have for the *determinant* of $a_{\text{inf}}$,

$$\det(a_{\text{inf}}) = +1. \tag{1.12}$$

Now for the *finite* $a$,

$$\det(\tilde{a}a) = \det(\tilde{a}) \cdot \det(a) = [\det(a)]^2,$$

while Eq. (1.7b) implies that

$$\det(\tilde{a}a) = 1.$$

Thus

$$\det(a) = \pm 1.$$

But Eqs. (1.11) and (1.12) pick out the plus sign,

$$\det(a) = \prod \det(a_{\text{inf}}) = +1, \tag{1.13}$$

for a proper Lorentz transformation.

In addition to these proper transformations, we consider *improper* ones, which can all be written as the product of a proper $a$ and one of the following:

$$a_P = \begin{pmatrix} -1 & 0 & 0 & 0 \\ 0 & -1 & 0 & 0 \\ 0 & 0 & -1 & 0 \\ 0 & 0 & 0 & 1 \end{pmatrix}, \tag{1.14a}$$

$$a_T = \begin{pmatrix} 1 & 0 & 0 & 0 \\ 0 & 1 & 0 & 0 \\ 0 & 0 & 1 & 0 \\ 0 & 0 & 0 & -1 \end{pmatrix}, \tag{1.14b}$$

$$a_R = \begin{pmatrix} -1 & 0 & 0 & 0 \\ 0 & -1 & 0 & 0 \\ 0 & 0 & -1 & 0 \\ 0 & 0 & 0 & -1 \end{pmatrix}. \tag{1.14c}$$

*Space inversion* is effected by $a_P$, since

$$x' = a_P x = (-\mathbf{r}, it),$$

while $a_T$ is the *time-reversal* transformation:

$$x' = a_T x = (\mathbf{r}, -it);$$

these both have $\det(a) = -1$. Clearly $a_R$ does both at once and has $\det(a) = +1$. We will see later that there exist quantities which, as far as *proper* Lorentz transformations go, are four-vectors [Eq. (1.8) is satisfied], but under $a_P$ satisfy

$$A' = -a_P A;$$

i.e.,

$$A' = (\mathbf{A}, -A_4).$$

These are known as *pseudo(four)-vectors*, and their transformation properties are summarized by the expression

$$A_\mu' = \det(a) a_{\mu\nu} A_\nu . \tag{1.15}$$

The product of a vector and a pseudovector is a *pseudoscalar*, while the product of two pseudovectors is again a scalar. For a pseudoscalar,

$$PS' = \det(a) PS, \tag{1.16}$$

instead of $S' = S$, which holds for a scalar.

## 1.2. Particle Kinematics

Fortunately, it is not much harder to describe relativistic motion than nonrelativistic motion.

### *a. Four-Momentum*

Suppose $x_\mu \equiv [r(t), it]$ describes the position of a particle. Then defining the (scalar) *proper time* interval $d\tau$ by

$$d\tau \equiv [-(dS)^2]^{1/2} = [-dx_\mu\, dx_\mu]^{1/2}, \tag{1.17}$$

we can form a new four-vector, the *four-velocity*, by setting

$$u_\mu \equiv dx_\mu/d\tau. \tag{1.18}$$

Now

$$u_\mu = \frac{dx_\mu}{dt}\frac{dt}{d\tau} = (\mathbf{v}, i)\frac{dt}{d\tau},$$

where $\mathbf{v} = d\mathbf{r}/dt$ is the ordinary velocity; then using

$$d\tau = [-\mathbf{v}^2(dt)^2 + dt^2]^{1/2} = dt/\gamma, \tag{1.19}$$

we have

$$u_\mu = (\gamma\mathbf{v}, i\gamma).$$

Multiplying by the (*rest-*) *mass* of the particle gives the *four-momentum*

$$p_\mu \equiv (\mathbf{p}, iE) \equiv mu_\mu = (\gamma m\mathbf{v}, i\gamma m), \tag{1.20}$$

which combines ordinary momentum and energy in one four-vector. Because it is covariant and reduces to the nonrelativistic expressions for small velocities, Eq. (1.20) must be the correct relativistic expression to use for energy and momentum conservation.

The fact that

$$p_\mu p_\mu = \mathbf{p}^2 - E^2$$

is a scalar invariant is extremely useful. In the rest system of the particle, $\mathbf{p} = 0$ and $E = m$, and $p_\mu p_\mu = -m^2$. Therefore in *any* system,

$$E^2 - \mathbf{p}^2 = m^2 \tag{1.21}$$

is the general relationship between energy and momentum. The equivalent expression

$$E = (\mathbf{p}^2 + m^2)^{1/2} \tag{1.21'}$$

can be expanded for small $\mathbf{p}^2$, giving

$$E \simeq m + \mathbf{p}^2/2m + \cdots,$$

thus reducing to the rest energy plus nonrelativistic kinetic energy for small velocities.

We will use the notation $p^2$ for the "square" of a four-vector $p_\mu$, defining

$$p^2 \equiv p_\mu p_\mu, \tag{1.22}$$

so that

$$p^2 = \mathbf{p}^2 - E^2 = -m^2.$$

### b. Two-Body Decay Kinematics

In the *rest system* of a parent particle, we can easily find how the energy is distributed in two-body decay (Fig. 1.2). From conservation of four-momentum,

$$P = p_1 + p_2 .$$

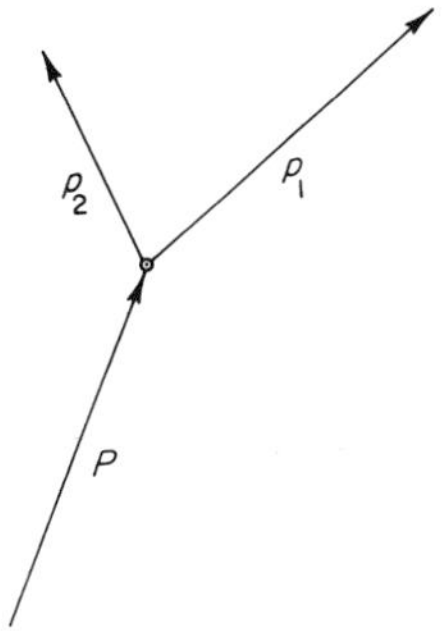


**Fig. 1.2.** Momenta in two-body decay.

Hence,

$$\begin{aligned} {p_2}^2 &= (P - p_1)^2 \\ &= -{m_2}^2 = p^2 + {p_1}^2 - 2P \cdot p_1 = -M^2 - {m_1}^2 + 2ME_1, \end{aligned}$$

since $\mathbf{p} = 0$. Thus

$$E_1 = (M^2 + {m_1}^2 - {m_2}^2)/2M, \tag{1.23a}$$

and similarly

$$E_2 = (M^2 + {m_2}^2 - {m_1}^2)/2M. \tag{1.23b}$$

Suppose that in the rest system the decay is isotropic, i.e., the decay products emerge with equal frequency in all directions. Then for decay in flight, the angular distribution is no longer isotropic but is distorted in the direction of flight. One can use the Lorentz transformation to find the energy and angle of emergence of each daughter in terms of angle in the rest system.

### c. Collision Kinematics. Mandelstam Variables

Here we will discuss the kinematics of scattering with two particles going in and two coming out, as indicated in Fig. 1.3. The experimentalist can never discover what is going on inside the circle, i.e., when the particles are very close together; the best he can do is measure the distribution in direction of the emerging particles. (If the outgoing particles have spins, he can try to

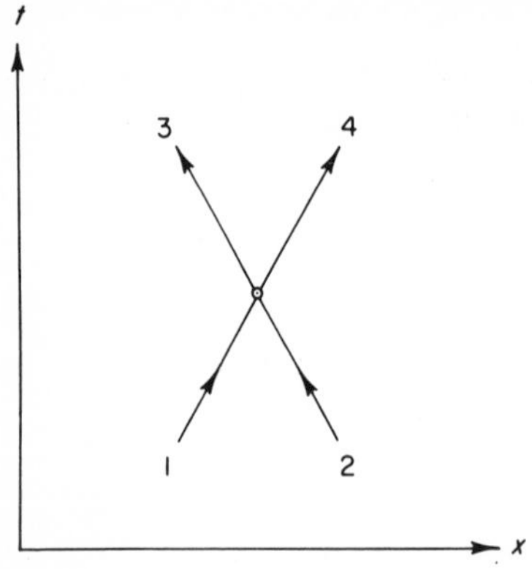


**Fig. 1.3.** Space–time diagram for a "two in, two out" collision.

measure them, too.) Furthermore, the experiments will usually use a target initially at rest. Suppose particle *2* is at rest; then

$$\mathbf{p}_2{}^{\mathrm{L}} \equiv 0$$

defines the *laboratory system.* On the other hand, theoretical analysis is often simpler in the reference frame in which the total momentum is zero; this is the so-called *center-of-mass* (more precisely, *center-of-momentum*) system, defined by the condition

$$\mathbf{p}_1^{\mathrm{CM}} + \mathbf{p}_2^{\mathrm{CM}} \equiv 0.$$

In either system, once the outgoing particles *3* and *4* are specified, the scattering is completely determined by the energy and the angle of scattering. We now want to show how the kinematic quantities (energy, momentum, angle) are expressible in terms of some relativistic invariants. We define

$$s \equiv -(p_1 + p_2)^2, \qquad t \equiv -(p_1 - p_3)^2, \qquad u \equiv -(p_1 - p_4)^2. \tag{1.24}$$

Because energy–momentum conservation requires

$$p_1 + p_2 = p_3 + p_4, \tag{1.25}$$

and each momentum must satisfy $p_1{}^2 = m_1{}^2$, etc., we can easily show that other combinations [like $(p_1 - p_2)^2$, etc.] add nothing new, and indeed that $s$, $t$, and $u$ are not independent but satisfy

$$s + t + u = m_1{}^2 + m_2{}^2 + m_3{}^2 + m_4{}^2. \tag{1.26}$$

Often $s$, $t$, and $u$ are known as the *Mandelstam variables.*

Let us examine these variables in the CM system. Then

$$s = -(p_1 + p_2)^2 = (E_1 + E_2)^2 - (\mathbf{p}_1 + \mathbf{p}_2)^2 = (E_1 + E_2)^2 \equiv W^2, \tag{1.27a}$$

so $s$ is the *energy* variable; $t$ and $u$, on the other hand, are *momentum transfer* variables. Letting

$$\mathbf{p} \equiv \mathbf{p}_1 = -\mathbf{p}_2 \qquad \text{and} \qquad \mathbf{q} \equiv \mathbf{p}_3 = -\mathbf{p}_4,$$

we can write

$$\begin{aligned} t &= -(p_1 - p_3)^2 = -p_1{}^2 - p_3{}^2 + 2p_1 \cdot p_3 \\ &= m_1{}^2 + m_3{}^2 - 2(E_1E_3 - \mathbf{p}\cdot\mathbf{q}), \end{aligned} \tag{1.27b}$$

and

$$\begin{aligned} u &= -(p_1 - p_4)^2 = -p_1{}^2 - p_4{}^2 + 2p_1 \cdot p_4 \\ &= m_1{}^2 + m_4{}^2 - 2(E_1E_4 + \mathbf{p}\cdot\mathbf{q}), \end{aligned}$$

where $t$ and $u$ depend on the *scattering angle* $\theta$ through the product

$$\mathbf{p}\cdot\mathbf{q} \equiv |\mathbf{p}|\ |\mathbf{q}| \cos\theta.$$

For the special case of *elastic scattering,*

$$m_1 = m_3 \qquad \text{and} \qquad m_2 = m_4.$$

Then

$$E_1 = E_3, \qquad E_2 = E_4, \qquad |\mathbf{p}| = |\mathbf{q}|,$$

and

$$t = 2m_1{}^2 - 2(E_1{}^2 - \mathbf{p}^2\cos\theta) = -2\mathbf{p}^2(1 - \cos\theta). \tag{1.27b$'$}$$

(The expression for $u$ is more complicated.) With the aid of Fig. 1.4, one can show that the change in momentum satisfies

$$|\Delta\mathbf{p}|^2 \equiv (\mathbf{p}_1 - \mathbf{p}_3)^2 = (\mathbf{p} - \mathbf{q})^2 = 4\mathbf{p}^2 \sin(\theta/2) = -t.$$

Since for an actual scattering process we must have

$$\mathbf{p}^2 > 0 \qquad \text{and} \qquad |\cos\theta| \leq 1,$$

the *physical region* for the elastic scattering $1 + 2 \to 3 + 4$ is defined by the relations[4]

$$\begin{aligned} &s > (m_1 + m_2)^2, \qquad -4\mathbf{p}^2 \leq t \leq 0, \\ &(E_1 - E_2) \geq u \geq -s + 2(m_1{}^2 + m_2{}^2). \end{aligned} \tag{1.28}$$

To give some practice in handling these variables, we now derive relations between energy, momentum, and angle and $s$, $t$, and $u$ for the general *unequal mass* case, first in the CM system and then in the laboratory system.

---

[4] The process $1 + 2 \to 3 + 4$ is known as the *s-channel*, because $s$ is the energy variable. We can also consider the *t-channel*, $1 + \bar{3} \to \bar{2} + 4$, where $t$ is the energy variable, or the *u-channel*, $1 + \bar{4} \to \bar{2} + 3$, where $u$ is the energy variable. The antiparticle of 3 is $\bar{3}$, etc. it turns out that these different processes are intimately related, as we shall see.

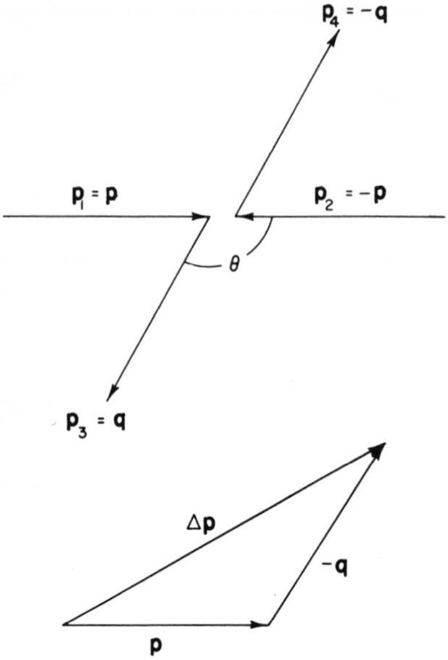


**Fig. 1.4.** Momenta and momentum transfer for the "two in, two out" collision in the CM system.

*CM System.* Since

$$\mathbf{p}_1 = -\mathbf{p}_2 \equiv \mathbf{p},$$
$$E_1{}^2 - E_2{}^2 = m_1{}^2 - m_2{}^2$$
$$= W(E_1 - E_2),$$

then

$$E_1 - E_2 = (m_1{}^2 - m_2{}^2)/W,$$

and using

$$E_1 + E_2 = W = \sqrt{s}$$

gives

$$E_1 = \frac{s + m_1{}^2 - m_2{}^2}{2\sqrt{s}} \tag{1.29a}$$

and

$$E_2 = \frac{s + m_2{}^2 - m_1{}^2}{2\sqrt{s}}. \tag{1.29b}$$

Therefore,

$$\mathbf{p}^2 = E_1{}^2 - m_1{}^2 = \frac{s^2 - 2s(m_1{}^2 + m_2{}^2) + (m_1{}^2 - m_2{}^2)^2}{4s}$$
$$= \frac{[s - (m_1 + m_2)^2][s - (m_1 - m_2)^2]}{4s}$$

or

$$\mathbf{p}^2 = \frac{\lambda(s, m_1^2, m_2^2)}{4s}, \tag{1.30}$$

where

$$\lambda(a, b, c) \equiv a^2 + b^2 + c^2 - 2(ab + ac + bc), \tag{1.31}$$

and $\lambda$ is clearly a symmetric function of its arguments. Here in the CM system, the energies and momenta of the outgoing particles must be given by exactly analogous expressions:

$$E_3 = \frac{s + m_3^2 - m_4^2}{2\sqrt{s}}, \qquad E_4 = \frac{s + m_4^2 - m_3^2}{2\sqrt{s}}, \tag{1.32}$$

$$\mathbf{q}^2 = \frac{\lambda(s, m_3^2, m_4^2)}{4s}. \tag{1.33}$$

To get the scattering angle, we write

$$\begin{aligned} t &= -(p_1 - p_2)^2 = m_1^2 + m_3^2 - 2(E_1 E_3 - \mathbf{p} \cdot \mathbf{q}) \\ &= 2\mathbf{p} \cdot \mathbf{q} - \frac{s^2 - s(m_1^2 + m_2^2 + m_3^2 + m_4^2) + (m_1^2 - m_2^2)(m_3^2 - m_4^2)}{2s}, \end{aligned}$$

or

$$2\,|\mathbf{p}|\,|\mathbf{q}|\cos\theta = t - m_1^2 - m_3^2 + \frac{(s + m_1^2 - m_2^2)(s + m_3^2 - m_4^2)}{2s},$$

so that

$$\begin{aligned} \cos\theta &= \frac{2s(t - m_1^2 - m_3^2) + (s + m_1^2 - m_2^2)(s + m_3^2 - m_4^2)}{[\lambda(s, m_1^2, m_2^2)\lambda(s, m_3^2, m_4^2)]^{1/2}} \\ &= \frac{s(t - u) + (m_1^2 - m_2^2)(m_3^2 - m_4^2)}{(\lambda_{s12}\lambda_{s34})^{1/2}}, \end{aligned} \tag{1.34}$$

where we have used $\lambda_{s12} \equiv \lambda(s, m_1^2, m_2^2)$, etc.

*Laboratory System.* Here

$$\mathbf{p}_2{}^{\mathrm{L}} = 0 \qquad \text{and} \qquad E_2{}^{\mathrm{L}} = m,$$

so that

$$\begin{aligned} s &= -p_1^2 - p_2^2 - 2p_1 \cdot p_2 \\ &= m_1^2 + m_2^2 + 2E_1{}^{\mathrm{L}} m_2, \end{aligned}$$

and

$$E_1{}^{\mathrm{L}} = (s - m_1^2 - m_2^2)/2m_2. \tag{1.35}$$

But

$$s = (E_1^{\mathrm{L}} + m_2)^2 - (\mathbf{p}_1^{\mathrm{L}})^2,$$

and, therefore,

$$(\mathbf{p}_1^{\mathrm{L}})^2 = \frac{s^2 - 2s(m_1^2 + m_2^2) + (m_1^2 - m_2^2)^2}{4m_2^2}$$

$$= \frac{\lambda(s, m_1^2, m_2^2)}{4m_2^2}. \tag{1.36}$$

Similarly

$$\begin{aligned} u &= -(p_2 - p_3)^2 = m_2^2 + m_3^2 - 2m_2 E_3^{\mathrm{L}}, \\ E_3^{\mathrm{L}} &= -(u - m_2^2 - m_3^2)/2m_2, \end{aligned} \tag{1.37}$$

and

$$u = (m_2 - E_3^{\mathrm{L}})^2 - (\mathbf{p}_3^{\mathrm{L}})^2,$$

tells us that

$$(\mathbf{p}_3^{\mathrm{L}})^2 = \lambda(u, m_2^2, m_3^2)/4m_2^2. \tag{1.38}$$

Then using

$$t = m_1^2 + m_3^2 - 2(E_1^{\mathrm{L}} E_3^{\mathrm{L}} - \mathbf{p}_1^{\mathrm{L}} \cdot \mathbf{p}_3^{\mathrm{L}}),$$

we find

$$\cos\theta_{\mathrm{L}} = \frac{2m_2^2(t - m_1^2 - m_3^2) - (s - m_1^2 - m_2^2)(u - m_2^2 - m_3^2)}{(\lambda_{s12}\lambda_{u23})^{1/2}}. \tag{1.39}$$

We can, if we wish, use these expressions to relate various kinematic quantities in the two systems without explicit use of the Lorentz transformation which connects them.

## Problems

**1.1.** Verify, from the Lorentz transformation equations [Eqs. (1.1)], that $S^2$ is invariant.

**1.2.** Given one event at (0, 0, 0; 0) and a second at $(0, 0, a; ib)$ with $b > 0$, find (and hence demonstrate the existence of) the Lorentz transformation which takes us to a frame for which the two events occur at (a) the same spatial point for $b^2 > a^2$ (*time-like* separation); (b) the same instant for $b^2 < a^2$ (*space-like* separation).

**1.3.** Show that the quantities

$$\partial_\alpha \equiv \partial/\partial x_\alpha$$

form a four-vector, and hence that $\partial_\mu \partial_\mu$ is invariant.

**1.4.** Show that $d_4x \equiv d^3\mathbf{x}\, dt$ is invariant under proper Lorentz transformations.

**1.5.** An unstable particle has lifetime $\tau$ and mass $m$ in its rest system. What is the observed lifetime in the lab system where it has kinetic energy $T$? Show how this follows from the Lorentz transformation.

**1.6.** (a) Find the $Q$-value (i.e., the kinetic energy released) in the decay $\pi^+ \to \mu^+\nu$. Find the kinetic energy carried off by each daughter particle in the $\pi^+$ rest system.

(b) Consider the decays in flight of a $\pi^+$ beam of kinetic energy $T_\pi = 10$ GeV. What are the maximum and minimum values of the neutrino energy $T_\nu$ ("forward" and "backward" neutrinos)?

**1.7.** (a) Use the Lorentz transformation equation to show that, for two-body decay of a moving particle, the laboratory angle of emergence (with respect to the direction of flight) of a daughter is related to the CM angle by

$$\tan\theta_L = \frac{\tan\theta_{CM}}{\gamma[1 + (v/v_1)\sec\theta_{CM}]},$$

where $v$ is the speed (lab system) of the parent, $v_1$ that of the daughter (CM system), and $\gamma \equiv 1(-v^2)^{-1/2}$.

(b) Suppose that the decay products emerge isotropically in the CM system. Discuss the resulting laboratory angular distributions for the two cases $v < v_1$, $v > v_1$. Apply this to the $\mu^+$ and $\nu$ distributions from the decay of $\pi$'s with 20-MeV kinetic energy.

**1.8.** Find the threshold in $\pi$ lab kinetic energy $T_\pi$ for the reaction $\pi^- p \to \Lambda K^0$.

**1.9.** What CM kinetic energy is available with a 200-GeV (kinetic energy) proton beam incident on a hydrogen target? Compare this with the CM kinetic energy available in a "clashing beam" arrangement, where each proton beam has 25-MeV kinetic energy.

**1.10.** Consider the decay in flight of a $\pi^0$ : $\pi^0 \to \gamma\gamma$. (a) Find maximum photon energy in terms of the $\pi^0$ kinetic energy. (b) If the two photons have equal energy, what is the "opening angle" between them?

**1.11.** Prove that $s + t + u = m_1^2 + m_2^2 + m_3^2 + m_4^2$.

**1.12.** Using Eq. (1.34) and $d\Omega = 2\pi\, d(\cos\theta)$, show that

$$d\Omega_{CM}/dt = 4\pi s/(\lambda_{s12}\lambda_{s34})^{1/2} = \pi/|\mathbf{p}|\,|\mathbf{q}|.$$

**1.13.** Prove that

$$v_{rel}\left(\equiv |\mathbf{v}_1 - \mathbf{v}_2| \equiv \left|\frac{\mathbf{p}_1}{E_1} - \frac{\mathbf{p}_2}{E_2}\right|\right)$$

satisfies

$$E_1 E_2 v_{\mathrm{rel}} = [-(p_1 p_2)^2 - m_1{}^2 m_2{}^2]^{1/2},$$

as long as $\mathbf{p}_1 \times \mathbf{p}_2 = 0$ (which holds in both the CM and lab systems).

## Bibliography

E. E. ANDERSON. *Moden Physics and Quantum Mechanics*, Chap. 1. Saunders, Philadelphia, Pennsylvania, 1971.

R. P. FEYNMAN, R. B. LEIGHTON, and M. SANDS. *The Feynman Lectures in Physics*. Vol. 1, Chaps. 15–17. Addison-Wesley, Reading, Massachusetts, 1963.

R. HAGEDORN. *Relativistic Kinematics*. Benjamin, New York, 1963.

R. K. RICHTMEYER, E. H. KENNARD, and J. N. COOPER. *Introduction to Modern Physics*. (6th ed.), Chaps. 2, 3. McGraw-Hill, New York, 1969.

E. F. TAYLOR and J. A. WHEELER. *Space-Time Physics*. Freeman, San Francisco, California, 1966.

CHAPTER TWO

# Relativistic Wave Equations

The wave nature of matter brings about the need for wave equations. In particle physics, where the matter is often moving at high speed, these equations must be relativistic.

## 2.1. Spin Zero

### *a. Klein–Gordon Equation*

Everyone knows how to write a wave equation for a nonrelativistic particle. Following deBroglie, one makes the association

$$\mathbf{p} \to -i\,\nabla, \qquad E \to i(\partial/\partial t). \tag{2.1}$$

(We choose units such that $\hbar = 1$ in addition to $c = 1$.) Then, since the nonrelativistic *Hamiltonian* is

$$H = \mathbf{p}^2/2m + V(\mathbf{r}),$$

the equation

$$H\psi = E\psi$$

implies

$$[-(1/2m)\ \nabla^2 + V(\mathbf{r})]\psi(\mathbf{r}, t) = i(\partial/\partial t)\psi(\mathbf{r}, t). \tag{2.2}$$

This is the (time-dependent) *Schrödinger equation*, and $\psi(\mathbf{r}, t)$ is the *wave function* or *position amplitude*. That is, $|\psi(\mathbf{r}, t)|^2$ is the probability density for finding the particle at point $\mathbf{r}$ at time $t$.

It seems reasonable to do the same for a relativistic particle. Thus for a free $[V(\mathbf{r}) \equiv 0]$ relativistic particle,

$$p^2 = \mathbf{p}^2 - E^2 = -m^2,$$

which leads to the *Klein–Gordon* equation (invented by Schrödinger in 1926, and independently by Gordon and by Klein, very slightly later):

$$[\nabla^2 - (\partial^2/\partial t^2)]\phi(r, t) = m^2\phi(r, t),$$

or

$$(\square - m^2)\phi(x) = 0. \tag{2.3}$$

Here we use the symbol $\square$ for the *d'Alembertian*:

$$\square \equiv \nabla^2 - \frac{\partial^2}{\partial t^2} \equiv \frac{\partial}{\partial x_\mu}\frac{\partial}{\partial x_\mu} \equiv \partial_\mu \partial_\mu .$$

The differential operation $(\square - m^2)$ is a Lorentz scalar. We also require $\phi(x)$ to be a scalar (or pseudoscalar), so that the Klein–Gordon equation is covariant. It describes particles of *zero spin*.

### b. Difficulties of the Klein–Gordon Equation

In contrast with the Schrödinger equation, however, here it is *not* possible to interpret $\phi(r, t)$ as a position amplitude. We only have an entirely satisfactory theory using the Klein–Gordon equation, or *any* relativistic wave equation, when $\phi$ is made a *field operator* which can create and destroy particles. (This is the method of *second quantization*, which we will discuss later.) This is because there are always *negative energy solutions* to such an equation; i.e.,

$$E = \pm(\mathbf{p}^2 + m^2)^{1/2} \tag{2.4}$$

satisfies $p^2 = -m^2$. When interaction is introduced (e.g., with the electromagnetic field), there will be nothing to prevent a particle from making a transition from a positive to a negative energy state, with the emission of some energy greater than $2m$.

Another related sickness of the *one-particle* Klein–Gordon equation (i.e., without second quantization) has to do with the probability current density which plays such an important role in the Schrödinger theory. There one has

$$\rho = \psi^*\psi \qquad \text{and} \qquad \mathbf{j} = (-i/2m)[\psi^* \nabla\psi - \psi \nabla\psi^*], \tag{2.5}$$

and, as a result of the Schrödinger equation for $\psi$ and $\psi^*$,

$$(\partial\rho/\partial t) + \nabla \cdot \mathbf{j} = 0. \tag{2.6}$$

This is the *equation of continuity* and means that the total probability is conserved. Note that the probability density $\rho$ given by Eq. (2.5) can never be negative.

For the Klein–Gordon case, the correct current density is found from the Klein–Gordon equation: Since

$$(\Box - m^2)\phi = 0$$

implies

$$(\Box - m^2)\phi^* = 0,$$

$$0 = \phi^*(\Box - m^2)\phi - \phi(\Box - m^2)\phi^* = -\partial_\mu(\phi^* \, \partial_\mu \phi - \phi \, \partial_\mu \phi^*).$$

Thus the appropriate conserved current density four-vector is

$$j_\mu \equiv (\mathbf{j}, i\rho) \equiv -i(\phi^* \, \partial_\mu \phi - \phi \, \partial_\mu \phi^*). \tag{2.7}$$

The space part is proportional to the nonrelativistic expression, Eq. (2.5), and the equation of continuity is satisfied:

$$\partial_\mu j_\mu = 0. \tag{2.8}$$

The trouble appears when we look at $\rho$:

$$\rho = -ij_4 = i\left(\phi^* \frac{\partial}{\partial t} \phi - \phi \frac{\partial \phi^*}{\partial t}\right). \tag{2.9}$$

We cannot interpret $\rho$ as the probability density because it is not *positive definite*. In fact, for energy eigenstates,

$$i(\partial/\partial t)\phi = E\phi \qquad \text{and} \qquad \rho = 2E\phi^*\phi,$$

so the sign of $\rho$ depends on the sign of the energy. As we will see later, $\rho$ can be interpreted as the *charge-density operator* in the quantized theory.

These problems caused the Klein–Gordon equation to be abandoned until its resurrection by Pauli and Weisskopf (in 1934) as a field operator equation.

## 2.2. Spin One-Half

### *a. Dirac Equation*

In an attempt to avoid the difficulties of the one-particle Klein–Gordon equation, Dirac (in 1928) introduced an equation *linear* instead of *quadratic* in $E$. It turns out that while the probability density $\rho$ is now all right, the negative energy difficulty persists. The *Dirac equation*, however, describes particles with spin *one-half*, in contrast to the Klein–Gordon equation which refers to spin zero.

Handling the Dirac equation is complicated by the presence of matrices, but since nature exhibits a great fondness for particles of spin one-half, we must face up to the difficulties. The Dirac wave function is not a single function but has several components which are organized into a column vector or *spinor*; naively, we might expect just *two* components corresponding to the two possibilities for the $z$-component of the spin, $S_z = \pm\frac{1}{2}$, but it turns out that relativity requires *four*. Thus,

$$\psi \equiv \begin{pmatrix} \psi_1 \\ \psi_2 \\ \psi_3 \\ \psi_4 \end{pmatrix}, \tag{2.10}$$

where each $\psi_i$ is a function of $x \equiv (\mathbf{x}, it)$. The equation satisfied by this four-component spinor will contain $4 \times 4$ matrices; the requirement that it be linear in $\mathbf{p}$ as well as $E$ leads to the combination

$$(\gamma_1 p_1 + \gamma_2 p_2 + \gamma_3 p_3 + i\gamma_4 E)\psi = im\,\psi,$$

where the $\gamma$'s are constant $4 \times 4$ matrices whose properties are determined by the requirements that we place on the equation. (In fact, they can be chosen in an infinite number of ways and there are, therefore, many conventions in use.) It can be rewritten as

$$i(\boldsymbol{\gamma} \cdot \mathbf{p} + i\gamma_4 E)\psi + m\psi = 0,$$

or

$$(i\gamma_\mu p_\mu + m)\psi = 0. \tag{2.11}$$

(The $m$ in this last equation really means $m \cdot \mathbf{1}$, where $\mathbf{1}$ is the unit $4 \times 4$ matrix.) Now since

$$E \to i(\partial/\partial t) \qquad \text{and} \qquad \mathbf{p} \to -i\,\boldsymbol{\nabla}, \tag{2.1}$$

is the same as

$$p_\mu \to -i\,\partial_\mu \equiv -i(\partial/\partial x_\mu), \tag{2.1$'$}$$

Eq. (2.11) becomes

$$\gamma_\mu\,\partial_\mu\,\psi(x) + m\psi(x) = 0. \tag{2.12}$$

Equations (2.11) and (2.12) are the Dirac equation in two of its guises.

Let us apply the operator $(i\gamma_\nu p_\nu - m)$ to Eq. (2.11):

$$\begin{aligned} 0 &= -[\gamma_\nu p_\nu \gamma_\mu p_\mu + m^2]\psi \\ &= -[\tfrac{1}{2}(\gamma_\mu\gamma_\nu + \gamma_\nu\gamma_\mu)p_\mu p_\nu + m^2]\psi. \end{aligned}$$

In order to recover

$$(p^2 + m^2)\psi = (-\Box + m^2)\psi = 0$$

from this equation (i.e., in order that $p^2 = -m^2$ hold for plane-wave solutions), we require that

$$\gamma_\mu \gamma_\nu + \gamma_\nu \gamma_\mu = 2\delta_{\mu\nu} \cdot \mathbf{1}. \tag{2.13}$$

This is the basic relation for the $\gamma$'s, and is more important than having an explicit *representation* for the $\gamma$'s—i.e., knowing all the entries of four matrices that satisfy Eq. (2.13). We choose to require that the $\gamma$'s be *self-adjoint* matrices; i.e., that

$$\gamma_\mu{}^\dagger \equiv (\tilde{\gamma}_\mu)^* = \gamma_\mu .$$

What about the current density? We need the equation for the row vector $\psi^\dagger$,

$$\psi^\dagger \equiv (\psi_1{}^*, \psi_2{}^*, \psi_3{}^*, \psi_4{}^*),$$

which is found by taking the *adjoint* (transpose, complex conjugate) of Eq. (2.12):

$$\partial_\mu{}^* \psi^\dagger \gamma_\mu = -m\psi^\dagger,$$

where

$$\partial_\mu{}^* \equiv (\mathbf{\nabla}, -\partial_4).$$

We can "repair" the minus sign in $\partial_\mu{}^*$ by multiplying on the right by $\gamma_4$ and using the anticommutation relations, Eq. (2.13):

$$\partial_\mu{}^* \psi^\dagger \gamma_\mu \gamma_4 = -\partial_\mu \psi^\dagger \gamma_4 \gamma_\mu = -m\psi^\dagger \gamma_4$$

or

$$\partial_\mu \bar{\psi} \gamma_\mu - m\bar{\psi} = 0, \tag{2.14}$$

where

$$\bar{\psi} \equiv \psi^\dagger \gamma_4 \tag{2.15}$$

is referred to as the *adjoint spinor.*

Multiplying this equation on the right by $\psi$, Eq. (2.12) on the left by $\psi$, and adding yields

$$\partial_\mu \bar{\psi} \gamma_\mu \psi + \bar{\psi} \gamma_\mu \partial_\mu \psi = 0$$

or

$$\partial_\mu (\bar{\psi} \gamma_\mu \psi) = 0. \tag{2.16}$$

Therefore,

$$j_\mu \equiv i\bar{\psi}\gamma_\mu\psi \tag{2.17}$$

satisfies the equation of continuity. As expected, the density

$$\rho = -ij_4 = \bar{\psi}\gamma_4\psi = \psi^\dagger\psi \tag{2.18}$$

has *no* time derivative and in fact is positive definite.

### *b. Solutions of the Dirac Equation*

It is instructive to examine the solutions to the Dirac equation, and for this purpose we must choose an explicit representation for the $\gamma_\mu$. We take

$$\gamma_4 = \begin{pmatrix} 1 & 0 & 0 & 0 \\ 0 & 1 & 0 & 0 \\ 0 & 0 & -1 & 0 \\ 0 & 0 & 0 & -1 \end{pmatrix} = \begin{pmatrix} \mathbf{\bar{1}} & 0 \\ 0 & -\mathbf{\bar{1}} \end{pmatrix}, \tag{2.19a}$$

$$\gamma_i = \begin{pmatrix} 0 & -i\sigma_i \\ i\sigma_i & 0 \end{pmatrix}, \quad \text{or} \quad \boldsymbol{\gamma} = \begin{pmatrix} 0 & -i\boldsymbol{\sigma} \\ i\boldsymbol{\sigma} & 0 \end{pmatrix}. \tag{2.19b}$$

Here,

$$\mathbf{\bar{1}} \equiv \begin{pmatrix} 1 & 0 \\ 0 & 1 \end{pmatrix}$$

and the $\sigma_i$ are the familiar $2 \times 2$ *Pauli matrices*, which satisfy

$$\sigma_x\sigma_y = i\sigma_z$$

and *cyclic* permutations of this equation. A particular representation for these is

$$\sigma_x = \begin{pmatrix} 0 & 1 \\ 1 & 0 \end{pmatrix}, \quad \sigma_y = \begin{pmatrix} 0 & -i \\ i & 0 \end{pmatrix}, \quad \sigma_z = \begin{pmatrix} 1 & 0 \\ 0 & -1 \end{pmatrix}. \tag{2.20}$$

The $\gamma$'s so defined are self-adjoint and satisfy Eq. (2.13).

Let us seek plane-wave solutions by setting

$$\psi(x) = e^{ipx}u(p) \equiv e^{i(\mathbf{p}\cdot\mathbf{x} - Et)}u(p). \tag{2.21}$$

The column vector $u(p)$ is also known as a spinor and must satisfy the Dirac equation in momentum space:

$$\begin{aligned} 0 &= (ip_\mu\gamma_\mu + m)u = (i\boldsymbol{\gamma}\cdot\mathbf{p} - \gamma_4 E + m)u \\ &= \left[\begin{pmatrix} 0 & \boldsymbol{\sigma}\cdot\mathbf{p} \\ -\boldsymbol{\sigma}\cdot\mathbf{p} & 0 \end{pmatrix} + \begin{pmatrix} m - E & 0 \\ 0 & m + E \end{pmatrix}\right]\begin{pmatrix} \chi \\ \eta \end{pmatrix}, \end{aligned}$$

or

$$\begin{pmatrix} m - E & \boldsymbol{\sigma} \cdot \mathbf{p} \\ -\boldsymbol{\sigma} \cdot \mathbf{p} & m + E \end{pmatrix} \begin{pmatrix} \chi \\ \eta \end{pmatrix} = 0. \tag{2.22}$$

Here we have written the *four-component spinor* $u$ in terms of the *two-spinors* $\chi$ and $\eta$:

$$u \equiv \begin{pmatrix} \chi \\ \eta \end{pmatrix} \equiv \begin{pmatrix} \chi_1 \\ \chi_2 \\ \eta_1 \\ \eta_2 \end{pmatrix}. \tag{2.23}$$

Thus the Dirac equation is equivalent to two coupled equations:

$$(m - E)\chi + \boldsymbol{\sigma} \cdot \mathbf{p}\eta = 0 \qquad \text{and} \qquad (m + E)\eta - \boldsymbol{\sigma} \cdot \mathbf{p}\chi = 0. \tag{2.24}$$

For each $\mathbf{p}$, there are *four* linearly independent solutions of the Dirac equation. For example, in the rest system ($\mathbf{p} = 0$), we can have

$$E = +m, \qquad \eta = 0, \qquad \chi = \begin{pmatrix} 1 \\ 0 \end{pmatrix} \quad \text{or} \quad \begin{pmatrix} 0 \\ 1 \end{pmatrix};$$

or

$$E = -m, \qquad \chi = 0, \qquad \eta = \begin{pmatrix} 1 \\ 0 \end{pmatrix} \quad \text{or} \quad \begin{pmatrix} 0 \\ 1 \end{pmatrix};$$

that is,

$$u = \begin{pmatrix} 1 \\ 0 \\ 0 \\ 0 \end{pmatrix}, \begin{pmatrix} 0 \\ 1 \\ 0 \\ 0 \end{pmatrix}, \begin{pmatrix} 0 \\ 0 \\ 1 \\ 0 \end{pmatrix}, \quad \text{or} \quad \begin{pmatrix} 0 \\ 0 \\ 0 \\ 1 \end{pmatrix}. \tag{2.25}$$

So, as in the Klein–Gordon case, the negative energy solutions appear on the same footing as the positive energy ones.

For any $\mathbf{p}$, we can write for the positive energy solution

$$\eta = \frac{\boldsymbol{\sigma} \cdot \mathbf{p}}{E_{\mathrm{p}} + m} \chi \tag{2.26}$$

and, therefore,

$$u = \begin{pmatrix} \chi \\ \dfrac{\boldsymbol{\sigma} \cdot \mathbf{p}}{E_{\mathrm{p}} + m} \chi \end{pmatrix}, \tag{2.27}$$

with $\chi$ *any* two spinor, and

$$E_{\mathrm{p}} \equiv +(\mathbf{p}^2 + m^2)^{1/2}.$$

For the negative energy solutions,

$$\chi = -\frac{\boldsymbol{\sigma}\cdot\mathbf{p}}{E_{\mathrm{p}}+m}\eta, \tag{2.28}$$

so that

$$u = \begin{pmatrix} -\dfrac{\boldsymbol{\sigma}\cdot\mathbf{p}}{E_{\mathrm{p}}+m}\eta \\ \eta \end{pmatrix}. \tag{2.29}$$

Thus we see that, in our representation, positive energy and small values of momentum mean that $\eta$ is small, while for negative energy (small momentum) $\chi$ is small. As we shall see later (Chapter 4), the negative energy solutions correspond to *antiparticles*, which have energy $E_{\mathrm{p}}$ and momentum $-\mathbf{p}$. Thus it is convenient for these negative energy spinors to replace $u$ by

$$v(p) \equiv u(-\mathbf{p}, -E_{\mathrm{p}}).$$

Furthermore, we henceforth prefer to use the symbol $\chi$ instead of $\eta$ for $v(p)$:

$$v(p) = \begin{pmatrix} \dfrac{\boldsymbol{\sigma}\cdot\mathbf{p}}{E_{\mathrm{p}}+m}\chi \\ \chi \end{pmatrix}. \tag{2.30}$$

We want to normalize these spinors by introducing a factor[1] $(E_{\mathrm{p}}+m/2m)^{1/2}$; this results in

$$\bar{u}u = 1 = -\bar{v}v. \tag{2.31}$$

This is a useful normalization since, as we shall see shortly, $\bar{u}u$ defined in this way is a scalar quantity. Putting this together and indicating the momentum $p$ and the spin state $i$ by subscripts, we have

$$u_{pi} = \left(\frac{E_p+m}{2m}\right)^{1/2}\begin{pmatrix} \chi_i \\ \dfrac{\boldsymbol{\sigma}\cdot\mathbf{p}}{E_p+m}\chi_i \end{pmatrix}, \qquad v_{pi} = \left(\frac{E_p+m}{2m}\right)^{1/2}\begin{pmatrix} \dfrac{\boldsymbol{\sigma}\cdot\mathbf{p}}{E_p+m}\chi_i \\ \chi_i \end{pmatrix}, \tag{2.32}$$

with $i = 1$ or $2$, $\chi_1 = \binom{1}{0}$, $\chi_2 = \binom{0}{1}$.

[1] Except for the $m = 0$ case, i.e. neutrinos, where one can normalize so that $u^\dagger u = 1$ instead, or, for convenience introduce a fictitious neutrino mass $m$, which is later set equal to 0.

### c. Spin Angular Momentum

To see that, indeed, the Dirac equation refers to particles with *spin angular momentum* one-half, we first revert to the original, noncovariant form. Multiplying Eq. (2.11) on the left by $\gamma_4$, we have

$$(i\gamma_4 \boldsymbol{\gamma} \cdot \mathbf{p} + \gamma_4 m)\psi = E\psi \qquad \text{or} \qquad (\boldsymbol{\alpha} \cdot \mathbf{p} + \beta m)\psi = E\psi, \tag{2.33}$$

with $\alpha_i \equiv i\gamma_4 \gamma_i$ and $\beta \equiv \gamma_4$. Therefore,

$$H \equiv \boldsymbol{\alpha} \cdot \mathbf{p} + \beta m \tag{2.34}$$

is the Hamiltonian. If we define

$$\boldsymbol{\Sigma} \equiv \begin{pmatrix} \boldsymbol{\sigma} & 0 \\ 0 & \boldsymbol{\sigma} \end{pmatrix}, \tag{2.35}$$

i.e., $\Sigma_x = -i\gamma_y \gamma_z$, etc., and introduce as *angular momentum operator*

$$\mathbf{J} \equiv \mathbf{r} \times \mathbf{p} + \tfrac{1}{2}\boldsymbol{\Sigma}, \tag{2.36}$$

we find that $[\mathbf{J}, H] = 0$. As we shall see later (Chapter 3, Section 3.2c), this implies angular momentum conservation. Now for a particle at rest, only the $\frac{1}{2}\boldsymbol{\Sigma}$ term contributes; hence this represents spin angular momentum. Since $\sigma_z$ has eigenvalues $\pm 1$, the $z$-component of spin can be $\pm\frac{1}{2}$.

### d. Behavior under Lorentz Transformations

What happens to the Dirac equation when we make a Lorentz transformation? In analogy with the behavior of vectors under ordinary rotations, we expect to get a new $\psi$ linearly related to the old one:

$$\psi(x) \to \psi'(x') = L\psi(x), \tag{2.37}$$

where $L$ is a $4 \times 4$ matrix. Covariance requires that

$$(\gamma_\mu \partial_\mu' + m)\psi'(x') = 0 \tag{2.38}$$

with the *same* $\gamma$'s as in the old system. Then using $\partial_\mu' = a_{\mu\nu} \partial_\nu$ and substituting Eq. (2.37) into (Eq. (2.38),

$$(\gamma_\mu a_{\mu\nu} \partial_\nu + m)L\psi(x) = 0.$$

Multiplying on the left by $L^{-1}$, we see that the new equation is equivalent to the old one if

$$L^{-1}\gamma_\mu L a_{\mu\nu} = \gamma_\nu,$$

or, using Eq. (1.7a),

$$L^{-1}\gamma_\alpha L = a_{\alpha\nu}\gamma_\nu. \tag{2.39}$$

This is a condition on the transformation matrix $L$. The adjoint equation (2.14) is equally covariant; so

$$\bar{\psi}'(x') = \bar{\psi}(x)\bar{L}$$

(defining $\bar{L}$) implies that

$$\bar{L}\gamma_\alpha \bar{L}^{-1} = a_{\alpha\nu}\gamma_\nu. \tag{2.40}$$

Comparison with Eq. (2.39) shows[2] that $\bar{L} = L^{-1}$, i.e.,

$$\bar{\psi}'(x') = \bar{\psi}(x)L^{-1}. \tag{2.41}$$

Equations (2.39) and (2.40) have some immediate and important consequences. For example,

$$\bar{\psi}'(x')\psi'(x') = \bar{\psi}(x)L^{-1}L\psi(x) = \bar{\psi}(x)\psi(x); \tag{2.42}$$

i.e., $\bar{\psi}\psi$ is a scalar quantity. Similarly,

$$\bar{\psi}'\gamma_\mu\psi' = \bar{\psi}L^{-1}\gamma_\mu L\psi = a_{\mu\nu}\bar{\psi}\gamma_\nu\psi; \tag{2.43}$$

so this is a four-vector. Quantities of the form $\bar{\psi}O\psi$, with $O$ some product of $\gamma$'s, are known as *bilinear covariants* and play a central role when one introduces interaction. Since $O$ is a $4 \times 4$ matrix, there are *16* linearly independent $O$'s, and they are conventionally chosen as shown in Table 2.1; the transformation properties are readily established. Here

$$\Sigma_{\mu\nu} \equiv -(i/2)(\gamma_\mu\gamma_\nu - \gamma_\nu\gamma_\mu), \tag{2.44}$$

and

$$\gamma_5 \equiv \gamma_1\gamma_2\gamma_3\gamma_4; \tag{2.45}$$

$\gamma_5$ has the interesting property of anticommuting with each $\gamma_\mu$. In the representation introduced above,

$$\gamma_5 = -\begin{pmatrix} 0 & 1 \\ 1 & 0 \end{pmatrix}. \tag{2.46}$$

### e. Hole Theory

Dirac proposed getting around the negative energy problem by assuming that in empty space all the negative energy states are filled; the Pauli exclusion principle then prevents transition from a positive to a negative energy state. The possibility of raising an electron from negative to positive energy then

---

[2] For *proper* Lorentz transformations, we can always work in terms of infinitesimal transformations, in which case the existence of $L^{-1}$ and the equality $\bar{L} = L^{-1}$ are very easy to establish.

**Table 2.1**

| Bilinear covariant | Transformation type | Number of components |
|---|---|---|
| $\bar{\psi}\psi$ | Scalar (S) | 1 |
| $\bar{\psi}\gamma_{\mu}\psi$ | Vector (V) | 4 |
| $\bar{\psi}\Sigma_{\mu\nu}\psi$ | Tensor (T) | 6 |
| $\bar{\psi}\gamma_{\mu}\gamma_5\psi$ | Pseudovector (A) | 4 |
| $\bar{\psi}\gamma_5\psi$ | Pseudoscalar (P) | 1 |

arises; the *hole* so created then exhibits positive energy and the same mass as the electron but opposite charge. The holes were, in fact, considered a defect of Dirac's theory until 1933 when Anderson discovered the positron, which has all the required properties. So the hole theory successfully deals with the negative energy-state problem, but since it calls for an infinite density of (unobservable) electrons everywhere, it can hardly be considered a one-particle theory.

In addition to spins zero and one-half, one can produce wave equations and field operators for spin one (*vector* particles), spin three-halves, etc. Here we will only discuss a very special example of spin one: the *electromagnetic* field and the wave equation for *photons*.

## 2.3. Electromagnetic Field

Here one must cope with the fact that the different components of the electric and magnetic fields are not independent but are related by Maxwell's equations. Rather than working with **E** and **B**, we use the scalar and vector potentials, which in fact form a four-vector $A$:

$$A_{\mu} = (\mathbf{A}, i\phi). \tag{2.47}$$

Then the standard relations

$$\mathbf{E} = -(\partial\mathbf{A}/\partial t) - \boldsymbol{\nabla}\phi, \qquad \mathbf{B} = \boldsymbol{\nabla} \times \mathbf{A},$$

are contained in the covariant equation

$$F_{\mu\nu} = \partial_{\mu} A_{\nu} - \partial_{\mu} A_{\nu}, \tag{2.48}$$

where

$$(F_{\mu\nu}) = \begin{pmatrix} 0 & B_3 & -B_2 & -iE_1 \\ -B_3 & 0 & B_1 & -iE_2 \\ B_2 & -B_1 & 0 & -iE_3 \\ iE_1 & iE_2 & iE_3 & 0 \end{pmatrix}. \tag{2.49}$$

Furthermore, the two Maxwell equations in the absence of sources

$$\nabla \times \mathbf{B} - (\partial E/\partial t) = 0, \qquad \nabla \cdot \mathbf{E} = 0,$$

are contained in the covariant field equation

$$\partial_\mu F_{\mu\nu} = 0. \tag{2.50}$$

However, different choices of $A_\mu$ can correspond to the same $F_{\mu\nu}$ and, therefore, to the same physical situation. The freedom to change $A_\mu$ without changing the physics is known as *gauge invariance*; the transformation

$$A_\mu(x) \to A_\mu'(x) = A_\mu(x) + \partial_\mu \Lambda(x), \tag{2.51}$$

which obviously leaves $F_{\mu\nu}$ unchanged, is known as a *gauge transformation*. It is often convenient to restrict ourselves to choices of $A_\mu$ such that

$$\partial_\mu A_\mu = 0. \tag{2.52}$$

This is known as the *Lorentz gauge*. Using this and field equation (2.50) gives us the required photon equation:

$$\square A_\mu = 0. \tag{2.53}$$

In momentum space, this equation corresponds to $p^2 = 0$; i.e., the photons have zero mass. Plane wave solutions of wave equation (2.53) have the form

$$A_\mu(x) = \varepsilon_\mu(p)e^{ipx}. \tag{2.54}$$

To conform with the Lorentz condition, Eq. (2.52), the *polarization vector* $\varepsilon_\mu(p)$ satisfies

$$p_\mu \varepsilon_\mu(p) = 0. \tag{2.55}$$

A gauge transformation will alter $\varepsilon_\mu(p)$:

$$\varepsilon_\mu(p) \to \varepsilon_\mu'(p) = \varepsilon_\mu(p) + ip_\mu \Lambda(p).$$

We can always choose the gauge so that, in a particular Lorentz frame, $\varepsilon_4 = 0$; Eq. (2.55) then implies that

$$\mathbf{p} \cdot \boldsymbol{\varepsilon}(p) = 0, \tag{2.56}$$

i.e., the polarization is *transverse* in that Lorentz frame. For $\hat{\mathbf{z}} \| \mathbf{p}$, linearly independent polarization vectors are then $\hat{\mathbf{e}}_x$ and $\hat{\mathbf{e}}_y$; the independent combinations

$$\boldsymbol{\varepsilon}^\pm \equiv 1/\sqrt{2}\,(\hat{\mathbf{e}}_x \pm i\hat{\mathbf{e}}_y) \tag{2.57}$$

correspond to a photon having $z$-component of angular momentum $\pm 1$.

Photons interact only by being absorbed or emitted; so it is natural to make $A_\mu$ a field operator to describe this. Indeed, second quantization was first invented for the electromagnetic case (by Dirac, 1927).

## Problems

**2.1.** Verify that the matrices given by Eqs. (2.19) satisfy the anticommutation relations, Eq. (2.13).

**2.2.** Verify the normalization of the spinors, Eq. (2.31).

**2.3.** Evaluate $u^\dagger u$, $v^\dagger v$.

**2.4.*** Show that $[H, \mathbf{L}] \neq 0$, but $[H, \mathbf{J}] = 0$, where $H$ is the Dirac Hamiltonian [Eq. (2.34)] and $\mathbf{J}$ is the total angular momentum [Eq. (2.36)].

**2.5.*** Show that, under an infinitesimal Lorentz transformation,

$$L^{-1}\gamma_5 L = \gamma_5 + 0(v^2),$$

which is enough to establish that

$$L^{-1}\gamma_5 L = \gamma_5$$

for a finite (proper) Lorentz transformation; and that

$$L^{-1}\gamma_5 L = -\gamma_5$$

for a space inversion. Use these results to establish the transformation properties for the bilinear covariants listed in Chapter 3, Table 3.1.

**2.6.** Show that the covariant equation

$$\partial_\mu F_{\mu\nu} = -j_\nu$$

[where $j_\nu = (\mathbf{j}, i\rho)$ is the *current density* four-vector] is half of Maxwell's equations.

### Bibliography

W. A. Blanpied. *Modern Physics*. Chap. 75. Holt, New York, 1971.

J. A. Eisele. *Modern Quantum Mechanics with Applications to Elementary Particle Physics*. Part II. Wiley, New York, 1969.

H. Muirhead. *Physics of Elementary Particles*. Pergamon, Oxford, 1965.

J. J. Sakurai. *Advanced Quantum Mechanics*. Addison-Wesley, Reading, Massachusetts, 1967.

CHAPTER THREE

# Quantum Mechanics and Scattering

*Cross sections* and *decay rates* play a central role in particle physics: They are the meeting ground of theory and experiment. Here we see how they are related to quantum-mechanical amplitudes. We begin by reviewing the abstract formalism of quantum mechanics.

## 3.1. States and Operators in Quantum Mechanics

### *a. Basic Properties of Hilbert Space*

The most definite specification that can be made of a quantum-mechanical system is to say that it is in a certain *state* (e.g., the ground state of a harmonic oscillator), which we can describe by writing down the wave function. Instead of talking about wave functions, it is often simpler and more direct to think of the state as represented by a *vector in Hilbert space*. This is by definition a complex vector space such that all the vectors have finite lengths (or *norms*).[1] If the Hilbert space has *n dimensions* (and *n* may be infinite), a set

[1] In addition, if a sequence of vectors in the space approaches a limit, that limit must also lie within the space.

of $n$ linearly independent basis vectors $\alpha_i$ can be chosen, and *any* state[2] $\psi$ can be written as a (complex) linear combination of these:

$$\psi = \sum_{i=1}^{n} c_i \alpha_i, \tag{3.1}$$

where the $c_i$ are complex numbers. We can distinguish different states by subscripts: $\psi_1$, $\psi_2, \ldots, \psi_a, \ldots$, and introduce *Dirac notation* to make it obvious that we are talking about Hilbert space vectors: $|\psi_a\rangle$. It is then unnecessary to include the $\psi$ symbol in this *ket vector*, so we (sometimes) write

$$|a\rangle \equiv |\psi_a\rangle.$$

The quantum-mechanical *principle of superposition* then follows from the linearity of any vector space:

$$\begin{aligned} |a_1\rangle + |a_2\rangle &= \sum_i C_i(a_1)|\alpha_i\rangle + \sum C_i(a_2)|\alpha_i\rangle \\ &= \sum_i [C_i(a_1) + C_i(a_2)]|\alpha_i\rangle \\ &= |a_3\rangle. \end{aligned} \tag{3.2}$$

We also want the generalization of the scalar product of two ordinary vectors. There

$$\begin{aligned} \mathbf{A} \cdot \mathbf{B} &\equiv (A_x \hat{\imath} + A_y \hat{\jmath} + A_z \hat{k}) \cdot (B_x \hat{\imath} + B_y \hat{\jmath} + B_z \hat{k}) \\ &= A_x B_x \hat{\imath} \cdot \hat{\imath} + A_x B_y \hat{\imath} \cdot \hat{\jmath} + \cdots \\ &= A_x B_x + A_y B_y + A_z B_z, \end{aligned}$$

since $\hat{\imath} \cdot \hat{\imath} = 1$, $\hat{\imath} \cdot \hat{\jmath} = 0$, etc. We proceed in the same way for Hilbert space vectors: Choosing the basis vectors $|\alpha_i\rangle$ to be *orthonormal* means that the *inner product* between any pair is given by $\delta_{ij}$. This is written

$$\langle \alpha_i | \alpha_j \rangle = \delta_{ij}, \tag{3.3}$$

where $\langle \alpha_i|$ is a new kind of vector, the *bra vector adjoint* to the ket vector $|\alpha_i\rangle$. For a general vector $|a\rangle$ $(\equiv \sum C_i |\alpha_i\rangle)$, the rule is

$$\langle a| \equiv \sum C_i^* \langle \alpha_i|,$$

and we have, in general,

$$\begin{aligned} \text{inner product} \equiv \langle a_1 | a_2 \rangle &= \sum_{i,\, j} C_i^*(a_1) C(a_2) \langle \alpha_i | \alpha_j \rangle \\ &= \sum_i C_i^*(a_1) C_i(a_2). \end{aligned} \tag{3.4}$$

---

[2] We will generally not bother to make the distinction between the states and the Hilbert space vector that represents them.

It should be clear that

$$\langle a|b\rangle^* = \langle b|a\rangle, \qquad \langle a|(|b\rangle + |c\rangle) = \langle a|b\rangle + \langle a|c\rangle,$$
$$\langle a|\lambda b\rangle = \lambda\langle a|b\rangle, \qquad \langle \lambda a|b\rangle = \lambda^*\langle a|b\rangle,$$

where $\lambda$ is a complex number and $|\lambda a\rangle$ means $\lambda|a\rangle$. The normalization of one of these *state vectors* is expressed by

$$\langle a|a\rangle = 1 = \sum_i |C_i|^2. \tag{3.5}$$

### b. Operators

The *observables* and *transformations* of quantum mechanics are represented by *linear operators* in Hilbert space. The operator $A$ can operate on both bras and kets:

$$A|a\rangle = |a'\rangle, \qquad \langle a|A = \langle a''|,$$

and give a new bra (left side of $A$) or ket (right side). We can also form

$$\begin{aligned}\langle a|A|b\rangle &= [\sum_i C_i^*(a)\langle\alpha_i|]\, A[\sum C_j(b)|\alpha_j\rangle] \\ &= \sum_{i,j} C_i^*(a)\langle\alpha_i|A|\alpha_j\rangle C_j(b),\end{aligned} \tag{3.6}$$

where we have used the linearity of $A$:

$$A(\lambda_1|a\rangle + \lambda_2|b\rangle) = \lambda_1 A|a\rangle + \lambda_2 A|b\rangle. \tag{3.7}$$

Equation (3.6) can be thought of as matrix multiplication, with the $\{C_i^*(a)\}$ forming a *row vector*, the $\{C_j(b)\}$ a *column vector*, and the $\langle\alpha_i|A|\alpha_j\rangle \equiv A_{ij}$ the entries of a matrix. Thus the $A_{ij}$ and even $\langle a|A|b\rangle$ are often referred to as *matrix elements.* The quantity $\langle a|A|a\rangle$ is called the *expectation value* of $A$ in the state $|a\rangle$.

The (Hilbert space) adjoint $A$ of the operator $A$ is defined by

$$\langle A^\dagger a|b\rangle = \langle a|A|b\rangle, \tag{3.8}$$

for any $\langle a|$ and $|b\rangle$. Here $\langle A^\dagger a|$ means the vector adjoint to $A^\dagger|a\rangle$. From Eq. (3.8), it follows that

$$(A^\dagger)_{ij} = A^*_{ji}. \tag{3.9}$$

If $A|a\rangle$ has the same *direction* as $|a\rangle$, i.e.,

$$A|a\rangle = \lambda|a\rangle \tag{3.10a}$$

with $\lambda$ a *number*, then $|a\rangle$ is called an *eigenvector*, and $\lambda$ an *eigenvalue*, of the operator $A$. Often the eigenvalue is used to label the state:

$$A|\lambda\rangle = \lambda|\lambda\rangle. \tag{3.10b}$$

An operator is said to be *Hermitian* if

$$A^\dagger = A. \tag{3.11}$$

Such operators play an important role, because observables always correspond to Hermitian operators. It turns out that the eigenvectors of *any* Hermitian operator form a *complete set*, i.e., *any* vector satisfies[3]

$$|\psi\rangle = \sum_\lambda C_\lambda |\lambda\rangle$$

when the $|\lambda\rangle$ are all of the eigenvectors of any Hermitian operator. We can always arrange that the $|\lambda\rangle$ be orthonormal and, therefore, that

$$C_\lambda = \langle\lambda|\psi\rangle,$$

so that Eq. (3.1′) becomes

$$|\psi\rangle = \sum_\lambda |\lambda\rangle\langle\lambda|\psi\rangle.$$

Since $|\psi\rangle$ is *any* vector, we can say that completeness implies

$$\sum_\lambda |\lambda\rangle\langle\lambda| = \bar{1}, \tag{3.12}$$

where $\bar{1}$ is the identity operator in Hilbert space:

$$\bar{1}|\psi\rangle = |\psi\rangle \qquad \text{for any} \quad |\psi\rangle.$$

### c. Amplitudes and Probabilities

If we take the expectation value of $A$,

$$\begin{aligned}\langle\psi|A|\psi\rangle &= \sum_{\lambda,\lambda'} \langle\psi|\lambda\rangle\langle\lambda|A|\lambda'\rangle\langle\lambda'|\psi\rangle \\ &= \sum_{\lambda,\lambda'} C_\lambda{}^*\lambda'\langle\lambda|\lambda'\rangle C_{\lambda'} = \sum_\lambda |C_\lambda|^2\lambda. \end{aligned} \tag{3.13}$$

Since $\sum_\lambda |C_\lambda|^2 = 1$, we are led to interpret $|C_\lambda|^2$ as the probability for finding the eigenvalue $\lambda$, and $C_\lambda$ as the corresponding *amplitude*. More generally, we say that the amplitude for finding the system in state $|a\rangle$, given that it is in state $|\psi\rangle$, is given by the *overlap*:

$$\text{ampl} = \langle a|\psi\rangle. \tag{3.14}$$

---

[3] It may be that more than one independent eigenvector belongs to the same eigenvalue; in that case, we must be sure to include *all* the independent eigenvectors in Eq. (3.1′). This situation is known as *degeneracy*.

For example, if $|\psi\rangle$ is a one-particle state and $|\mathbf{r}\rangle$ the eigenvector of the position operator, then

$$\langle\mathbf{r}|\psi\rangle \equiv \psi(\mathbf{r}) \tag{3.15}$$

is the amplitude for finding the particle at position $\mathbf{r}$, i.e., the wave function.

### *d. Continuum States*

At this point we must confess that if we have a *continuous* range of eigenvalues, all of the sums over states given above will be partly or wholly integrations. Thus by

$$\sum_{\lambda} |\lambda\rangle\langle\lambda|,$$

we may really mean

$$\int |\lambda\rangle\, d\lambda \langle\lambda|.$$

We normalize these states in the *continuous spectrum* by

$$\int_{\delta\lambda} \langle\lambda|\lambda'\rangle\, d\lambda' = \begin{cases} 1 & \text{if} \quad \delta\lambda \text{ contains } \lambda, \\ 0 & \text{otherwise.} \end{cases} \tag{3.16}$$

This is exactly like the relation

$$\sum_{\lambda} \langle\lambda|\lambda'\rangle = \begin{cases} 1 \\ 0 \end{cases}$$

which clearly holds for the *discrete spectrum*. With this understanding, none of the above formulas containing $\sum_{\lambda}$, etc., need to be changed.

The normalization condition (3.16) can be rewritten as

$$\langle\lambda|\lambda'\rangle = \delta(\lambda - \lambda') \tag{3.17}$$

where $\delta(\lambda - \lambda')$ is the *Dirac delta function*, which satisfies

$$\delta(x) = 0, \qquad \text{for} \quad x \neq 0, \tag{3.18}$$

and $\int \delta(x)\, dx = 1$, if the region of integration contains $x = 0$. Equation (3.17) is like the orthonormality relation for the discrete case, Eq. (3.3).

### *e. Unitary Operators*

*Unitary operators* play an important role in quantum mechanics. A unitary operator $U$ satisfies

$$U^{\dagger}U = UU^{\dagger} = \bar{1}; \tag{3.19}$$

i.e., $U^{\dagger} = U^{-1}$. If we use $U$ to transform all of the vectors,

$$|a'\rangle = U|a\rangle; \tag{3.20}$$

then all inner products are unchanged:

$$\langle a' | b' \rangle = \langle a | U^\dagger U | b \rangle = \langle a | b \rangle. \tag{3.21}$$

Sometimes these unitary operators are used merely to provide alternative descriptions of the same physical system; however, we will more often be interested in unitary operators that bring about a transformation from one physical system to another.

The unitary operators corresponding to *infinitesimal transformations* are very useful. These have the form

$$U = \bar{1} + i\varepsilon G, \tag{3.22}$$

with $\varepsilon$ a small real number. Unitarity requires that

$$\bar{1} = U^\dagger U = \bar{1} + i\varepsilon(G - G^\dagger),$$

neglecting the $O(\varepsilon^2)$ term. Therefore $G$ must be Hermitian. We can make *finite* transformations by repeated application of infinitesimal ones. Thus

$$(\bar{1} + i\varepsilon G)^n = \bar{1} + in\varepsilon G + \frac{n(n-1)}{2!}(i\varepsilon)^2 G^2 + \cdots. \tag{3.23}$$

Now if we put $\varepsilon = \lambda/n$ and let $n \to \infty$ while keeping $\lambda$ fixed, we get an exact expression for the *finite* transformation:

$$U(\lambda) = \lim_{n\to\infty}\left(\bar{1} + i\frac{\lambda}{n}G\right)^n = \bar{1} + i\lambda G + \frac{(i\lambda G)^2}{2!} + \cdots$$
$$\equiv e^{i\lambda G}. \tag{3.24}$$

This transformation is *continuous*, and $G$ is called the *generator* of the transformation. Note that since $G$ is Hermitian, $U(\lambda)$ is automatically unitary:

$$U(\lambda)^\dagger U(\lambda) = e^{-i\lambda G} e^{i\lambda G} = \bar{1}.$$

There also exist *discrete* transformations which *cannot* be put into the form of Eq. (3.26); these are analogous to the improper Lorentz transformations discussed in Chapter 1, Section 1.1c.

## 3.2. Equations of Motion

It is important for us to be able to describe the time development of quantum-mechanical systems. There are two main ways of doing this: the *Schrödinger representation* and the *Heisenberg representation*. In the former, the time development is borne by the states, while the operators corresponding to observables (like $p$, $q$, etc.) are time independent; in the latter, the states are independent of time while the operators change with time.

### *a. Schrödinger Representation*

We start with a state $\psi_S$ at time $t_0$ and inquire about a later time $t$. We expect that a *linear operator* relates the states at the two times:

$$|\psi_S(t)\rangle = U(t, t_0)|\psi_S(t_0)\rangle, \tag{3.25}$$

and since we demand

$$\langle\psi_S(t)|\psi_S(t)\rangle = \langle\psi_S(t_0)|\psi_S(t_0)\rangle = 1,$$

$U$ must be a unitary operator. We clearly want

$$U(t_0, t_0) = \bar{1}, \tag{3.26}$$

and

$$|\psi_S(t)\rangle = U(t, t_1)|\psi_S(t_1)\rangle = U(t, t_1)U(t_1, t_0)|\psi_S(t_0)\rangle. \tag{3.27}$$

Then just homogeneity in time requires that $U(t, t_0)$ can be written in the form

$$U(t, t_0) = U(t - t_0) = e^{-iH(t-t_0)}. \tag{3.28}$$

The operator $H$ so defined is called the *Hamiltonian*, and is independent of time.[4] Thus

$$|\psi_S(t)\rangle = e^{-iHt}|\psi_S(0)\rangle \tag{3.29}$$

gives the time development; then

$$i\,(\partial/\partial t)|\psi_S(t)\rangle = H|\psi_S(t)\rangle \tag{3.30}$$

is the *Schrödinger equation* [compare Eq. (2.2)].

### *b. Heisenberg Representation*

We choose the (time-independent) *Heisenberg states* to coincide with the *Schrödinger states* at $t = 0$;

$$|\psi_H\rangle = |\psi_S(0)\rangle. \tag{3.31}$$

We require all expectation values of operators to be the same in both representations:

$$\langle\psi_S(t)|A_S(t)|\psi_S(t)\rangle = \langle\psi_H|A_H(t)|\psi_H\rangle. \tag{3.32}$$

---

[4] There are also systems with time-dependent Hamiltonians, but these lack the homogeneity in time assumed in obtaining Eq. (3.28).

Using Eq. (3.29), we must have

$$A_{\mathrm{H}}(t) = e^{iHt} A_{\mathrm{S}}(t) e^{-iHt} \tag{3.33}$$

for the time development of *Heisenberg operators*. Suppose $A_{\mathrm{S}}$ is time dependent (e.g., dynamic variables like $p$, $q$, etc.); then

$$i\,(d/dt)A_{\mathrm{H}}(t) = [A_{\mathrm{H}}(t), H] \equiv A_{\mathrm{H}}(t),H, - HA_{\mathrm{H}}(t) \tag{3.34}$$

is the *Heisenberg equation of motion* for the operators. It has some immediate and important consequences which we now discuss.

### c. Invariance and Conservation Laws

Suppose we transform the (Heisenberg) states with some unitary operator $U$. If the Hamiltonian is *invariant* under this transformation, i.e., if

$$H' = U^{-1}HU = H,$$

then the *commutator*

$$[U, H] = 0 \tag{3.35}$$

and

$$dU/dt = 0. \tag{3.36}$$

Therefore the expectation values of $U$ are *constants of the motion*. Thus we see that the *invariance* of $H$ under $U$ gives rise to a *conservation law*.

Very often $U$ is a continuous transformation; in this case we express the conservation law in terms of the generator:

$$U = e^{i\lambda G}.$$

Then

$$[U, H] = 0$$

implies

$$[G, H] = 0, \tag{3.35$'$}$$

and, therefore,

$$dG/dt = 0. \tag{3.36$'$}$$

We can also express Eqs. (3.36) and (3.36′) as conservation laws for the expectation values:

$$(d/dt)\langle U\rangle = 0 \qquad \text{or} \qquad (d/dt)\langle G\rangle = 0.$$

For the expectation values, of course, it does not matter whether the Heisenberg or the Schrödinger representation is used. Clearly, these conservation laws imply equality of the conserved quantity before and after scattering or decay. We discuss two very important examples now and many others later on.

### *d. Translation Invariance and Momentum Conservation*

Suppose we consider the operation of *translation* of the coordinate axes by a distance **a**:

$$U_{\mathbf{a}}|\psi\rangle = |\psi'\rangle.$$

The generator of this translation is called the *momentum operator* **P**:

$$U_{\mathbf{a}} = e^{i\mathbf{a}\cdot\mathbf{P}}. \tag{3.37}$$

Invariance of $H$ under translation implies

$$[\mathbf{P}, H] = 0 \tag{3.38}$$

and thus *momentum conservation.*

In terms of a coordinate space wave function,

$$\begin{aligned}\psi'(\mathbf{r}) &= U_{\mathbf{a}}\psi(\mathbf{r}) = \psi(\mathbf{r} + \mathbf{a})\\ &= \psi(\mathbf{r}) + \mathbf{a}\cdot\nabla\psi(\mathbf{r}) + \cdots,\end{aligned}$$

so for an infinitesimal translation,

$$\psi'(\mathbf{r}) = (1 + \mathbf{a}\cdot\nabla)\psi(\mathbf{r}) = (1 + i\mathbf{a}\cdot\mathbf{P})\psi(\mathbf{r}), \tag{3.39}$$

using Eq. (3.37). Hence we have found the usual representation for the momentum operator

$$\mathbf{P} = -i\nabla \tag{3.40}$$

(single-particle coordinate-space representation). For several (noninteracting) particles, the same reasoning gives

$$\mathbf{P} = -i\sum_i \nabla_i = \sum_i \mathbf{p}_i.$$

### *e. Rotation Invariance and Angular Momentum Conservation*

Here the generators are the angular momentum operators $J_x$, $J_y$, $J_z$:

$$U_{\theta_z} = e^{i\theta J_z} \tag{3.41}$$

for a rotation (of the coordinate axes) of $\theta$ around the $z$ direction, etc. Invariance of $H$ under rotations implies

$$[J, H] = 0 \tag{3.42}$$

and thus *angular momentum conservation.* The eigenvalues of angular momentum are very useful for classifying states.

Consider wave function $\psi(\mathbf{r})$ again:

$$\begin{aligned}\psi'(\mathbf{r}) &= U_{\theta_z}\psi(\mathbf{r}) = \psi(\mathbf{r}') = \psi(\mathbf{r} + \Delta\mathbf{r})\\ &= \psi(\mathbf{r}) + \Delta\mathbf{r}\cdot\nabla\psi(\mathbf{r}) + \cdots;\end{aligned}$$

so for an infinitesimal rotation of $\varepsilon$ around $z$

$$\psi'(\mathbf{r}) = \psi(\mathbf{r}) + \mathbf{\Delta r} \cdot \nabla\psi(\mathbf{r}),$$

where $\mathbf{\Delta r}$ depends upon $\bar{r}$:

$$\mathbf{\Delta r} = \varepsilon\hat{\mathbf{z}} \times \mathbf{r} = \varepsilon(x\hat{\mathbf{y}} - y\hat{\mathbf{x}}). \tag{3.43}$$

Then

$$\begin{aligned}\psi'(\mathbf{r}) &= \psi(\mathbf{r}) + \varepsilon\left(x\frac{\partial}{\partial y} - y\frac{\partial}{\partial x}\right)\psi(\mathbf{r}) \\ &= (1 + i\varepsilon L_z)\psi(\mathbf{r}).\end{aligned} \tag{3.44}$$

Here

$$L_z \equiv -i(\mathbf{r} \times \nabla)_z = (\mathbf{r} \times \mathbf{p})_z, \tag{3.45}$$

and similarly, rotations around $x$ and $y$ are generated by $L_x$ and $L_y$. These, of course, are the *orbital angular momentum* operators. If the wave function describes several particles,

$$\mathbf{L} = \mathbf{L}_1 + \mathbf{L}_2 + \cdots = \sum_i \mathbf{r}_i \times \mathbf{p}_i.$$

The situation is more complicated if the wave function has components which get mixed up under rotations. A familiar example of such a function is a vector $\mathbf{A}$. Under an infinitesimal rotation of $\varepsilon$ around $z$, a *constant* $\mathbf{A}$ becomes

$$\mathbf{A}' = \begin{pmatrix} A_x' \\ A_y' \\ A_z' \end{pmatrix} = \begin{pmatrix} A_x + \varepsilon A_y \\ A_y - \varepsilon A_x \\ A_z \end{pmatrix} = R\mathbf{A}; \tag{3.46}$$

so

$$R = \bar{1} + \varepsilon\begin{pmatrix} 0 & 1 & 0 \\ -1 & 0 & 0 \\ 0 & 0 & 0 \end{pmatrix} = \bar{1} + i\varepsilon S_z, \tag{3.47}$$

with generator

$$S_z = \begin{pmatrix} 0 & -i & 0 \\ i & 0 & 0 \\ 0 & 0 & 0 \end{pmatrix}, \tag{3.48a}$$

Similarly,

$$S_x = \begin{pmatrix} 0 & 0 & 0 \\ 0 & 0 & -i \\ 0 & i & 0 \end{pmatrix}, \tag{3.48b}$$

and

$$S_y = \begin{pmatrix} 0 & 0 & i \\ 0 & 0 & 0 \\ -i & 0 & 0 \end{pmatrix}. \tag{3.48c}$$

This is *spin angular momentum*; there is no orbital angular momentum for this example because there is no spatial variation of **A**.

Let us relax this condition and investigate the case of a vector *function* $\mathbf{A}(\mathbf{r})$ Under our usual infinitesimal rotation,

$$\begin{aligned}\mathbf{A}(\mathbf{r}) \to \mathbf{A}'(\mathbf{r}) &= R\mathbf{A}(\mathbf{r}') \\ &= R[\mathbf{A}(\mathbf{r}) + (\Delta\mathbf{r} \cdot \nabla)\mathbf{A}(\mathbf{r})],\end{aligned}$$

or

$$\mathbf{A}'(\mathbf{r}) = \{1 + i\varepsilon[S_z + \bar{1}(\mathbf{r} \times \mathbf{p})_z]\}\mathbf{A}(\mathbf{r}), \tag{3.49}$$

when we use Eq. (3.43) and discard terms $O(\varepsilon^2)$ Thus, for this example

$$\mathbf{J} = \mathbf{L} + \mathbf{S}; \tag{3.50}$$

the *total* angular momentum **J** is made up of orbital plus spin parts. Rotational invariance means that **J** is conserved, but not, in general, **L** or **S** separately

Unlike translations, successive rotations do not commute:

$$U_{\theta z} U_{\phi x} \neq U_{\phi x} U_{\theta z}.$$

If we are clever at geometry, we can evaluate the effect of the rotations

$$U_{\theta z}^{-1} U_{\phi x}^{-1} U_{\theta z} U_{\phi x}$$

directly; it comes out to be a $U_{\varepsilon_y}$.[5] The result is most easily expressed in terms of the generators **J**,

$$[\mathbf{J}_x, \mathbf{J}_y] = i\mathbf{J}_z, \tag{3.51}$$

and cyclic permutations, or

$$[J_i, J_j] = i\varepsilon_{ijk} J_k \qquad \text{or} \qquad \mathbf{J} \times \mathbf{J} = i\mathbf{J}.$$

($\varepsilon_{ijk} = \pm 1$, if $ijk$ is an $\left\{{\text{even} \atop \text{odd}}\right\}$ permutation of $xyz$, 0 otherwise.) An easier way to get the result is to realize that the **J** commutation relations must be just the same as the **L** or **S** commutation relations;

$$[S_i, S_j] = i_{ijk} S \tag{3 52}$$

follows directly from the matrices (3.48).

---

[5] Compare, e.g., Sakurai, 1964.

The **L** and **S** we have found here are particular *representations* of the commutation relations (3.51). We quote some of the standard results of angular momentum theory[6]: The commutation relations require that the allowed eigenvalues of $\mathbf{J}^2$ be $j(j+1)$, where $j = 0, \frac{1}{2}, 1, \frac{3}{2}, \ldots$, and those of $J_z$ be $j, j-1, \ldots, -j$. Sensible behavior for the spatial wave functions limit the allowed eigenvalues of $\mathbf{L}^2$ to $l(l+1)$, with $l = 0, 1, 2, \ldots$. The allowed eigenvalues $s(s+1)$ of $\mathbf{S}^2$ with $s = 0, \frac{1}{2}, 1, \frac{3}{2}, \ldots$, correspond to the $S_x, \ldots$ being $(2s+1) \times (2s+1)$ matrices; our vector example thus means *spin one.* Particles have definite values of $s$; those with $s = 0, 1, 2, \ldots$ are called *bosons,* while $s = \frac{1}{2}, \frac{3}{2}, \ldots$ correspond to *fermions.* The rules for the addition of angular momenta then limit a state with an odd number of fermions to $j = \frac{1}{2}, \frac{3}{2}, \ldots$, while an even number of fermions means $j = 0, 1, 2, \ldots$.

## 3.3. Scattering and Cross Sections. Method of Partial Waves

About the best an experimentalist can do to investigate the structure and interactions of elementary particles is to perform *scattering experiments.* That is, he aims a beam of particles at a target (or causes two beams to intersect) and observes what comes out, measuring the frequency of finding various scattered particles emerging at various angles to the incident beam direction. To express the results of such experiments in a manner independent of the number of particles in the beam and in the target, we introduce the idea of scattering *cross section.*

### *a. Cross Sections*

Suppose a single target is located somewhere within an area $A$ and a single projectile is made to pass through that area; then the probability of scattering $P$ is proportional to $A^{-1}$. Writing

$$P = \sigma / A$$

defines the cross section $\sigma$ for the particular kind of scattering in question. We can think of $\sigma$ as representing a small area within $A$; if the projectile happens to go through that area, we have a scattering. It is clear from this definition that $\sigma$ is independent of whether we view the scattering in the CM or the lab, since both $P$ and $A$ are the same in both systems. For a target with many scatterers and a beam of projectiles, the *rate* $R$ of scattering will be

$$R = FAN_t P,$$

---

[6] See any quantum mechanics text.

where $F$ is the *flux* of incident particles (number arriving per unit area per unit time), and $N_t$ the number of target particles. Introducing the number density of incident particles $n_i$ and the relative velocity $v_{rel}$, we can write

$$F = n_i v_{rel},$$

so that

$$R = n_i N_t v_{rel} \sigma. \tag{3.53}$$

We may want to focus on scattering into a particular solid angle $\Delta\Omega$ at an angle $\theta$ to the incident beam. Then we expect that

$$\Delta\sigma \propto \Delta\Omega,$$

so that we are led to consider the *differential cross section*, $d\sigma(\theta)/d\Omega$. Unlike $\sigma$, $d\sigma/d\Omega$ *does* depend on the reference frame because $d\Omega$ depends on it.

### b. Partial Waves. Elastic Scattering

We now wish to discuss how the scattering cross section depends upon the quantum-mechanical amplitude. We do this first by the method of *partial waves*, which is especially simple for the scattering of spinless particles, and begin by considering the case of *purely elastic* scattering. This will hold exactly for scattering of a spinless particle by a central potential, or the scattering of two spinless particles below the lowest *inelastic threshold* (i.e., the energy required to produce new particles). We are interested in an energy eigenstate (steady-state solution), so putting

$$\psi(\mathbf{r}, t) = \psi(\mathbf{r})e^{-iEt}$$

in the (time-dependent) Schrödinger equation (2.2) gives the *time-independent* Schrödinger equation:

$$[-(1/2m)\,\nabla^2 + V(\mathbf{r})]\psi(\mathbf{r}) = E\psi(\mathbf{r}). \tag{2.2'}$$

For large separation $r$ (and assuming that the potential falls off faster than $1/r$ for large $r$) this becomes the *free-wave equation*,

$$(\nabla^2 + k^2)\psi(\mathbf{r}) = 0, \tag{2.2''}$$

where $k^2 \equiv 2mE$. Suppose the incident beam is a plane wave traveling in $+z$ direction; this means an $e^{ikz}$ term in the wave function. In addition, there is the scattering caused by the potential, represented at large $r$ by a spherical wave propagating (when the time dependence is included) *outward*:

$$f(\theta)e^{ikr}/r.$$

Thus we are after an asymptotic solution of Eq. (2.2″) of the form

$$\psi(\mathbf{r}) \simeq e^{ikz} + f(\theta)e^{ikr}/r \tag{3.54}$$

(cf. Problem 3.6), where $f(\theta)$ is the scattering amplitude; there can be no dependence on azimuthal angle $\phi$ because the $z$-component of angular momentum is zero. The scattered wave must have a $1/r$ dependence in order that the rate of scattered particles passing through a large spherical surface (centered at the origin) be independent of its radius. This rate is given in terms of the probability density of scattered wave $|\psi_{sc}|^2$ by

$$R = \int |\psi_{sc}|^2 v_{rel} r^2 \, d\Omega = v_{rel} \int |f(\theta)|^2 \, d\Omega.$$

Since the probability density of incident wave is

$$|\psi_{incident}|^2 = n_p = |e^{ikz}|^2 = 1$$

and $N_t = 1$ (we are considering a single scattering center), Eq. (3.53) tells us that

$$\sigma = \int |f(\theta)|^2 \, d\Omega, \tag{3.55a}$$

and

$$d\sigma/d\Omega = |f(\theta)|^2, \tag{3.55b}$$

for the total and differential elastic scattering cross sections.

We now introduce the *partial wave expansion* for $f(\theta)$:

$$f(\theta) = \sum_{l=0}^{\infty} (2l+1) \frac{a_l}{2ik} P_l(\cos\theta), \tag{3.56}$$

and for $e^{ikz}$:

$$\begin{aligned} e^{ikz} = e^{ikr\cos\theta} &= \sum_{l=0}^{\infty} (2l+1) i^l j_l(kr) P_l(\cos\theta) \\ &\underset{r\to\infty}{\simeq} \sum (2l+1) i \frac{\sin(kr - l\pi/2)}{kr} P_l(\cos\theta) \\ &= \sum (2l+1) \frac{e^{ikr} - (-1)^l e^{-ikr}}{2ikr} P_l(\cos\theta). \end{aligned} \tag{3.57}$$

Here the $P_l(\cos\theta)$ are the *Legendre functions*, and $j_l(kr)$ the spherical *Bessel functions*[7]; $l$ is the orbital angular momentum quantum number and the complex number $a_l$ is the *partial wave amplitude*. Putting Eqs. (3.56) and (3.57) in Eq. (3.54) yields

$$\psi(r) \simeq \frac{1}{2ikr} \sum (2l+1)\{(1+a_l)e^{ikr} - (-1)^l e^{-ikr}\} P_l(\cos\theta). \tag{3.58}$$

[7] Compare, e.g., Blanpied, 1971.

Each $e^{ikr}$ term is an *outgoing spherical wave*, while each $e^{-ikr}$ term is an *incoming* one; thus

$$\psi \simeq \psi_{\text{out}} + \psi_{\text{in}}. \tag{3.59}$$

Now since conservation of probability requires that $\text{rate}_{\text{out}} = \text{rate}_{\text{in}}$, we must have

$$\int |\psi_{\text{out}}|^2 \, d\Omega = \int |\psi_{\text{in}}|^2 \, d\Omega. \tag{3.60}$$

This must hold *separately* for *each* partial wave because of angular momentum conservation. Thus the scattering can only change the *phase* of each outgoing wave, not its modulus. Hence

$$S_l \equiv 1 + a_l \tag{3.61}$$

satisfies $|S_l| = 1$, and we can set

$$S_l = e^{2i\delta_l}, \tag{3.62}$$

with the *phase shift* $\delta_l$ so defined being real. Then, since

$$a_l = e^{2i\delta_l} - 1,$$

we have

$$\begin{aligned} f(\theta) &= (1/2ik) \sum (2l + 1)(e^{2i\delta_l} - 1) P_l(\cos\theta) \\ &= (1/k) \sum (2l + 1) e^{i\delta_l} \sin\delta_l \, P_l(\cos\theta). \end{aligned} \tag{3.63}$$

For potential scattering, the phase shifts [which, along with $f(\theta)$, $S_l$, and $a_l$, are functions of $k$] can easily be found, e.g., by numerically integrating the Schrödinger equation for each partial wave. One can also show that, if $R$ is the *range* of the potential, $\delta_l$ is appreciably different from zero only if $l \lesssim kR$.

The resulting expression for the *differential* cross section,

$$d\sigma/d\Omega = |f(\theta)|^2,$$

includes interference between different partial waves, but that for the *total* cross section does not, since the different $P_l(\cos\theta)$ are orthogonal:

$$\sigma = 2\pi \int |f(\theta)|^2 \, d(\cos\theta) = (4\pi/k^2) \sum (2l + 1) \sin^2 \delta_l, \tag{3.64}$$

upon using

$$\int P_l^2 \, d(\cos\theta) = 2/(2l + 1)$$

and Eq. (3.63).

From Eqs. (3.63) and (3.64) [and $P_l(1) = 1$], we derive a general relationship:

$$\operatorname{Im} f(0) = (1/k) \sum (2l + 1) \sin^2 \delta_l,$$

and therefore

$$\operatorname{Im} f(0) = (k/4\pi)\sigma. \tag{3.65}$$

This relation is known as the *optical theorem*; it says if there is any scattering at all, there must be some scattering in the forward direction. We will see that it holds quite generally.

### c. Inelastic Scattering

Let us now drop our requirement of purely elastic scattering and consider *inelastic* processes to be present as well. We retain $f(\theta)$ as the *elastic* scattering amplitude and note that all the relations are unchanged down through Eq. (3.59). We have the elastic scattering part of the wave function

$$\psi^{\text{el}} \simeq \psi_{\text{out}}^{\text{el}} + \psi_{\text{in}}, \tag{3.59'}$$

and since now $\text{rate}_{\text{out}}^{\text{el}} \leq \text{rate}_{\text{in}}$ and hence

$$\int |\psi_{\text{out}}^{\text{el}}|^2 \, d\Omega \leq \int |\psi_{\text{in}}|^2 \, d\Omega, \tag{3.60'}$$

we must have $|S_l| \leq 1$. An additional parameter is now needed to specify $S_l$:

$$S_l = \eta_l \, e^{2i\delta_l}, \tag{3.62'}$$

with $\eta_l$ and $\delta_l$ real and $1 \geq \eta_l \geq 0$; $\eta_l$ is the *inelasticity parameter*. As before

$$f(\theta) = (1/2ik) \sum (2l + 1)(S_l - 1) P_l(\cos \theta), \tag{3.63'}$$

and for the *total elastic* cross section,

$$\sigma_{\text{el}} = (\pi/k^2) \sum (2l + 1) \, |S_l - 1|^2. \tag{3.64'}$$

We can also define a *reaction* cross section by

$$\begin{aligned} \sigma_r \equiv \frac{\text{rate}_{\text{in}} - \text{rate}_{\text{out}}^{\text{el}}}{v_{\text{rel}}} &= \int \{|\psi_{\text{in}}|^2 - |\psi_{\text{out}}^{\text{el}}|^2\} r^2 \, d\Omega \\ &= (\pi/k)^2 \sum (2l + 1)(1 - |S_l|^2). \end{aligned} \tag{3.66}$$

The *total* cross section is then

$$\sigma_{\text{T}} = \sigma_{\text{el}} + \sigma_r = (2\pi/k^2) \sum (2l + 1)(1 - \operatorname{Re} S_l). \tag{3.67}$$

If we now look back at Eq. (3.63′), we see that

$$\begin{aligned} \operatorname{Im} f(0) &= (1/2k) \sum (2l + 1)(1 - \operatorname{Re} S_l) \\ &= (k/4\pi)\sigma_{\text{T}}. \end{aligned} \tag{3.65'}$$

So the optical theorem relates the forward *elastic* scattering amplitude to the *total* cross section. This makes sense, since *any* kind of scattering means depletion of the original beam, which implies some elastic scattered wave in the forward direction to cancel part of the original plane wave.

## 3.4. Scattering. The *S*-Matrix

We now turn to a more general way of describing scattering. The *S-matrix* concept is a powerful one and is used very extensively in particle physics. The quantity $S$ relates "after collision" to "before collision," or "after decay" to "before decay." That is, suppose $|\mathrm{i}\rangle$ is an initial (Schrödinger) state in the remote past ($t$ at $-\infty$) when the particles are well separated and therefore not interacting. As time proceeds, the particles approach, interact (at $t \simeq 0$), and then they, or other particles, recede and are detected in the remote future ($t = +\infty$). The evolved state at time $t$ is $U(t, -\infty)|\mathrm{i}\rangle$, and at $t = \infty$,

$$U(\infty, -\infty)|\mathrm{i}\rangle \equiv S|\mathrm{i}\rangle.$$

This is a superposition of various kinds of final states; the amplitude for finding a particular final state $|\mathrm{f}\rangle$ is given by the overlap $\langle \mathrm{f}|S|\mathrm{i}\rangle$. Thus the *S-matrix element*

$$S_{\mathrm{fi}} \equiv \langle \mathrm{f}|S|\mathrm{i}\rangle \tag{3.68}$$

is the *amplitude* for the *transition* $\mathrm{i} \to \mathrm{f}$. From its definition, $S$ is a unitary operator.

### *a. Invariance*

We now discuss the relation between invariance and conservation laws, developed above, in terms of $S$ instead of in terms of $H$; this is just another way of saying the same thing, since the Schrödinger equation (3.30) tells us that the invariance of $H$ implies the invariance of $U(\infty, -\infty) = S$. Consider a transformation of the states given by some unitary operator $U$. Then if all transition amplitudes are *unchanged* by this transformation,

$$\langle \mathrm{f}|S|\mathrm{i}\rangle = \langle \mathrm{f}'|S|\mathrm{i}'\rangle = \langle \mathrm{f}|U^{-1}SU|\mathrm{i}\rangle,$$

we conclude that

$$U^{-1}SU = S \qquad \text{or} \qquad [S, U] = 0. \tag{3.69}$$

Conversely, if $S$ commutes with $U$, the transition amplitudes are invariant. If $U$ is a continuous operator generated by $G$ we can replace Eq. (3.69) by

$$[S, G] = 0. \tag{3.69$'$}$$

Now if the states $|\mathrm{i}\rangle$ and $|\mathrm{f}\rangle$ are eigenstates of $G$ with eigenvalues $g_\mathrm{i}$ and $g_\mathrm{f}$,

$$0 = \langle \mathrm{f}|[S, G]|\mathrm{i}\rangle = \langle \mathrm{f}|S|\mathrm{i}\rangle (g_\mathrm{i} - g_\mathrm{f}), \tag{3.70}$$

so either $g_\mathrm{i} = g_{\mathrm{fi}}$ or $S_{\mathrm{fi}} = 0$. That is, either the eigenvalue is unchanged, or the transition will not take place.

For example, rotation invariance implies that if we choose i and f to be angular momentum eigenstates, we must have

$$\langle J'J_z'|S|JJ_z\rangle = \delta_{JJ'}\delta_{J_zJ_z'}S_J. \tag{3.70'}$$

Hence for spinless particles, $S$ is diagonal in $l$; for elastic scattering this and the unitary property of $S$ bring us back to the partial wave expression, Eq. (3.62).

In contrast to $H$, which singles out time and therefore is *not* invariant under Lorentz transformations, $S$ *is* invariant under proper Lorentz transformations; this expresses the *isotropy* of space–time. In addition (and like $H$) $S$ is also invariant under space–time translations, which expresses the *homogeneity* of space–time. For infinitesimal space–time translations [see Eq. (3.37)],

$$U = \bar{1} + ia_\lambda P_\lambda,$$

so that translation invariance implies

$$[S, P_\lambda] = 0,$$

and therefore for momentum eigenstates

$$S_{\mathrm{fi}}(P_{\mathrm{i}\lambda} - P_{\mathrm{f}\lambda}) = 0. \tag{3.71}$$

($P_\mathrm{i}$ and $P_\mathrm{f}$ are initial and final *total* four-momenta.) This expresses energy momentum conservation for collisions and leads us to suspect that we might be able to factor a four-momentum $\delta$-function out of $S_{\mathrm{fi}}$.

### *b. Transition Rates and Cross Sections*

We now want to establish the connection between the $S$-matrix elements and *transition rates* and cross sections. It is already clear that the transition rate from state i to state f and hence the cross section $\sigma_{\mathrm{fi}}$ are $\propto |S_{\mathrm{fi}}|^2$; the kinematic factors of proportionality can be derived from scattering theory or the simplest time-dependent perturbation theory. We first give the answer and then show how it comes about. One finds

$$\sigma_{\mathrm{fi}} = \frac{(2\pi)^6}{v_{\mathrm{rel}}4E_aE_b}\int \delta(P_\mathrm{i} - P_\mathrm{f})|T_{\mathrm{fi}}|^2 \prod_{k=1}^{n} \frac{d^3\mathbf{p}_k}{(2\pi)^3 2E_k}. \tag{3.72}$$

In this expression, the energy momentum $\delta$-function,

$$\delta(P_\mathrm{i} - P_\mathrm{f}) \equiv \delta(P_{\mathrm{i}x} - P_{\mathrm{f}x})\delta(P_{\mathrm{i}y} - P_{\mathrm{f}y})\delta(P_{\mathrm{i}z} - P_{\mathrm{f}z})\delta(E_\mathrm{i} - E_\mathrm{f}),$$

enforces energy–momentum conservation in the integrations over all the possible final state momenta of the $n$ final state particles. The energies $E_k$ are given by

$$E_k = (\mathbf{p}_k{}^2 + m_k{}^2)^{1/2},$$

and $E_a$ and $E_b$ are the energies of the two initial state particles. The *invariant scattering amplitude* $T_{\mathrm{fi}}$ is essentially $S_{\mathrm{fi}}$ with a $\delta(P_{\mathrm{i}} - P_{\mathrm{f}})$ factored out.

We now want to derive this result and show how the form of the expression for $\sigma_{\mathrm{fi}}$ depends on how the states are normalized. We use a method that is unconvincing at one point but gets to the correct answer quickly. To begin it is convenient to think in terms of discrete states with normalization

$$\langle a | a \rangle = 1.$$

This is accomplished by imagining the system to be enclosed in a box (volume $V = L^3$) and imposing periodic boundary conditions, so that the allowed momenta satisfy

$$p_x = (2\pi/L)n_x, \qquad p_y = (2\pi/L)n_y, \qquad p_z = (2\pi/L)n_z, \tag{3.73}$$

with $n_x$, $n_y$, $n_z$ integers. The allowed momentum eigenstates thus have spatial wave functions

$$\langle x | p \rangle = (1/\sqrt{V})e^{i\mathbf{p}\cdot\mathbf{x}}. \tag{3.74}$$

We always have in mind taking the limit $V \to \infty$.

First note that if there is *no* interaction, we must have no transitions:

$$S^0_{\mathrm{fi}} = \delta_{\mathrm{fi}}.$$

Since this cannot contribute to scattering, we remove it by defining the *reactance* operator

$$iR \equiv S - \bar{1}. \tag{3.75}$$

But Eq. (3.71) implies that $R_{\mathrm{fi}} = 0$ for $P_{\mathrm{f}} \neq P_{\mathrm{i}}$, so that it is convenient to define a new matrix element $M_{\mathrm{fi}}$ such that (in the $V \to \infty$ limit)

$$R_{\mathrm{fi}} = (2\pi)^4 \delta(P_{\mathrm{i}} - P_{\mathrm{f}}) M_{\mathrm{fi}},$$

and

$$S_{\mathrm{fi}} = \delta_{\mathrm{fi}} + i(2\pi)^4 \delta(P_{\mathrm{i}} - P_{\mathrm{f}}) M_{\mathrm{fi}}. \tag{3.76}$$

The probability for a transition ($\mathrm{f} \neq \mathrm{i}$) is given by

$$|S_{\mathrm{fi}}|^2 = (2\pi)^8 [\delta(P_{\mathrm{i}} - P_{\mathrm{f}})]^2 |M_{\mathrm{fi}}|^2. \tag{3.77}$$

To handle the $[\delta(P_{\mathrm{i}} - P_{\mathrm{f}})]^2$, it is helpful to know that

$$2\pi\delta(x - x') = \int_{-\infty}^{\infty} dk\, e^{ik(x-x')},$$

which follows from the properties of Fourier transforms:

$$f(x) = \frac{1}{2\pi}\int_{-\infty}^{\infty} dk\, e^{ikx}\left[\int_{-\infty}^{\infty} e^{-ikx'} f(x')\, dx'\right]$$

$$= \frac{1}{2\pi}\int_{-\infty}^{\infty} dx'\, f(x') \int_{-\infty}^{\infty} dk\, e^{ik(x-x')}$$

$$= \int_{-\infty}^{\infty} dx'\, f(x')\, \delta(x-x') = f(x)\int_{-\infty}^{\infty} \delta(x-x')\, dx'.$$

Then writing

$$(2\pi)^4\, \delta(P_i - P_f) = \int_{-\infty}^{\infty} \exp[i(P_i - P_f)\cdot x]\, d^4x$$

$$= \lim_{V,\, T\to\infty} \int_{VT} \exp[i(P_i - P_f)x]\, d^4x$$

($d^4x \equiv dx\, dy\, dz\, dt$), we can put

$$(2\pi)^8[\delta(\quad)]^2 = (2\pi)^4\, \delta(P_i - P_f) \lim_{V,T\to\infty} \int_{VT} \exp[i(P_i - P_f)x]\, d^4x$$

$$= (2\pi)^4\, \delta(P_i - P_f) \lim_{V,\, T\to\infty} \int_{VT} d^4x$$

$$= (2\pi)^4\, \delta(P_i - P_f) \lim_{V,\, T\to\infty} VT. \tag{3.78}$$

This result can also be derived in a slightly less mysterious, although lengthier manner, simply by being careful to square $S_{fi}$ *before* taking the limit $V, T \to \infty$.

[Thus for a finite time interval $T$, the $2\pi\delta(E_i - E_f)$ is actually

$$\int_{-T/2}^{T/2} \exp[i(E_f - E_i)t]\, dt = 2i\, \frac{\sin[(E_f - E_i)T/2]}{i(E_f - E_i)}.$$

Then instead of $|2\pi\delta(E_i - E_f)|^2$, we have

$$4\,\frac{\sin^2[(E_f - E_i)T/2]}{(E_f - E_i)^2} \equiv A(E_i - E_f).$$

Now because $A(E)$ is very sharply peaked around $E = 0$, we can write

$$\int_{E<0}^{E>0} f(E)A(E)\, dE \simeq f(0)\int_{-\infty}^{\infty} A(E)\, dE = f(0)\cdot 2\pi T,$$

i.e.,

$$A(E) \xrightarrow[T\to\infty]{} T\cdot 2\pi\, \delta(E),$$

$$\left|\int_{-T/2}^{T/2} e^{i(E_f - E_i)t}\, dt\right|^2 \xrightarrow[T\to\infty]{} T\cdot 2\pi\, \delta(E_f - E_i).$$

Furthermore, for finite $V$, $2\pi\delta(p_{ix} - p_{fx})$ is actually

$$\int_{-L/2}^{L/2} e^{i(p_{ix}-p_{fx})x}\, dx = \int_{-L/2}^{L/2} e^{i(2\pi/L)(n-n')x}\, dx = L\delta_{nn'};$$

so instead of $|2\pi\delta(p_{ix} - p_{fx})|^2$, we have

$$L^2(\delta_{nn'})^2 = L^2\,\delta_{nn'} \xrightarrow[L\to\infty]{} L \cdot 2\pi\,\delta(p_{ix} - p_{fx}).$$

Putting these results together then gives Eq. (3.78).]

Now Eq. (3.77) gives the transition probability during a time interval $T$, so the *transition rate* is, using (3.78),

$$\text{rate}_{fi} = |S_{fi}|^2/T = (2\pi)^4\,\delta(P_i - P_f)V\,|M_{fi}|^2. \tag{3.79}$$

For a particular set of final state particles, there are a large number of states that are very much alike; recall that the number of states for a particle in a box of volume $V$ is given by the *phase volume*:

$$dN = d^3\mathbf{p}V/h^3 = d^3\mathbf{p}V/(2\pi)^3, \tag{3.80}$$

which in fact follows directly from Eqs. (3.73), since

$$dN = dn_x\,dn_y\,dn_z = (L/2\pi)^3\,dp_x\,dp_y\,dp_z.$$

One of these factors will appear for *each* particle in the final state. Then summing over these final states means integrating over the phase volume, so that if there are $n$ final state particles,

$$\sum \text{rate}_{fi} = (2\pi)^4 \int \delta(P_i - P_f)\,|M_{fi}|^2 \prod_{k=1}^{n} \frac{d^3\mathbf{p}_k}{(2\pi)^3}\, V^{n+1}. \tag{3.81}$$

Now, from Eq. (3.53)

$$\sum \text{rate}_{fi} = n_p v_{rel} N_t \sigma_{fi} = (1/V)v_{rel}\,\sigma_{fi},$$

and combining this with Eq. (3.81) gives

$$\sigma_{fi} = \frac{V^{n+2}}{v_{rel}} (2\pi)^4 \int \delta(P_i - P_f)\,|M_{fi}|^2 \prod_k \frac{d^3\mathbf{p}_k}{(2\pi)^3} \tag{3.82}$$

for the i $\to$ f cross section.

Since the cross section must be independent of the normalization volume $V$, the $V^{n+2}$ in Eq. (3.82) must be canceled by the $V$ dependence of the matrix element $M_{fi}$. We can get rid of the $V$'s completely by returning to *δ-function normalization*:

$$\langle \mathbf{p} | \mathbf{p}' \rangle = \delta(\mathbf{p} - \mathbf{p}'). \tag{3.83}$$

Here the single-particle wave functions for momentum eigenstates are

$$\langle \mathbf{x} | \mathbf{p} \rangle = \frac{1}{(2\pi)^{3/2}}\, e^{i\mathbf{p}\cdot\mathbf{x}}. \tag{3.84}$$

This is normalization to fixed density $[=(2\pi)^{-3}]$ rather than to fixed number. To convert from

$$\langle \mathbf{p} | \mathbf{p} \rangle = 1 \qquad \text{to} \qquad \langle \mathbf{p} | \mathbf{p}' \rangle = \delta(\mathbf{p} - \mathbf{p}'),$$

we simply multiply an amplitude by a factor of $[V/(2\pi)^3]^{1/2}$ for each particle in or out [compare Eqs. (3.74) and (3.84)]. Since there are two particles going in and $n$ coming out, the matrix element in the new normalization is

$$\overline{M}_{\mathrm{fi}} \equiv \left(\frac{V}{(2\pi)^3}\right)^{(n+2)/2} M_{\mathrm{fi}}, \tag{3.85}$$

and

$$\sigma_{\mathrm{fi}} = \frac{(2\pi)^6}{v_{\mathrm{rel}}} (2\pi)^4 \int \delta(P_{\mathrm{i}} - P_{\mathrm{f}}) \, |\overline{M}_{\mathrm{fi}}|^2 \prod_k d^3\mathbf{p}_k . \tag{3.86}$$

*Invariant normalization* is another possibility; its virtue is that if operator $O$ is Lorentz invariant then so is $\langle a | O | b \rangle$. Here we put

$$\langle \mathbf{p} | \mathbf{p}' \rangle = 2E_p (2\pi)^3 \, \delta(\mathbf{p} - \mathbf{p}'). \tag{3.87}$$

The momentum wave functions are then

$$\langle \mathbf{x} | \mathbf{p} \rangle = (2E_p)^{1/2} \, e^{i\mathbf{p}\cdot\mathbf{x}}, \tag{3.84'}$$

which corresponds to density $2E_p$. The *invariant matrix element* is then

$$T_{\mathrm{fi}} \equiv N M_{\mathrm{fi}}, \tag{3.88}$$

with

$$N \equiv \left[ (2E_a V)(2E_b V) \prod_{k=1}^{n} (2E_k V) \right]^{1/2}, \tag{3.89}$$

and

$$\sigma_{\mathrm{fi}} = \frac{(2\pi)^4}{v_{\mathrm{rel}} 4E_a E_b} \int \delta(P_{\mathrm{i}} - P_{\mathrm{f}}) \, |T_{\mathrm{fi}}|^2 \prod_k \frac{d^3\mathbf{p}_k}{(2\pi)^3 2E_p}, \tag{3.72}$$

where $a$ and $b$ refer to the incoming particles.

[Note that the phase space factor in the sum over final states must always match the normalization. That is

$$\langle \mathbf{p} | \mathbf{p}' \rangle = N' \, \delta(\mathbf{p} - \mathbf{p}')$$

implies phase space factor

$$d^3\mathbf{p}/N'.$$

This is so that

$$\sum_p \langle \mathbf{p} | \mathbf{p}' \rangle \equiv \int \langle \mathbf{p} | \mathbf{p}' \rangle \, (d^3\mathbf{p}/N') = 1$$

is satisfied.]

The expression

$$d^3\mathbf{p}_k/2E_k$$

is known as the *invariant phase space*; we can demonstrate its invariance by establishing the equality

$$\int \delta(p^2 + m^2)\theta(E)\, d^4p = \int d^3\mathbf{p}/2E_p. \tag{3.90}$$

The integral on the left-hand side of this equation is invariant because each factor under the integral is $\delta(p^2 + m^2)$ because its argument is; the *step function*

$$\theta(E) \equiv \begin{cases} 1, & E > 0, \\ 0, & E < 0, \end{cases}$$

because a proper Lorentz transformation cannot change the sign of $E$; and $d^4p$ ($\equiv d^3\mathbf{p}\, dE$) because the *Jacobian*

$$\partial(p)/\partial(p') = 1$$

if $p'$ is related to $p$ by a Lorentz transformation. To demonstrate Eq. (3.90), consider

$$I \equiv \int dE\, \delta(p^2 + m^2)\theta(E).$$

Because

$$\int \delta[f(x)]\, dx = \int \frac{d[f(x)]\, df(x)}{|df(x)/dx|} = \sum \left|\frac{df(x_i)}{dx}\right|^{-1},$$

where $f(x_i)$ are the *zeros* of $f(x)$,

$$\delta(p^2 + m^2) = \delta(\mathbf{p}^2 + m^2 - E^2) = \frac{1}{2E_p}\,[\delta(E - E_p) + \delta(E + E_p)]$$

with $E_p \equiv (\mathbf{p}^2 + m^2)^{1/2}$. Therefore,

$$I = 1/2E_p \int \delta(E - E_p)\, dE = 1/2E_p.$$

Equation (3.90) follows immediately.

Introducing

$$\delta_+(p^2 + m^2) \equiv \delta(p^2 + m^2)\theta(E),$$

we can now rewrite Eq. (3.72) in the form

$$\sigma_{fi} = (v_{rel} \cdot 4E_a E_b)^{-1} \int (2\pi)^4\, \delta(P_i - P_f)\, |T_{fi}|^2 \cdot \prod 2\pi\delta_+(p_k{}^2 + m_k{}^2)\, \frac{d^4p_k}{(2\pi)^4}. \tag{3.72$'$}$$

Each (one-dimensional) $\delta$-function is multiplied by a factor of $(2\pi)$, and each $d^4p$ is divided by $(2\pi)^4$.

We can obtain a similarly elegant expression when we consider a decay instead of a collision, that is, one particle in the initial state instead of two. Starting from Eq. (3.81) and multiplying $M_{fi}$ by $(2E_p V)^{1/2}$ for each particle to convert to invariant normalization, we find the *decay rate*

$$\Gamma_{fi} = (2E_i)^{-1} \int (2\pi)^4 \, \delta(P_i - P_f) \, | T_{fi} |^2 \prod_k 2\pi \, \delta_+(p_k{}^2 + m_k{}^2) \frac{d^4 p_k}{(2\pi)^4}. \tag{3.91}$$

If we want the *lifetime* $\tau$ of the particle, we need the *total* decay rate; i.e., we must sum over *all possible* decay channels and, if the final state particles have *spins*, over all these spin states as well (the summation already made in Eq. (3.81) ff being only over the *momenta* of a particular set of particles). Thus, denoting the sums over final spins by $\sum_{\mathrm{sp_f}}$,

$$\Gamma = \sum_{f} \sum_{\mathrm{sp_f}} \Gamma_{fi}, \tag{3.92}$$

and $\tau = \Gamma^{-1}$.

If the particles involved in a collision have spins, then Eq. (3.72) and those preceding it refer to particular initial and final spin states. On the other hand, experiments most often deal with unpolarized beams and targets and do not determine the final state spins.[8] Thus all possible spin states are present, and we must *sum* over the *final* state spins and *average* over the *initial* state spins. Since there are $(2s_a + 1)(2s_b + 1)$ initial state spins and they all have the same statistical weight, the average over initial spins will be accomplished by

$$\bar{\sum_{\mathrm{sp_i}}} \equiv \frac{1}{(2s_a + 1)(2s_b + 1)} \sum_{\mathrm{sp_i}}. \tag{3.93}$$

We denote the combined spin averaging and summation by

$$\sum_{\mathrm{sp}} \equiv \bar{\sum_{\mathrm{sp_i}}} \sum_{\mathrm{sp_f}}. \tag{3.93'}$$

Then to convert Eq. (3.72′) to an expression for the *spin-averaged* cross section, we merely insert $\sum_{\mathrm{sp}}$ in front of the $|T_{fi}|^2$.

The factor $v_{\mathrm{rel}} E_a E_b$ appearing in Eq. (3.72) is easily evaluated. Since [cf. Eq. (1.20)]

$$v_{\mathrm{rel}} = \left| \frac{\mathbf{p}_a}{E_a} - \frac{\mathbf{p}_b}{E_b} \right|,$$

[8] If the beam and/or target is *polarized* (i.e., unequal population for different spin states), one introduces the *density matrix* to describe it. (See, e.g., Muirhead, 1965.)

we have

$$\text{CM system:}\quad v_{\text{rel}} E_a E_b = |\mathbf{p}|(E_a + E_b) = |\mathbf{p}|\sqrt{s}, \tag{3.94a}$$

$$\text{lab system:}\quad v_{\text{rel}} E_a E_b = |\mathbf{p}_a| m_b. \tag{3.94b}$$

(In fact,

$$|\mathbf{p}|\sqrt{s} = |\mathbf{p}_a| m_b = [-(p_a \cdot p_b)^2 - m_a{}^2 m_b{}^2]^{1/2} = [\lambda_{sab}/2]^{1/2};$$

remember, $\mathbf{v}_a$ is parallel to $\mathbf{v}_b$ in these systems.)

### c. Phase Space Integrals

We now want to evaluate the phase space factor for two and three particles in the final state. We start with the *two*-particle case; the process i → f is now 1 + 2 → 3 + 4.

$$I_2 = \int \delta(P_{\text{i}} - P_{\text{f}})\delta_+(p_3{}^2 + m_3{}^2)\delta_+(p_4{}^2 + m^2)\, d^4p_1\, d^4p_2$$

$$= \int \delta(P_{\text{i}} - P_{\text{f}}) \frac{d^3\mathbf{p}_3\, d^3\mathbf{p}_4}{2E_3\ 2E_4} = \int \delta(W - E_3 - E_4) \frac{d^3\mathbf{p}_3}{4E_3 E_4},$$

where now $E_3$ and $E_4$ are both functions of $\mathbf{p}_3$. We write

$$I_2 = \int \frac{\delta(W - E_3 - E_4)}{4E_3 E_4} \frac{d^3\mathbf{p}_3}{dW}\, dW = \int \frac{\mathbf{p}_3{}^2}{4E_3 E_4}\, d\Omega \frac{d|\mathbf{p}_3|}{dW}.$$

The $\delta$-functions are now used up, so we cannot do any more integrations until $|T_{\text{fi}}|^2$ is inserted. To finish the calculation, we note that

$$\left(\frac{d|\mathbf{p}_3|}{dW}\right)^{-1} = \frac{dW}{d|\mathbf{p}_3|} = \frac{dE_3}{d|\mathbf{p}_3|} + \frac{dE_4}{d|\mathbf{p}_3|},$$

and use the *CM system*, where $|\mathbf{p}_3| \equiv |\mathbf{q}| = |\mathbf{p}_4|$. Thus

$$\frac{dW}{d|\mathbf{p}_3|} = |\mathbf{q}|\left(\frac{1}{E_3} + \frac{1}{E_4}\right) = \frac{|\mathbf{q}|\sqrt{s}}{E_3 E_4},$$

so that

$$I_2 = |\mathbf{q}| \int d\Omega/4\sqrt{s}. \tag{3.95}$$

Putting this result and Eq. (3.94a) into Eq. (3.72) gives us

$$\sigma_{\text{fi}} = \frac{1}{16(2\pi)^2} \frac{|\mathbf{q}|}{|\mathbf{p}|s} \int |T_{\text{fi}}|^2\, d\Omega, \tag{3.96a}$$

or

$$\frac{d\sigma_{\mathrm{fi}}}{d\Omega} = \frac{1}{64\pi^2}\frac{|\mathbf{q}|}{|\mathbf{p}|s}\,|T_{\mathrm{fi}}|^2. \tag{3.96b}$$

for the process $1 + 2 \rightarrow 3 + 4$.

In the case of elastic scattering of spinless particles, we can compare this result with Eq. (3.56),

$$d\sigma(\theta)/d\Omega = |f(\theta)|^2,$$

and conclude that, within a phase factor,

$$f(\theta) = T_{\mathrm{fi}}(\theta)/8\pi\sqrt{s}. \tag{3.97}$$

The phase space factor for a *three*-particle final state has *five* integrations left after the δ-functions are integrated out, and the form of the result depends on how we choose the independent variables; our particular choice will appear in a moment. Labeling the final state particles by 1, 2, and 3, and working in the CM system, we have

$$\begin{aligned} I_3 &= \int \delta(E)\delta(p_1 + p_2 + p_3)\frac{d^3p_1}{2E_1}\frac{d^3p_3}{2E_2}\frac{d^3p_3}{2E_3} \\ &= \int \delta(E)\,\frac{p_1{}^2\,dp_1\,d\Omega_1}{2E_1}\,\frac{p_2{}^2\,dp_2\,d\Omega_2}{2E_2}\cdot\frac{1}{2E_3}, \end{aligned}$$

where for this paragraph $p_1 \equiv |\mathbf{p}_1|$, etc., and $E \equiv W - E_1 - E_2 - E_3$. Measuring the direction of $\mathbf{p}_2$ from that of $\mathbf{p}_1$, we replace $d\Omega_2$ by $d\Omega_{12}$. Then using

$$dp_1 = (E_1/p_1)\,dE_1, \quad \text{etc.,}$$

and

$$p_3{}^2 = p_1{}^2 + p_2{}^2 - 2p_1p_2\cos\theta_{12},$$

so that

$$2E_3\,dE_3 = dE_3{}^2 = dp_3{}^2 = -2p_1p_2\,d\cos\theta_{12}$$

($E_1$ and $E_2$ are being held fixed here), we are left with

$$I_3 = \tfrac{1}{8}\int \delta(E)\,dE_1\,dE_2\,dE_3\,d\Omega_1\,d\phi_{12}. \tag{3.98}$$

Therefore, for the collision $a + b \rightarrow 1 + 2 + 3$, we can rewrite Eq. (3.72) in the form

$$\sigma = \frac{1}{8(2\pi)^5}\frac{1}{4|\mathbf{p}|\sqrt{s}}\int |T_{\mathrm{fi}}|^2\cdot\delta(E)\,dE_1\,dE_2\,dE_3\,d\Omega_1\,d\phi_{12}. \tag{3.99}$$

### d. Unitarity

We close this section by showing how the *unitarity* of the $S$-matrix implies the optical theorem.

$$S^{\dagger}S = \bar{1} \qquad \text{and} \qquad S = \bar{1} + iR$$

imply

$$i(R^{\dagger} - R) = R^{\dagger}R. \tag{3.100}$$

Sandwiching this relation between states i and f yields

$$i(\langle \mathrm{f}|\, R^{\dagger}\,|\mathrm{i}\rangle - \langle \mathrm{f}|\, R\,|\mathrm{i}\rangle) = \sum_{\mathrm{n}} \langle \mathrm{f}|\, R^{\dagger}\,|\mathrm{n}\rangle \langle \mathrm{n}|\, R\,|\mathrm{i}\rangle,$$

or

$$i(R^{*}_{\mathrm{if}} - R_{\mathrm{fi}}) = \sum_{\mathrm{n}} R^{*}_{\mathrm{nf}} R_{\mathrm{ni}},$$

where the $\sum_{\mathrm{n}}$ is the sum over a complete set of states. Then removing the δ-functions brings us to the $M_{\mathrm{fi}}$:

$$i(2\pi)^4\delta(P_{\mathrm{i}} - P_{\mathrm{f}})(M^{*}_{\mathrm{if}} - M_{\mathrm{fi}}) = (2\pi)^8 \sum_{\mathrm{n}} \delta(P_{\mathrm{n}} - P_{\mathrm{f}})\delta(P_{\mathrm{n}} - P_{\mathrm{i}})M^{*}_{\mathrm{nf}} M_{\mathrm{ni}}.$$

But

$$\delta(P_{\mathrm{n}} - P_{\mathrm{f}})\delta(P_{\mathrm{n}} - P_{\mathrm{i}}) = \delta(P_{\mathrm{i}} - P_{\mathrm{f}})\delta(P_{\mathrm{n}} - P_{\mathrm{i}}),$$

and therefore

$$i(M^{*}_{\mathrm{if}} - M_{\mathrm{fi}}) = (2\pi)^4 \sum_{\mathrm{n}} \delta(P_{\mathrm{n}} - P_{\mathrm{i}})M^{*}_{\mathrm{nf}} M_{\mathrm{ni}}.$$

Converting to the invariant $T_{\mathrm{fi}}$ [cf. Eq. (3.88)] gives

$$i(T^{*}_{\mathrm{if}} - T_{\mathrm{fi}})/N = (2\pi)^4 \sum_{\mathrm{n}} \delta(P_{\mathrm{n}} - P_{\mathrm{i}})T^{*}_{\mathrm{nf}} T_{\mathrm{ni}}/N'N''.$$

The factor $N'N''$ takes care of the factors of $V$ in the sum over intermediate states [cf. Eq. (3.80)] and affects the conversion to invariant phase space:

$$\sum_{\mathrm{n}} \delta(P_{\mathrm{n}} - P_{\mathrm{f}}) \to \int \delta(P_{\mathrm{n}} - P_{\mathrm{i}}) \prod \left(\frac{d^3\mathbf{p}_k}{(2\pi)^3 2E_k}\right).$$

The remaining summation is over the different spin states and kinds of particles in the intermediate states; the $\delta(P_{\mathrm{n}} - P_{\mathrm{i}})$ insures that we include only states that conserve energy momentum in the sum.

Then taking f = i (forward elastic scattering), we have

$$i(T^{*}_{\mathrm{ii}} - T_{\mathrm{ii}}) = (2\pi)^4 \sum \int \delta(P_{\mathrm{n}} - P_{\mathrm{i}})\, |T_{\mathrm{ni}}|^2 \prod \left(\frac{d^3\mathbf{p}_k}{(2\pi)^3 2E_k}\right). \tag{3.101}$$

Comparing this with Eq. (3.72) shows us that

$$2\,\mathrm{Im}\,T_{\mathrm{ii}} = 4E_a E_b v_{\mathrm{rel}}\,\sigma_{\mathrm{T}},$$

where $\sigma_{\mathrm{T}}$ is the *total* cross section:

$$\sigma_{\mathrm{T}} = \sum_{\mathrm{f}} \sigma_{\mathrm{if}}\,;$$

in the CM system,

$$\mathrm{Im}\,T_{\mathrm{ii}} = 2\,|\mathbf{p}|\,\sqrt{s}\,\sigma_{\mathrm{T}}, \tag{3.102}$$

and hence

$$\mathrm{Im}\,T_{\mathrm{ii}} = (\lambda_{sab})^{1/2}\,\sigma_{\mathrm{T}}. \tag{3.102'}$$

This version of the optical theorem is completely general.

Comparing with Eq. (3.65′), we see that

$$\mathrm{Im}\,f(0) = \frac{|\mathbf{p}|}{4\pi}\,\sigma_{\mathrm{T}} = \frac{1}{8\pi\sqrt{s}}\,\mathrm{Im}\,T_{\mathrm{ii}},$$

in agreement with Eq. (3.97).

## 3.5. On Finding *S* from *H*

Since the time evolution of a system is controlled by the Hamiltonian $H$, it is clear that $H$ determines the $T_{\mathrm{fi}}$. How is $H$ determined? By more or less educated guessing. We assume a certain form for $H$; this defines a theory or model. We then attempt to use $H$ to predict the $T_{\mathrm{fi}}$ and therefore the cross sections (or decay rates). It seems that only by using a perturbation expansion can we obtain explicit expressions for the $T_{\mathrm{fi}}$. This is useful for the electromagnetic and weak interactions, but not for the strong interactions where the expansion parameter is large.

### *a. Interaction Representation*

To develop this perturbation method (which, in fact, is just a different way of doing the same *time-dependent perturbation theory* familiar from introductory quantum mechanics), we first introduce the *interaction representation.* In this representation, the state vectors have a time dependence that comes only from the *interaction*, while the operators have the same time dependence as Heisenberg operators with *no* interaction; if the interaction is "turned off," there is no scattering and the interaction representation becomes the Heisenberg representation. Thus the interaction representation state vectors are supposed to have only the time dependence that comes from scattering.

The trick of using the interaction representation, as we shall see, yields an equation for the time development operator $U(t, t_0)$:

$$i(\partial/\partial t)U(t, t_0) = H_{\mathrm{I}}U(t, t_0), \tag{3.103}$$

where $H_I$ is the interaction part of the Hamiltonian "in the interaction representation":

$$H_I = e^{iH_0(t-t_0)} H' e^{-iH_0(t-t_0)}, \tag{3.104}$$

and the total Hamiltonian $H$ is written as a sum of *free particle* and *interaction* parts:

$$H = H_0 + H'. \tag{3.105}$$

Then $S = U(\infty, -\infty)$ gives us $S$ and hence $S_{fi}$ and $T_{fi}$.

To see how this comes about, we start from the Schrödinger representation and the Schrödinger equation (3.30):

$$i(\partial/\partial t) \,|\, \psi_S(t)\rangle = H \,|\, \psi_S(t)\rangle, \tag{3.30}$$

and write, as above,

$$H = H_0 + H'. \tag{3.105}$$

We introduce an operator $R$ that represents the time development induced by $H_0$ alone:

$$i(\partial/\partial t)R(t, t_0) = H_0 R(t, t_0), \tag{3.106}$$

so that

$$R = \exp[-iH_0(t - t_0)], \tag{3.107}$$

since we also require

$$R(t_0, t_0) = \bar{1}.$$

We now define the interaction state vector $|\, \psi_I(t)\rangle$ by

$$|\, \psi_S(t)\rangle = R(t, t_0) \,|\, \psi_I(t)\rangle. \tag{3.108}$$

The two representations coincide at $t = t_0$. Introducing for clarity a slightly more concise notation,

$$|\, \psi_S(t)\rangle \equiv |\, S\rangle \qquad \text{and} \qquad |\, \psi_I(t)\rangle \equiv |\, I\rangle,$$

Eq.(3.108) reads

$$|\, S\rangle = R \,|\, I\rangle. \tag{3.108$'$}$$

Inserting this into the Schrödinger equation, we have

$$i(\partial/\partial t) \,|\, S\rangle = H \,|\, S\rangle, \tag{3.30$'$}$$

$$i(\partial/\partial t)(R \,|\, I\rangle) = HR \,|\, I\rangle = (H_0 + H')R \,|\, I\rangle; \tag{3.109}$$

but

$$i(\partial/\partial t)(R \,|\, I\rangle) = H_0 R \,|\, I\rangle + iR(\partial/\partial t) \,|\, I\rangle; \tag{3.110}$$

so comparing Eqs. (3.109) and (3.110) yields

$$i(\partial/\partial t)\,|\,\mathrm{I}\rangle = R^{-1}H'R\,|\,\mathrm{I}\rangle$$
$$= H_{\mathrm{I}}\,|\,\mathrm{I}\rangle, \tag{3.111}$$

where

$$H_{\mathrm{I}} \equiv R^{-1}H'R \tag{3.104$'$}$$

is the interaction Hamiltonian in the new representation, as mentioned above. Equation (3.111) is the required equation of motion of $|\,\mathrm{I}\rangle$; it looks like the Schrödinger equation, but with $H$ replaced by $H_{\mathrm{I}}$.

Since the matrix elements are the same in all representations, an operator $A$ in the interaction representation is found by writing

$$\langle \mathrm{I}'|A_{\mathrm{I}}|\mathrm{I}\rangle = \langle \mathrm{S}'|A_{\mathrm{S}}|\mathrm{S}\rangle = \langle \mathrm{I}'|R^{-1}A_{\mathrm{S}}R|\mathrm{I}\rangle;$$

therefore

$$A_{\mathrm{I}} = R^{-1}A_{\mathrm{S}}R. \tag{3.112}$$

In view of Eq. (3.107), we see that $A_{\mathrm{I}}$ has the same time dependence as $A_{\mathrm{H}}$ in the *absence of interaction.*

To proceed, we express the time development of $|\mathrm{I}\rangle$ in terms of the operator $U \equiv U(t, t_0)$:

$$|\mathrm{I}\rangle = U|\mathrm{I}_0\rangle, \tag{3.113}$$

where

$$|\mathrm{I}_0\rangle \equiv |\psi_{\mathrm{I}}(t_0)\rangle. \tag{3.114}$$

Then Eq. (3.111) can be written as

$$i(\partial/\partial t)U = H_{\mathrm{I}}\,U. \tag{3.115}$$

Furthermore $U$ can be shown to be unitary.[9] We now take $t_0 \to -\infty$ and $t \to +\infty$. The matrix elements of the resulting $U(\infty, -\infty)$ are just the transition amplitudes [as in Eq. (3.68)], so that $S \equiv U(\infty, -\infty)$ is essentially the same operator as introduced there. [The difference being that $S$ here belongs between interaction representation states, while $S$ of Eq. (3.68) belongs between Schrödinger representation

---

[9] Equation (3.115) and the fact that $H_{\mathrm{I}}$ is Hermitian are enough to insure that

$$U^{\dagger}U = \bar{\mathrm{I}}. \tag{A}$$

Since there must exist some operator (call it $U^{-1}$) that relates $|\mathrm{I}_0\rangle$ to $|\mathrm{I}\rangle$,

$$U^{-1}U = UU^{-1} = \bar{\mathrm{I}};$$

then multiplying Eq. (A) on the right by $U^{-1}$ tells us that $U^{\dagger} = U^{-1}$. Q.E.D.

states.] Thus, as stated above, the connection between $H$ and $S$ is given by the two equations:

$$\begin{aligned} S &= U(\infty, -\infty), \\ i(\partial/\partial t)U(t, t_0) &= H_I(t)U(t, t_0). \end{aligned} \tag{3.116}$$

### b. Perturbation Expansion

We next want to show how perturbation theory takes us from Eqs. (3.116) to an explicit expression for $S$:

$$S = \bar{1} + \sum_{n=1}^{\infty} \frac{(-i)^n}{n!} \int_{-\infty}^{\infty} \cdots \int_{-\infty}^{\infty} dt_n \cdots dt_1 \, T[H_I(t_n) \cdots H_I(t_1)]. \tag{3.117}$$

Here $T[\ \ ]$ indicates *time ordering*, i.e., the operators $H_I$ inside the $[\ \ ]$ must be arranged so that each has only "earlier" operators to its right. This ordering is important because, in general, $H_I(t)$ and $H_I(t')$ do not commute for $t \neq t'$.

We now assume that $H_I$ is small ($\propto \varepsilon$) and that $U$ can be expanded in a power series in $\varepsilon$,

$$U = \sum_{n=0}^{\infty} U_n,$$

where $U_n \propto \varepsilon^n$; then by equating, in Eq. (3.115), terms of equal order, we find

$$i(\partial/\partial t)U_{n+1} = H_I U_n. \tag{3.118}$$

Since $U_0 = \bar{1}$, we have

$$i(\partial/\partial t)U_1 = H_I, \qquad i(\partial/\partial t)U_2 = H_I U_1, \quad \text{etc.}$$

and, therefore,

$$\begin{aligned} U_1 &= -i \int_{t_0}^{t} H_I(t')\, dt', \\ U_2 &= -i \int_{t_0}^{t} H_I(t'')U_1(t'')\, dt'' \\ &= (-i)^2 \int_{t_0}^{t} dt'' \int_{t_0}^{t''} dt' H_I(t'')H_I(t'), \\ U_n &= (-i)^n \int_{t_0}^{t} dt_n \int_{t_0}^{t_n} dt_{n-1} \cdots \int_{t_0}^{t_2} dt_1 \cdot H_I(t_n)H_I(t_{n-1}) \cdots H_I(t_1). \end{aligned} \tag{3.119}$$

Notice that the products of the $H_I$'s in this equation are always time ordered; that is, the time argument of each $H_I$ is *later* than that of the $H_I$ that appears to the right of it. One can use this fact to extend the regions of the $t$-integrations:

$$U_n = \frac{(-i)^n}{n!} \int_{t_0}^{t} dt_n \int_{t_0}^{t} dt_{n-1} \cdots \int_{t_0}^{t} dt_1 \, T[H_I(t_n)H_I(t_{n-1}) \cdots H_I(t_1)], \tag{3.120}$$

where the *time-ordering operator* $T$ puts its arguments into the proper time order:

$$T[A(t_1)B(t_2)] \equiv \begin{cases} A(t_1)B(t_2), & t_1 > t_2, \\ B(t_2)A(t_1), & t_2 > t_1, \end{cases} \qquad (3.121)$$
$$\equiv \theta(t_1 - t_2)A(t_1)B(t_2) + \theta(t_2 - t_1)B(t_2)A(t_1),$$

etc. The $1/n!$ appears in Eq. (3.120) because extending the range of integrations means that we are doing the original integral $n!$ times over, renaming the variables of integration each time. (Figure 3.1 shows the regions of integration for $n = 2$.)

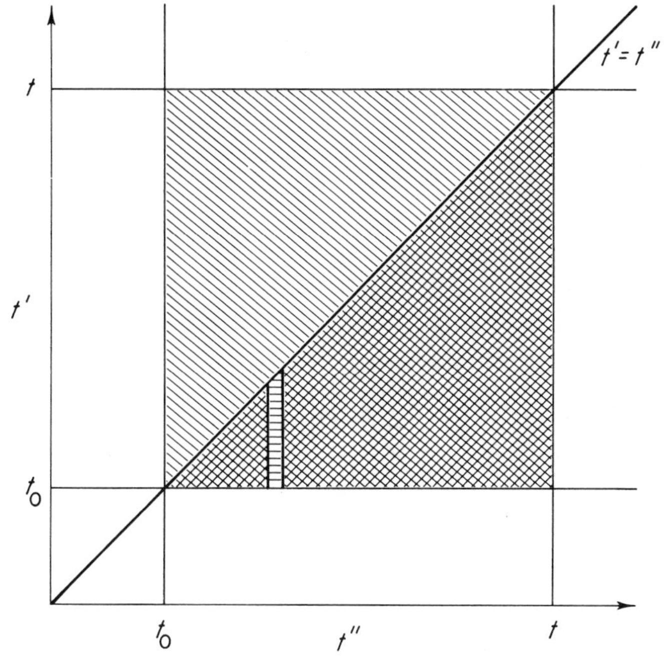


**Fig. 3.1.** Integration regions for Eq. (3.119) (triangle) and for Eq. (3.120) (square).

Thus our perturbation expansion for $U$ is

$$U = \bar{1} + \sum_{n=1}^{\infty} \frac{(-i)^n}{n!} \int \cdots \int_{t_0}^{t} dt_n \cdots dt_1 \, T[H_I(t_n) \cdots H_I(t_1)] \qquad (3.122)$$

which we can write *formally* as

$$U = T \exp\left[ -i \int_{t_0}^{t} H_I(t_1)\, dt_1 \right]. \qquad (3.122')$$

If we now take the limits $t \to +\infty$, $t_0 \to -\infty$, we have the explicit (perturbation theory) expression of $S$ in terms of $H$ quoted above:

$$S = \bar{1} + \sum_{n=1}^{\infty} \frac{(-i)^n}{n!} \int_{-\infty}^{\infty} \cdots \int dt_n \cdots dt_1 \, T[H_\text{I}(t_n) \cdots H_\text{I}(t_1)]$$

$$= T \exp\left[-i \int_{-\infty}^{\infty} H_\text{I}(t)\, dt\right]. \tag{3.117}$$

Finally, we remark that $H_\text{I}(t)$ can often be written as the integral of a scalar invariant[10] *Hamiltonian density* $\mathcal{H}_\text{I}(x)$:

$$H_\text{I}(t) = \int d^3\mathbf{x}\, \mathcal{H}_\text{I}(x), \tag{3.123}$$

so that Eq. (3.117) can be expressed in the covariant form ($d^4x \equiv d^3\mathbf{x}\, dt$),

$$S = T \exp\left[-i \int \mathcal{H}_\text{I}(x)\, d^4x\right]. \tag{3.124}$$

[Note that since a Lorentz transformation can change the temporal order of two space–time points if their separation is spacelike, this can result in changing the order of operators inside a time-ordering bracket. This does not matter, however, since we are always going to require that our operators commute for spacelike separations:

$$[A(x), B(x')] = 0 \qquad \text{if} \quad |\mathbf{x} - \mathbf{x}'| > |t - t'|.$$

This condition is known in the trade as *microcausality*; a measurement at $x$ can affect a measurement at $x'$ only if $x'$ is in the *absolute future* of $x$; i.e., a light signal, or something slower, can travel from $x$ to $x'$.]

One can show that Eq. (3.124) is equivalent to ordinary time-dependent perturbation theory. Hence we merely point out that the *Born approximation* results from discarding everything after the first-order term in Eq. (3.124):

$$S_\text{Born} = \bar{1} - i \int \mathcal{H}_\text{I}(x)\, d^4x. \tag{3.125}$$

In order to get an idea of how Eq. (3.124) or (3.125) is used for electromagnetic or weak interactions, we have to turn to how $\mathcal{H}_\text{I}(x)$ is formed out of *field operators*. [Following standard usage, we will henceforth put 1 instead of $\bar{1}$ for the unit operator.]

## Problems

**3.1.** Prove that

$$\langle a|b\rangle^* = \langle b|a\rangle, \qquad \langle a|\lambda b\rangle = \lambda\langle a|b\rangle,$$

$$\langle \lambda a|b\rangle = \lambda^*\langle a|b\rangle, \qquad (A^\dagger)_{ij} = A^*_{ji},$$

$$(ABC\cdots)^\dagger = \cdots C^\dagger B^\dagger A^\dagger, \qquad (ABC\cdots)^{-1} = \cdots C^{-1}B^{-1}A^{-1}.$$

---

[10] For some kinds of coupling, there are some noncovariant terms in $\mathcal{H}_\text{I}$, but these do not contribute to $S_{\text{fi}}$.

**3.2.** Prove that the *eigenvalues* of a Hermitian operator are real; that the *expectation values* of a Hermitian operator are real; and that if $\lambda_1$ and $\lambda_2$ are eigenvalues of a Hermitian operator and $\lambda_1 \neq \lambda_2$, then the corresponding eigenvectors are orthogonal.

**3.3.** Show that if an infinitesimal unitary transformation brings about a change $\delta X$ in an operator, then

$$\delta X = -i\varepsilon[X, G],$$

where $G$ is the generator of the transformation. What does this give when the transformation is time evolution?

**3.4.** Prove Eq. (3.28) using infinitesimal $\Delta t$'s.

**3.5.** Verify that the spin matrices [Eqs. (3.48)] satisfy the angular momentum commutation relations [Eq. (3.52)].

**3.6.** Show that Eq. (3.54) satisfies Eq. (2.2″) asymptotically (i.e., throwing away terms that fall off faster than $1/r$, for $r \to \infty$).

**3.7.*** (a) Show that

$$\mathbf{L}^2 \equiv L_x{}^2 + L_y{}^2 + L_z{}^2 = -\frac{1}{\sin\theta}\frac{\partial}{\partial\theta}\left(\sin\theta\frac{\partial}{\partial\theta}\right) - \frac{1}{\sin^2\theta}\frac{\partial^2}{\partial\phi^2}.$$

(b) Using the fact that $P_l(x)$ satisfies Legendres' equation,

$$\frac{d}{dx}(1 - x^2)\frac{dP_l(x)}{dx} + l(l+1)P_l(x) = 0,$$

show that

$$\mathbf{L}^2 P_l(\cos\theta) = l(l+1)P_l(\cos\theta),$$

thus identifying $l$ with orbital angular momentum.

**3.8.** Find the total elastic cross section for scattering from a large black sphere (radius $R$) using the approximation that $\eta_l = 0$ for $l < kR$, and $\eta_l = 1$, $\delta_l = 0$ for $l > kR$.

**3.9.** For the scattering of a single nonrelativistic particle from a fixed, static potential, we write, instead of Eq. (3.75),

$$S_{\mathrm{fi}} = \delta_{\mathrm{fi}} + i2\pi\,\delta(E_{\mathrm{i}} - E_{\mathrm{f}})M_{\mathrm{fi}},$$

with no momentum $\delta$-function. Carry through the argument to find *Fermi's Golden Rule 2*:

$$\sum_{\mathrm{f}} \mathrm{rate}_{\mathrm{fi}} = 2\pi \sum_{\mathrm{f}} \delta(E_{\mathrm{i}} - E_{\mathrm{f}})|M_{\mathrm{fi}}|^2 = 2\pi|M_{\mathrm{fi}}|^2\,(dN/dE),$$

and show that ($\delta$-function normalization)

$$\sigma_{\mathrm{fi}} = (2\pi)^4 m^2 \int |\overline{M}_{\mathrm{fi}}|^2 \, d\Omega,$$

where $m$ is the particle mass.

**3.10.** Show that Eq. (3.96b) can be rewritten as

$$d\sigma_{\mathrm{fi}}/dt = (16\pi\lambda_{s12})^{-1} |T_{\mathrm{fi}}|^2.$$

(Hint: See Chapter 1, Problem 1.12.)

**3.11.** Show that, as an alternative to Eq. (3.98), $I_3$ can be expressed as

$$I_3 = \tfrac{1}{8} \int (p_1 E_2/E_3) \, d\Omega_1 \, d\Omega_2 \, dE_2 .$$

**3.12.** For a single particle scattering from a fixed external potential $V(\mathbf{r})$, use the Born approximation [Eq. (3.125)] to evaluate $\overline{M}_{\mathrm{fi}}$.

## Bibliography

W. A. Blanpied. *Modern Physics*. Holt, New York, 1971.

P. T. Matthews. *Introduction to Quantum Mechanics*. McGraw-Hill, New York, 1968.

H. Muirhead. *Physics of Elementary Particles*. Chaps, 6, 7. Pergamon, Oxford, 1965.

J. L. Powell and B. Crasemann. *Quantum Mechanics*. Chap. 8. Addison-Wesley, Reading, Massachusetts, 1961.

J. J. Sakurai. *Invariance Principles and Elementary Particles*. Chap. 2. Princeton Univ. Press, Princeton, New Jersey, 1964.

K. Ziock. *Basic Quantum Mechanics*. Chap. 13. Wiley, New York, 1969.

CHAPTER FOUR

# Quantum Field Theory

## 4.1. Second Quantization

In particle physics, it is quite common to have particles created or destroyed; so it seems very natural (and necessary) to introduce *creation*- and *destruction-operators* to describe these processes. This procedure, called *second quantization*, is also very useful in other areas of quantum physics, because of the great resulting simplification in the handling of several (or many) identical particles. The resulting theory is known as *quantum field theory*; the particles are the *quanta* of the field(s) in the same way that the photons are the quanta of the electromagnetic field. The goal of the theory, as far as particle physics is concerned, is to predict, from a few basic principles, $S$-matrix elements and hence cross sections which can be compared with experiment.

In this chapter we attempt to give the reader a brief, skeletal introduction to quantum field theory, mainly so that he will understand what a field operator is and how the scheme for calculating $S$-matrix elements emerges. He is warned that anything like mastery of the subject will require much further study.

### *a. Creation and Destruction Operators*

The basic step is to invent operators $a_i^\dagger$ and $a_i$ which, respectively, *create* and *destroy* particles in Heisenberg state *i*. For bosons we postulate the commutation relations

$$[a_i, a_j^\dagger] = \delta_{ij},$$
$$[a_i, a_j] = [a_i^\dagger, a_j^\dagger] = 0. \tag{4.1}$$

These commutation relations and the *number operator*

$$N_i \equiv a_i^\dagger a_i \tag{4.2}$$

lead to the creation and destruction interpretation of $a_i^\dagger$ and $a_i$.

To see this, consider a Heisenberg state of identical free (i.e., noninteracting) bosons, with $n_i$ particles in each *single-particle state* $i$ : $|n_1 \cdots n_i \cdots\rangle$. This state must be an eigenstate of each number operator $N_i$:

$$N_i|n_1 \cdots n_i \cdots\rangle = n_i|n_1 \cdots n_i \cdots\rangle. \tag{4.3}$$

Now Eqs. (4.1) and (4.2) imply

$$[N_i, a_i^\dagger] = a_i^\dagger,$$

and

$$[N_i, a_i] = -a_i. \tag{4.4}$$

Hence

$$N_i(a_i^\dagger|n_1 \cdots n_i \cdots\rangle) = (a_i^\dagger N_i + a_i^\dagger)|n_1 \cdots n_i \cdots\rangle$$
$$= (n_i + 1)(a_i^\dagger|n_1 \cdots n_i \cdots\rangle);$$

that is, $a_i^\dagger|n_1 \cdots n_i \cdots\rangle$ is an eigenstate of $N_i$, with eigenvalue $n_i + 1$; so we see that $a_i^\dagger$ increases $n_i$ by one. Similarly,

$$N_i(a_i|n_1 \cdots n_i \cdots\rangle) = (n_i - 1)|n_1 \cdots n_i \cdots\rangle;$$

so $a_i$ decreases $n_i$ by one. Denoted by $|0\rangle$, the state with

$$n_1 = \cdots = n_i = 0$$

is called the *vacuum state*. Since

$$0 = \langle 0|N_i|0\rangle = (a_i|0\rangle)^\dagger a_i|0\rangle,$$

it must be true that

$$a_i|0\rangle = 0,$$

and hence that the lowest eigenvalue of $N_i$ is zero.

### b. Field Operators. Neutral Spin Zero

Suppose we are interested in neutral spin zero particles (mass $m$); hence the Klein–Gordon equation (Chapter 2, Section 2.1) is appropriate. We choose the single particle states to be momentum eigenstates $|\mathbf{k}\rangle$ so that the index $i$ becomes $\mathbf{k}$ and the $\delta_{ij}$ becomes $\delta(\mathbf{k} - \mathbf{k}')$:

$$[a_{\mathbf{k}}, a_{\mathbf{k}'}^{\dagger}] = \delta(\mathbf{k} - \mathbf{k}'),$$

$$[a_{\mathbf{k}}, a_{\mathbf{k}'}] = [a_{\mathbf{k}}{}^{\dagger}, a_{\mathbf{k}'}^{\dagger}] = 0. \tag{4.1'}$$

By inserting the appropriate space and time dependence, we can construct a (Heisenberg) field operator $\phi(x)$ which satisfies the Klein–Gordon equation:

$$\phi(x) \equiv \phi(\mathbf{x}, t) = (2\pi)^{-3/2} \sum_{\mathbf{k}} [1/(2\omega_{\mathbf{k}})^{1/2}]\{a_{\mathbf{k}} e^{ikx} + a_{\mathbf{k}}{}^{\dagger} e^{-ikx}\}, \tag{4.5}$$

and $(\Box - m^2)\phi(x) = 0$. Here $kx \equiv \mathbf{k} \cdot \mathbf{x} - \omega_{\mathbf{k}} t$, with $\omega_{\mathbf{k}} = (\mathbf{k}^2 + m^2)^{1/2}$, $\sum_{\mathbf{k}}$ really means $\int d^3\mathbf{k}$, and $(3\pi)^{-3/2}(2\omega_{\mathbf{k}})^{-1/2}$ is a normalizing factor. Note that we have put *both* creation and destruction operators into $\phi(x)$, in order that it be Hermitian; the time dependence previously associated with negative energies, $e^{i\omega_k t}$, now multiplies the creation operators $a_{\mathbf{k}}{}^{\dagger}$. We think of $\phi(x)$ as an operator which, at time $t$, creates and destroys particles at position $\mathbf{x}$.

Now the Klein–Gordon equation can always be derived from the appropriate Hamiltonian by use of the field generalization of *Hamilton's equations*:

$$\dot{p} = -(\partial H/\partial q), \qquad \dot{q} = +(\partial H/\partial p).$$

The dynamical variables in the Klein–Gordon field case are $\phi(x)$ and

$$\pi(x) \equiv \partial\phi(x)/\partial t; \tag{4.6}$$

so

$$\dot{\pi}(x) = -[\partial H/\partial\phi(x)], \qquad \dot{\phi}(x) = [\partial H/\partial\pi(x)]. \tag{4.7}$$

The Hamiltonian is given in terms of a *Hamiltonian density*, $\mathcal{H}(x)$:

$$H = \int d^3\mathbf{x}\, \mathcal{H}(x), \tag{4.8}$$

and

$$\mathcal{H}(x) = \tfrac{1}{2}\{(\nabla\phi)^2 + m^2\phi^2 + \pi^2\}. \tag{4.9}$$

Equations (4.7)–(4.9) imply the Klein–Gordon equation for $\phi(x)$.

By using the expansion for $\phi(x)$, Eq. (4.5), in Eqs. (4.6), (4.8), and (4.9), one can verify that[1]

$$H = \sum_{\mathbf{k}} (N_{\mathbf{k}} + \tfrac{1}{2})\omega_k . \tag{4.10}$$

Because the eigenvalues of $N_{\mathbf{k}}$ are never negative, this equation insures that there are no negative energy states; hence the negative energy difficulty is cured once $\phi$ is a field operator.

Equation (4.10) implies that the vacuum state has an infinite *zero-point energy*:

$$\langle 0|H|0\rangle = \sum_{\mathbf{k}} \tfrac{1}{2}\omega_k .$$

But since only energy *differences* are significant, this is no problem. In fact, we can make the vacuum energy *zero* by putting the operators in Eq. (4.9) into *normal order*. This means that we divide $\phi$ into its creation and destruction parts:

$$\phi = \phi_+ + \phi_-,$$

with

$$\phi_+ = (2\pi)^{-3/2} \sum_{\mathbf{k}} [1/(2\omega_k)^{1/2}] a_{\mathbf{k}}^{\dagger} e^{-ikx},$$

and

$$\phi_- = (2\pi)^{-3/2} \sum_{\mathbf{k}} [1/(2\omega_k)^{1/2}] a_{\mathbf{k}} e^{ikx}, \tag{4.11}$$

and rearrange the pieces of any product like $\phi^2$ so that any destruction part appears to the right of any creation part. Using : : to indicate normal order, we have

$$:\phi^2: = \phi_+\phi_+ + \phi_-\phi_- + 2\phi_+\phi_- . \tag{4.12}$$

But this implies

$$\langle 0| :\phi^2: |0\rangle = 0,$$

and hence

$$\langle 0|H|0\rangle = 0.$$

[1] The alert reader will object that, since we are actually using $\delta$-function normalization, the extra term in Eq. (4.10) should really be $\frac{1}{2}\,\delta(0)\omega_k$. By resorting to unit normalization in a box, we can see that this merely says that the vacuum energy is proportional to the volume $V$. Of course, we do not really care how big this term is.

### c. Charged Scalar Particles

Now we drop the requirement that $\phi$ be Hermitian, and so instead of Eq. (4.5) we write

$$\phi(x) = (2\pi)^{-3/2} \sum_{\mathbf{k}} [1/(2\omega_k)^{1/2}]\{a_{\mathbf{k}} e^{ikx} + b_{\mathbf{k}}^{\dagger} e^{-ikx}\}. \tag{4.13}$$

Here $b_{\mathbf{k}}^{\dagger}$ is the creation operator for the *antiparticle* of the particle created by $a_{\mathbf{k}}^{\dagger}$; $b_{\mathbf{k}}$ is the antiparticle destruction operator. In addition to Eqs. (4.1′), we require

$$[b_{\mathbf{k}}, b_{\mathbf{k}'}^{\dagger}] = \delta(\mathbf{k} - \mathbf{k}'),$$

$$[b_{\mathbf{k}}, b_{\mathbf{k}'}] = [b_{\mathbf{k}}^{\dagger}, b_{\mathbf{k}'}^{\dagger}] = 0,$$

and

$$[a_{\mathbf{k}}, b_{\mathbf{k}'}] = [a_{\mathbf{k}}, b_{\mathbf{k}'}^{\dagger}] = \cdots = 0. \tag{4.14}$$

Equation (4.13) implies that $\phi$ destroys particles and creates antiparticles and, correspondingly, that $\phi^{\dagger}$ creates particles and destroys antiparticles.

The Hamiltonian density is now

$$\mathcal{H} = \nabla\phi^{\dagger}\cdot\nabla\phi + m^2\phi^{\dagger}\phi + \pi^{\dagger}\pi, \tag{4.15}$$

which leads to

$$H = \sum_{\mathbf{k}} (N_{\mathbf{k}}^{(a)} + N_{\mathbf{k}}^{(b)} + 1)\omega_k. \tag{4.16}$$

Here $N_{\mathbf{k}}^{(a)} \equiv a_{\mathbf{k}}^{\dagger} a_{\mathbf{k}}$ is the number operator for the particles, $N_{\mathbf{k}}^{(b)} \equiv b_{\mathbf{k}}^{\dagger} b_{\mathbf{k}}$ that for the antiparticles. As before, the zero point energy can be removed by putting the operators in $\mathcal{H}$ into normal order.

To compute the charge $Q$ (except for the factor of $e$), we use Eq. (2.9), and find

$$Q = \int \rho\, d^3\mathbf{x} = \sum_{\mathbf{k}} [N_{\mathbf{k}}^{(a)} - N_{\mathbf{k}}^{(b)}]. \tag{4.17}$$

Thus the particles have positive charge and the antiparticles negative charge. Hence $\phi$ lowers the charge by one, while $\phi^{\dagger}$ raises it by one; the combination $\phi^{\dagger}\phi$ conserves charge.

### d. Dirac Particles

Analogous to $\phi(x)$, the Dirac spinor $\psi(x)$ can be made into a field operator and can be expanded in terms of creation and destruction operators and plane wave solutions to the Dirac equation:

$$\psi(x) = (2\pi)^{-3/2} \sum_{\mathbf{k}} (m/E_k)^{1/2}\{c_{\mathbf{k}\sigma} u(\mathbf{k}, \sigma) e^{ikx} + d_{\mathbf{k}\sigma}^{\dagger} v(\mathbf{k}, \sigma) e^{-ikx}\}. \tag{4.18}$$

Here $c_{\mathbf{k}\sigma}$ is the destruction operator for the (spin one-half) fermion with momentum $\mathbf{k}$ and spin $\sigma$, described by the spinor $u(\mathbf{k}, \sigma)$; $d^\dagger_{\mathbf{k}\sigma}$ is the creation operator for the *antifermion*, described by the negative energy spinor $v(\mathbf{k}, \sigma)$. Correspondingly, $\bar{\psi}(x)$ creates fermions and destroys antifermions:

$$\bar{\psi}(x) = (2\pi)^{-3/2} \sum_{\mathbf{k}} (m/E_k)^{1/2} \{c^\dagger_{\mathbf{k}\sigma} \bar{u}(\mathbf{k}, \sigma) e^{-ikx} + d_{\mathbf{k}\sigma} \bar{v}(\mathbf{k}, \sigma) e^{ikx}\}. \tag{4.19}$$

The Dirac equation for these field operators can be derived from the Hamiltonian [cf. Eq. (2.33)]:

$$H = \int \mathcal{H}\, d^3\mathbf{x} = \int \{\bar{\psi}(x)(\gamma \cdot \nabla + m)(\psi x)\, d^3\mathbf{x}. \tag{4.20}$$

However, there is a vital difference from the spin-zero formalism. If one imposes commutation relation like Eqs. (4.1) and (4.14), one finds (among other difficulties) that the Hamiltonian [Eq. (4.20)] is *not* positive definite. To avoid this, we require *anticommutation relations*

$$\begin{aligned} \{c_{\mathbf{k}\sigma}, c^\dagger_{\mathbf{k}'\sigma'}\} &= \delta(\mathbf{k} - \mathbf{k}')\delta_{\sigma\sigma'}, \\ \{d_{\mathbf{k}\sigma}, d^\dagger_{\mathbf{k}'\sigma'}\} &= \delta(\mathbf{k} - \mathbf{k}')\delta_{\sigma\sigma'}, \end{aligned} \tag{4.21}$$

all other *anticommutators* vanishing. On the other hand, the *commutation* relations between the number operators and the creation and destruction operators are the familiar ones:

$$\begin{aligned} N^{(c)}_{\mathbf{k}\sigma} &\equiv c^\dagger_{\mathbf{k}\sigma} c_{\mathbf{k}\sigma}, \\ [N^{(c)}_{\mathbf{k},\sigma}, c_{\mathbf{k}\sigma}] &= -c_{\mathbf{k}\sigma}, \end{aligned} \tag{4.22}$$

etc., using Eqs. (4.21). Hence the interpretation of the operators is just the same.

Since the anticommutation relations imply

$$(c^\dagger_{\mathbf{k}\sigma})^2 = 0,$$

a given single-particle state cannot be occupied by more than one particle; this is the *Pauli exclusion principle*. Furthermore, for $\mathbf{k}\sigma \neq \mathbf{k}'\sigma'$ the relation

$$c^\dagger_{\mathbf{k}\sigma} c^\dagger_{\mathbf{k}'\sigma'} = -c^\dagger_{\mathbf{k}'\sigma'} c^\dagger_{\mathbf{k}\sigma}$$

implies *antisymmetry* under exchange of the identical fermion labels, i.e., *Fermi–Dirac statistics*. This is in contrast to the spin-zero case discussed earlier, where for $\mathbf{k} \neq \mathbf{k}'$ we have

$$c_{\mathbf{k}}^\dagger c^\dagger_{\mathbf{k}'} = +c^\dagger_{\mathbf{k}'} c_{\mathbf{k}}^\dagger,$$

which means *symmetry* under exchange of the identical boson labels, i.e., *Bose–Einstein statistics*. In fact the relation between spin and statistics:

$$\begin{aligned} \text{integral spin} &\Leftrightarrow \text{Bose–Einstein statistics}, \\ \text{odd half-integral spin} &\Leftrightarrow \text{Fermi–Dirac statistics}, \end{aligned}$$

is characteristic of *relativistic* quantum field theory and has been proved very generally.

Using these anticommutation relations, then, and dropping the infinite vacuum terms (i.e., specifying normal ordering[2] for $H$ and $Q$), one finds

$$H = \sum_{\mathbf{k},\sigma} (N_{\mathbf{k}\sigma}^{(c)} + N_{\mathbf{k}\sigma}^{(d)}) E_k, \tag{4.23}$$

and

$$Q = \sum_{\bar{\mathbf{k}}\sigma} (N_{\bar{\mathbf{k}}\sigma}^{(c)} - N_{\bar{\mathbf{k}}\sigma}^{(d)}). \tag{4.24}$$

The latter relation shows that the antifermions have opposite charge from the fermions, as we expect.

### e. Electromagnetic Field

The case of the electromagnetic field is more difficult than those we have already discussed; here we merely mention some of the main points. Because the four components of the vector potential $A_\mu$ cannot represent independent degrees of freedom (for a given $\mathbf{k}$ there are only *two* independent polarizations for a photon), in quantizing the electromagnetic field one must either give up explicit covariance, or else introduce spurious degrees of freedom.

As discussed in Chapter 2, Section 2.3, the polarization information is carried by polarization unit vectors $\varepsilon_\mu(k)$, which because of the Lorentz condition satisfy

$$k_\mu \varepsilon_\mu(k) = 0, \tag{2.55}$$

and we can also require

$$\mathbf{k} \cdot \boldsymbol{\varepsilon} = 0 \tag{2.56}$$

in any particular reference frame. In that frame, the expansion in terms of photon creation and annihilation operators that we need is

$$\mathbf{A}(x) = (2\pi)^{-3/2} \sum_{\mathbf{k},i} \frac{1}{(2\omega_k)^{1/2}} \boldsymbol{\varepsilon}_i(k)\{a_{\mathbf{k}i} e^{ikx} + a_{\mathbf{k}i}^\dagger e^{-ikx}\}. \tag{4.25}$$

(Because the classical vector potential is real, the field operator $\mathbf{A}$ must be Hermitian; hence the photon is its own antiparticle.) Here the $a_{\mathbf{k}i}$ and $a_{\mathbf{k}i}^\dagger$ satisfy commutation relations

$$\begin{aligned} [a_{\mathbf{k}i}, a_{\mathbf{k}j}^\dagger] &= \delta_{ij}\, \delta(\mathbf{k} - \mathbf{k}'), \\ [a_{\mathbf{k}i}, a_{\mathbf{k}j}] &= [a_{\mathbf{k}i}^\dagger, a_{\mathbf{k}j}^\dagger] = 0. \end{aligned} \tag{4.26}$$

Also since $\Box A_\mu = 0$, $\omega_k = |\mathbf{k}|$.

---

[2] In the normal ordering, there is an extra minus sign for every odd permutation of Dirac operators, to go with the *anticommutation* relations.

## 4.2. Interaction and Scattering

### *a. Interaction Terms*

In all of the foregoing discussion, the most interesting aspect, the *interactions between particles*, has been left out, and we have been dealing with *free particles* and *free-field equations.* Interaction is introduced by inserting additional *interaction terms* in the Hamiltonian. Thus the *total* Hamiltonian, when there are several different kinds of particles present, is the sum of free particle Hamiltonians *plus* the interaction part:

$$H = H_1 + H_2 + \cdots + H' = H_0 + H'.$$

The interaction Hamiltonian $H'$ can be written in terms of the Lorentz-invariant[3] interaction Hamiltonian density $\mathcal{H}'$:

$$H' = \int \mathcal{H}' \, d^3\mathbf{x}.$$

In order to calculate $S$-matrix elements from a given $\mathcal{H}'$, we use the perturbation theory developed in Chapter 3, Section 3.5. This means working in the interaction representation, where the operators have the same time dependence as the corresponding Heisenberg operators in the absence of interaction, i.e., the same time dependence as the free-field operators. Hence the interaction representation field operators satisfy the free-field equations of motion, i.e., Klein–Gordon equation, Dirac equation, etc. Furthermore, one can readily verify that commutation (anticommutation) relations are also the same for the interaction representation creation and destruction operators; so the interpretation is unchanged. In addition, operators corresponding to *different* particles commute (or anticommute if both correspond to fermions). In the following, and whenever we are doing perturbation theory, the field operators and the creation and destruction operators should be understood to be interaction representation operators.

Thus, because the interaction $\mathcal{H}'(x)$ is, in general, a product of field operators with the *same* space–time argument $x$, in going over to the interaction representation,

---

[3] Actually, it is a closely related function, the *Lagrangian density* $\mathcal{L}'$, that is always invariant under Lorentz transformations. Under certain circumstances, $\mathcal{H}'$ *can* contain some noninvariant terms, but these do *not* contribute to the $S$-matrix elements; so we can simply ignore them.

$$\begin{aligned}
\mathcal{H}' &\to \mathcal{H}_{\mathrm{I}}, \\
\phi_1(x)\phi_2(x)\cdots &\to e^{iH_0t}\phi_1(x)\phi_2(x)\cdots e^{-iH_0t} \\
&= [e^{iH_0t}\phi_1(x)e^{-iH_0t}][e^{iH_0t}\phi_2(x)e^{-iH_0t}]\cdots \\
&= \phi_{1_{\mathrm{I}}}(x)\phi_{2_{\mathrm{I}}}(x)\cdots,
\end{aligned}$$

which is what we get simply by making the substitutions $\phi \to \phi_{\mathrm{I}}$ in $\mathcal{H}'$. We then drop the subscript I, leaving *interaction representation* to be understood.

In general, we will be concerned with interaction terms that are either *trilinear* in the field operators,

$$\mathcal{H}' = g\phi_1(x)\phi_2(x)\phi_3(x), \tag{4.27}$$

or *quadrilinear*,

$$\mathcal{H}' = g\phi_1(x)\phi_2(x)\phi_3(x)\phi_4(x), \tag{4.28}$$

where $\phi_1$, etc., stands for any type of field operator and $g$ is called a *coupling constant*. Since each field operator contains both creation and destruction parts,

$$\phi = \phi_+ + \phi_-,$$

the trilinear interaction term produces *trilinear vertices* of the type shown in Fig. 4.1. (In this diagram time $t$ increases upward and the abscissa is a spatial

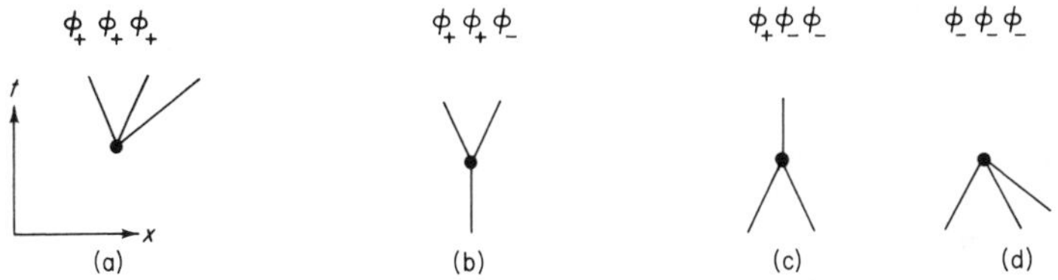


**Fig. 4.1.** Vertices implied by a *trilinear interaction*. A line emerging *upward* from the dot indicates a particle *created* at that point, *downward* indicates a particle *destroyed* at that point.

coordinate.) The vertex with no particles destroyed and three created at $x$ [Fig. 4.1(a)] comes from the $\phi_+\phi_+\phi_+$ product, one particle destroyed and two created [Fig. 4.1(b)] from the $\phi_+\phi_+\phi_-$ product, etc. In general, of course, it is necessary to include the labels which distinguish the different fields (particles). Similarly, from the quadrilinear interaction we get the *quadrilinear vertices* shown in Fig. 4.2.

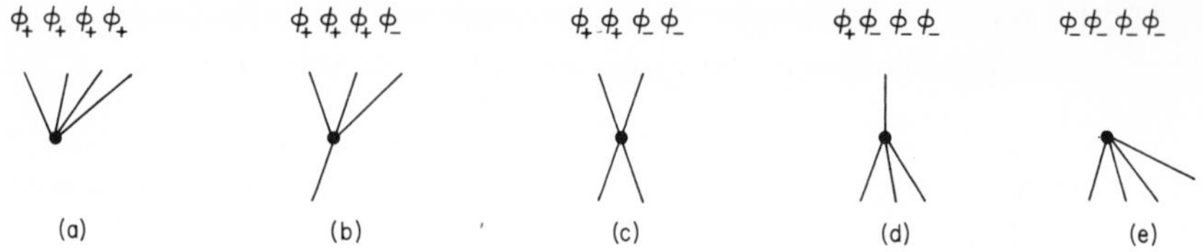


**Fig. 4.2.** Vertices implied by a *quadrilinear interaction.*

If we are dealing with Dirac particles, then to take care of the spinor indices and have an invariant $\mathcal{H}'$, we must use the bilinear covariant combinations $\bar{\psi}O\psi$. Since $\psi$ destroys fermions and creates antifermions while $\bar{\psi}$ does the opposite, this combination will always leave the *fermion number*, which is the number of fermions *minus* the number of antifermions, unchanged.

## *b. Calculating an S-matrix Element. Feynman Rules*

Let us now work through a specific example to see how one goes from $\mathcal{H}'$ to an $S$-matrix element using perturbation theory. Consider a (fictitious) system of interacting scalar mesons, which are all neutral but distinct; and are labeled $a$, $b$, and $c$. They are described by Klein–Gordon operators $\phi^a$, $\phi^b$, $\phi^c$, with a trilinear interaction:

$$\mathcal{H}'(x) = g\phi^a\phi^b\phi^c.$$

To make the vacuum energy vanish, we really mean the operators to be in normal order:

$$\begin{aligned}\mathcal{H}' &= g\!:\!\phi^a\phi^b\phi^c\!: = g\!:\!(\phi_+{}^a + \phi_-{}^a)(\phi_+{}^b + \phi_-{}^b)(\phi_+{}^c + \phi_-{}^c)\!:\\ &= g\{\phi_+{}^a\phi_+{}^b\phi_+{}^c + \phi_+{}^a\phi_+{}^b\phi_-{}^c + \phi_+{}^b\phi_+{}^c\phi_-{}^a + \phi_+{}^c\phi_+{}^a\phi_-{}^b\\ &\quad + \phi_+{}^a\phi_-{}^b\phi_-{}^c + \phi_+{}^b\phi_-{}^c\phi_-{}^a + \phi_+{}^c\phi_-{}^a\phi_-{}^b + \phi_-{}^a\phi_-{}^b\phi_-{}^c\},\end{aligned} \tag{4.29}$$

while $\mathcal{H}_{\mathrm{I}}$ has exactly the same form with all the operators understood to be interaction representation operators. To proceed, we use the perturbation expansion for $S$ of Chapter 3, Section 3.5:

$$S = 1 + \sum_{n=1}^{\infty} \frac{(-i)^n}{n!} \int_{-\infty}^{\infty} \cdots \int_{-\infty}^{\infty} dt_n \cdots dt_1 T\{H_{\mathrm{I}}(t_n) \cdots H_{\mathrm{I}}(t_1)\} \tag{3.117}$$

$$= 1 + \sum_{n=1}^{\infty} \frac{(-i)^n}{n!} \int \cdots \int d^4x_n \cdots d^4x_1 T\{\mathcal{H}_{\mathrm{I}}(x_n) \cdots \mathcal{H}_{\mathrm{I}}(x_1)\}$$

$$S \equiv 1 + \sum_{n=1}^{\infty} S_n. \tag{4.30}$$

At this point it is convenient to decide what process we are interested in. Suppose we consider elastic scattering, $ab \to ab$. Then automatically

$$\langle \mathrm{f} | S_1 | \mathrm{i} \rangle = 0;$$

so the lowest nontrivial term in the perturbation expansion for the $S$-matrix element is $\langle \mathrm{f} | S_2 | \mathrm{i} \rangle$.

In order to evaluate this matrix element, we must rearrange the field operators into normal order. First we write

$$T\{\mathcal{H}_\mathrm{I}(x)\mathcal{H}_\mathrm{I}(y)\} = \theta(x-y)\mathcal{H}_\mathrm{I}(x)\mathcal{H}_\mathrm{I}(y) + \theta(y-x)\mathcal{H}_\mathrm{I}(y)\mathcal{H}_\mathrm{I}(x), \qquad (4.31)$$

where

$$\theta(x-y) \equiv \theta(t_x - t_y),$$
$$x \equiv (\mathbf{x}, it_x), \quad \text{etc.},$$

and $\theta(t)$ is the *unit step function*:

$$\theta(t) = \begin{cases} 1, & t > 0, \\ 0, & t < 0. \end{cases} \qquad (4.32)$$

Next we note that all of the operators commute except for the pairs $\phi_+{}^a$, $\phi_-{}^a$, etc., and for these we have, using Eq. (4.11),

$$[\phi_-(x), \phi_+(y)] = (2\pi)^{-3} \sum_{\mathbf{k}, \mathbf{k}'} \frac{1}{(2\omega_k 2\omega_{k'})^{1/2}} [a_\mathbf{k}, a_{\mathbf{k}'}^\dagger] e^{i(kx-k'y)}$$
$$= \int \frac{d^3\mathbf{k}}{(2\pi)^3 2\omega_k} e^{ik(x-y)} \equiv i\Delta^{(+)}(x-y), \qquad (4.33)$$

so that

$$[\phi_-{}^a, \phi_+{}^a] = i\Delta_a^{(+)}, \quad \text{etc.}$$

$[k \equiv (\mathbf{k}, i\omega_k)]$. It is important to note that $i\Delta_a^{(+)}$ is *not* an operator, but simply a function of $x - y$. Thus taking the commutator gets rid of a pair of operators; this is known as *contraction*.

By using this and the obvious relation

$$AB = BA + [A, B],$$

we can write the product $\mathcal{H}_\mathrm{I}(x)\ \mathcal{H}_\mathrm{I}(y)$ in normal order. The result has many terms, and rather than writing them all out, we save just those which will contribute to the $ab \to ab$ matrix element; these *must* have the form $\phi_+{}^a\phi_+{}^b\phi_-{}^a\phi_-{}^b$, with the $\phi_+{}^c\phi_-{}^c$ having disappeared because of the $[\phi_-{}^c(x), \phi_+{}^c(y)]$ commutator. The only possible terms are

$$\begin{aligned}\langle \mathrm{f} | \mathcal{H}_\mathrm{I}(x)\mathcal{H}_\mathrm{I}(y) | \mathrm{i} \rangle = g^2 i\Delta_c^{(+)}(x-y) \langle \mathrm{f} | \{ &\phi_+{}^a(x)\phi_+{}^b(x)\phi_-{}^a(y)\phi_-{}^b(y) \\ &+ \phi_+{}^a(y)\phi_+{}^b(y)\phi_-{}^a(x)\phi_-{}^b(x) \\ &+ \phi_+{}^a(x)\phi_+{}^b(y)\phi_-{}^a(y)\phi_-{}^b(x) \\ &+ \phi_+{}^a(y)\phi_+{}^b(x)\phi_-{}^a(x)\phi_-{}^b(y)\} | \mathrm{i} \rangle.\end{aligned}$$

The operators in the { } can be pictured, respectively, as in Fig. 4.3, with the internal line being the $i\Delta_c^{(+)}(x-y)$ function. The internal line corresponds to a *virtual particle* of type $c$ going from $y$ to $x$.

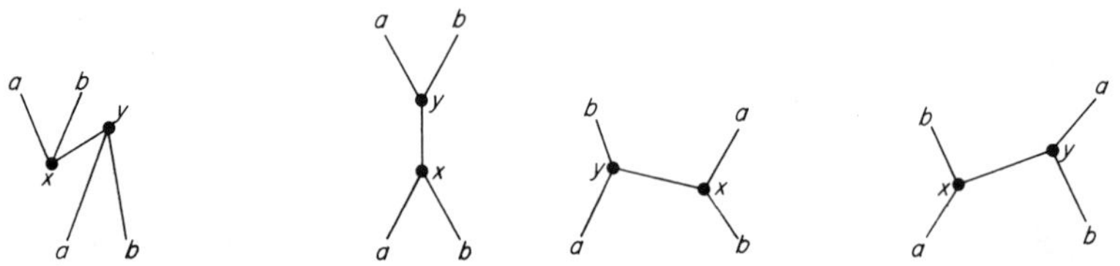


**Fig. 4.3.** The terms of Eq. (4.33′)($t_x < t_y$). Distinction is made between *internal lines* which represent contractions, and external lines (all the others) which will correspond to incoming and outgoing particles.

Now the *second* term of Eq. (4.31) is just the same except that $x \rightleftarrows y$; since the { } expression is unchanged under this switch,

$$\langle \mathrm{f} \,|\, \mathcal{H}_{\mathrm{I}}(y)\mathcal{H}_{\mathrm{I}}(x) \,|\, \mathrm{i}\rangle = g^2 i\Delta_c^{(+)}(y-x)\langle \mathrm{f} \,|\, \{\ \} \,|\, \mathrm{i}\rangle. \tag{4.34}$$

Putting this together as in Eq. (4.31), we have

$$\begin{aligned}\langle \mathrm{f} \,|\, T\{\mathcal{H}_{\mathrm{I}}(x)\mathcal{H}_{\mathrm{I}}(y)\} \,|\, \mathrm{i}\rangle = ig^2[&\theta(x-y)\Delta_c^{(+)}(x-y)\\ &+\theta(y-x)\Delta_c^{(+)}(y-x)]\langle \mathrm{f} \,|\{\ \}|\, \mathrm{i}\rangle.\end{aligned} \tag{4.35}$$

The function $i[\ ]$ occurs *whenever* we have an interaction coming from the exchange of particle $c$, and is known as the *Feynman propagator* $\Delta_{\mathrm{F}}{}^c$:

$$\begin{aligned}\Delta_{\mathrm{F}}{}^c(x-y) &\equiv i[\theta(x-y)\Delta_c^{(+)}(x-y)+\theta(y-x)\Delta_c^{(+)}(y-x)]\\ &= \theta(x-y)[\phi_-{}^c(x), \phi_+{}^c(y)] + \theta(y-x)[\phi_-{}^c(y), \phi_+{}^c(x)].\end{aligned} \tag{4.36}$$

Note that *both* time orderings are included in the Feynman propagator; hence diagrams with both time orderings, as in Fig. 4.3, are automatically being included when we use a $\Delta_{\mathrm{F}}(x-y)$.

Any $\Delta_{\mathrm{F}}(x-y)$ can also be written as the Fourier transform of the corresponding *momentum–space Feynman propagator* $\Delta_{\mathrm{F}}(k)$:

$$\Delta_{\mathrm{F}}(x) = (2\pi)^{-4}\int d^4k e^{ikx}\Delta_{\mathrm{F}}(k), \tag{4.37}$$

with

$$\Delta_{\mathrm{F}}(k) \equiv -i(k^2+m^2-i\varepsilon)^{-1}; \tag{4.38}$$

here $\varepsilon$ is a small, positive constant that is set to zero at the end of the calculation, $d^4k = d^3\mathbf{k}\,d\omega$ and $k = (\mathbf{k}, i\omega)$.

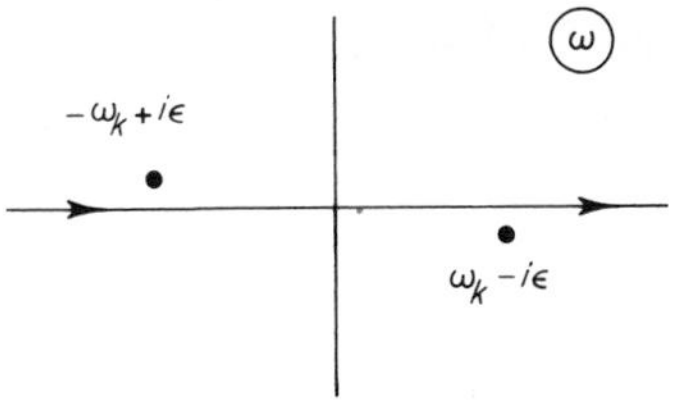


**Fig. 4.4.** Integration contour in the complex $\omega$ plane. The two poles lie just above and below the contour.

To see this, we do the energy integral in Eq. (4.37) first. In the *complex $\omega$-plane* the contour of integration lies along the real axis. Since

$$k^2 + m^2 = (\omega_k - \omega)(\omega_k + \omega)$$

$[\omega_k \equiv (\mathbf{k}^2 + m^2)^{1/2}]$, the integrand of Eq. (4.37) has *poles* slightly above and below the real axis, as shown in Fig. 4.4. Now

$$e^{ikx} = e^{i\mathbf{k}\cdot\mathbf{x} - i\omega t};$$

so for $t > 0$ the contour can be closed by a large semicircle in the *lower* half-plane. (The contribution of the semicircle $\to 0$ as the radius $\to \infty$.) This contour includes the lower, *positive energy pole*, and the *Cauchy residue theorem* says that

$$\int = + \frac{2\pi i(-i)}{(2\pi)^4} \int \frac{d^3\mathbf{k}}{2\omega_k} e^{i(\mathbf{k}\cdot\mathbf{x} - \omega_k t)} = i\Delta^+(x).$$

For $t < 0$ on the other hand, we must close the contour in the *upper* half-plane, enclosing the *negative energy pole*, and

$$\int = \frac{-2\pi i(-i)}{(2\pi)^4} \int \frac{d^3\mathbf{k}}{-2\omega_k} e^{i(\mathbf{k}\cdot\mathbf{x} + \omega_k t)}$$

$$= +(2\pi)^{-3} \int \frac{d^3\mathbf{k}}{2\omega_k} e^{-i(\mathbf{k}\cdot\mathbf{x} - \omega_k t)} = i\Delta^+(-x).$$

These results agree with Eq. (4.36).

The next step is to do the $x$ and $y$ integrations; here we notice that since $x$ and $y$ are now *dummy variables*, we can write

$$\begin{aligned} 1/2! \iint d^4x\, d^4y\, T\{\mathcal{H}_I(x)\mathcal{H}_I(y)\} \\ = ig^2 \iint d^4x\, d^4y\, \Delta_F{}^c(x-y) \langle f | \{\phi_+{}^a(x)\phi_+{}^b(x)\phi_-{}^a(y)\phi_-{}^b(y) \\ + \phi_+{}^a(x)\phi_+{}^b(y)\phi_-{}^a(y)\phi_-{}^b(x)\} | i \rangle. \end{aligned} \tag{4.39}$$

(In fact the $1/n!$ in $S_n$ will always disappear in just the same way.) At this point we introduce the expansions of the field operators, Eqs. (4.11), and label the momenta of the initial and final states $\mathbf{k}_a$, $\mathbf{k}_b$ and $\mathbf{k}_a'$, $\mathbf{k}_b'$, respectively. The surviving contribution is

$$\begin{aligned}\langle \mathrm{f}|\{\ \}|\mathrm{i}\rangle \equiv \langle \mathbf{k}_a\mathbf{k}_b'|\{\ \}|\mathbf{k}_a\mathbf{k}_b\rangle &= [(2\pi)^{-6}(16\omega_{k_a}\omega_{k_b}\omega_{k_a'}\omega_{k_b'})^{-1/2}] \\ &\times \{\exp[-i(k_a' + k_b')x + i(k_a + k_b)y] \\ &+ \exp[-i(k_a'x + k_b'y - k_a y - k_b x)]\}. \end{aligned} \tag{4.40}$$

When we use Eq. (4.37) for $\Delta_F{}^c(x - y)$, the $x$- and $y$-integrations become trivial, and just give $\delta$-functions:

$$\begin{aligned}\langle \mathbf{k}_a'\mathbf{k}_b'|S_2|\mathbf{k}_a\mathbf{k}_b\rangle &= (-i)^2 ig^2[(2\pi)^{-6}(16\omega_{k_a'}\omega_{k_b'}\omega_{k_a}\omega_{k_b})^{-1/2}] \\ &\times (2\pi)^{-4}\int d^4k\, \Delta_F{}^c(k)\{(2\pi)^4\,\delta(k - k_a' - k_b') \\ &\times (2\pi)^4\,\delta(k - k_a - k_b) + (2\pi)^4\,\delta(k - k_a' + k_b) \\ &\times (2\pi)^4\,\delta(k_a - k_b' - k)\}. \end{aligned} \tag{4.41}$$

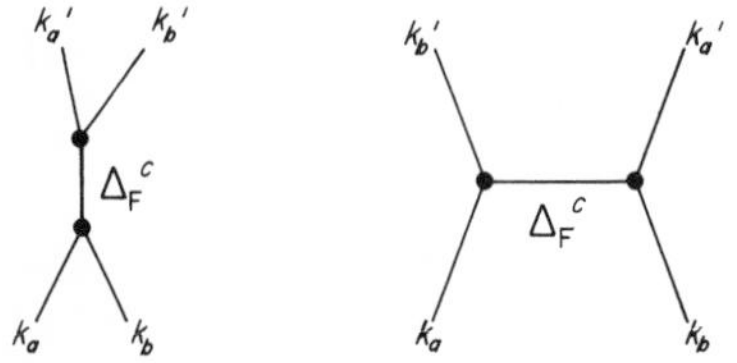


**Fig. 4.5.** *Feynman diagrams*, corresponding to the two contributions to the $S$-matrix element. Internal lines stand for Feynman propagators, external lines for incoming and outgoing particles.

This sum of two terms corresponds to the two diagrams of Fig. 4.5. While there remains the integration over the internal line four-momenta $k$, the $\delta$-function at each vertex imposes *energy-momentum conservation at each vertex*. Using the $\delta$-functions to evaluate the $\int d^4k$, we have

$$\begin{aligned}\langle \mathbf{k}_a'\mathbf{k}_b'|S_2|\mathbf{k}_a\mathbf{k}_b\rangle &= -ig^2[(2\pi)^{-6}(16\omega_{k_a'}\omega_{k_b'}\omega_{k_a}\omega_{k_b})^{-1/2}] \\ &\times (2\pi)^4\,\delta(k_a' + k_b' - k_a - k_b)\{\Delta_F{}^c(k_a + k_b) \\ &+ \Delta_F{}^c(k_a - k_b')\}. \end{aligned} \tag{4.42}$$

(The factor $[(2\pi)^{-6}(16\omega_{k_a'}\omega_{k_b'}\omega_{k_a}\omega_{k_b})^{-1/2}]$ is associated with the *$\delta$-function normalization* that we have been using; if instead we use *invariant normalization* of the states [see Eq. (3.87)], so that we are calculating the contribution to $i(2\pi)^4\,\delta(P_f - P_i)T_{fi}$, the [ ] factor disappears entirely.) This exhibits the overall energy-momentum $\delta$-function appropriate to $S$-matrix elements, and each (momentum–space) Feynman propagator has momentum argument appropriate to the internal line of the corresponding diagram of Fig. 4.5.

These diagrams, called *Feynman diagrams* or *Feynman graphs*, greatly facilitate writing down matrix elements of $S_n$ for *any* sort of interaction and *any* kind of particles. This is simply because the steps that take us from Eq.

(4.30) to the final matrix elements are essentially unchanged, except perhaps in degree of complexity. Thus to find a given matrix element of $S_n$, we first put down all of the appropriate Feynman diagrams, then write out the terms that correspond to each and sum. In general, the *vertex factors* depend on the form of the interaction $\mathcal{H}'$, and the Feynman propagators on the kind of particle being exchanged (i.e., the kind of field operators being contracted). For the external lines (initial- and final-state particles), there are factors of Dirac spinors, photon polarization vectors, etc.

The Dirac particle Feynman propagator comes from using the relation

$$\psi_{\alpha-}(x)\bar{\psi}_{\beta+}(y) = -\bar{\psi}_{\beta+}(y)\psi_{\alpha-}(x) + \{\psi_{\alpha-}(x), \bar{\psi}_{\beta+}(y)\} \tag{4.43}$$

to get the Dirac operators into normal order ($\alpha$ and $\beta$ indicate spinor components, and $+$ and $-$ subscripts indicate creation and destruction parts of the field operator). Because of the anticommutation relations, Eqs. (4.21), the *anticommutator* in Eq. (4.43) is a *function* and not an *operator*. One finds for the momentum–space propagator

$$S_F(k) = -i\,\frac{i\gamma_\mu k_\mu - m}{k^2 + m^2 - i\varepsilon} \equiv -i\,\frac{i\not{k} - m}{k^2 + m^2 - i\varepsilon}, \tag{4.44}$$

and

$$S_F(x-y) = (2\pi)^{-4} \int d^4k e^{ik(x-y)} S_F(k). \tag{4.45}$$

($\not{k} \equiv \gamma_\mu k_\mu$.) In an analogous manner, one finds for the photon propagator

$$[D_F(k)]_{\mu\nu} = -i\delta_{\mu\nu}/(k^2 - i\varepsilon)\,, \tag{4.46}$$

and

$$D_F(x-y) = (2\pi)^{-4} \int d^4k e^{ik(x-y)} D_F(k). \tag{4.47}$$

The procedure for getting the perturbation theory matrix elements is embodied in the *Feynman rules*, shown in Table 4.1. We will go through a number of examples illustrating some of the simpler aspects of the derivation and application of these rules.

Let us now use our result, Eq. (4.42), to calculate the *ab* elastic scattering cross section (to lowest order of perturbation theory). Writing

$$T_{fi} = -g^2\{\Delta_F{}^c(k_a + k_b) + \Delta_F{}^c(k_a - k_b')\},$$

and introducing the Mandelstam variables of Chapter 1, Section 1.2c,

$$T_{fi} = -ig^2\{(s - m^2 + i\varepsilon)^{-1} + (u - m^2 + i\varepsilon)^{-1}\}. \tag{4.48}$$

Thus, from Eq. (3.26b),

$$d\sigma/d\Omega = (g^4/64\pi^2 s)[(s - m^2)^{-1} + (u - m^2)^{-1}]^2. \tag{4.49}$$

**Table 4.1**

*Relativistic Feynman Rules for S-Matrix Elements*[a]

1. Write down all possible Feynman diagrams to a given order of the interaction for the process in question

For each diagram

2. Include a factor of $(2\pi)^4 g$ for each vertex (with momentum factors for derivative coupling and Dirac operators for fermions)
3. Include a factor $(2\pi)^{-4}\,\Delta_F(k)$ [or $S_F(k)$ or $D_F(k)$] for each internal line, remembering energy momentum conservation at each vertex
4. Include one "extra" δ-function for each unconnected part of a diagram (e.g., Fig. 4.6 has two unconnected parts)
5. Integrate over the four momentum of any closed loop, and include an extra minus sign for each closed fermion loop
6. Add the contributions of each diagram
7. Include a factor $(-i)^n$ ($n$ being the order of perturbation theory), and for invariant normalization $(2m)^{1/2}$ for each external fermion line[b]
8. Include external fermion line spinors and external photon line polarization vectors

[a] To compute $T_{fi}$, divide the corresponding $S_{fi}$ (invariant normalization) by $i(2\pi)^4\delta(P_i - P_f)$.

[b] For δ-function normalization, on the other hand, we need a factor of $(2\pi)^{-3/2}(2\omega_k)^{-1/2}$ for each external boson line, and a factor

$$(2\pi)^{-3/2}\,(m/E_k)^{1/2}$$

for each external fermion line [and, of course, the $(-i)^n$]. We are using the spinor normalization $\bar{u}u = 1$, for even massless fermions; the $m$ appearing in rule (7) will cancel the one in the $(E_k + m/2m)^{1/2}$ spinor normalization factor [cf. Eq. (2.32)].

**Fig. 4.6.** A Feynman diagram with two unconnected parts.

An expression like this can be compared with the experimental data for $d\sigma/d\Omega$, to evaluate $g^2$ and as a test of our assumption about the form of the interaction.

### c. *Minimal Electromagnetic Coupling*

As a start consider the interaction of electrons (or muons) with the electromagnetic field; this can be derived from the principle of *minimal electromagnetic coupling*. This results in an electric current operator

$$J_\mu^{\,e} \equiv ie\bar{\psi}\gamma_\mu\psi \tag{4.50}$$

which interacts with the four vector potential $A_\mu$:

$$\mathcal{H}'(x) = -A_\mu(x)J_\mu^{e}(x) = \phi\rho^{e} - \mathbf{A}\cdot\mathbf{j}^{e} \tag{4.51}$$

(just as for classical physics). Notice that the resulting vertices are of the trilinear type, but with $\gamma_\mu$'s appearing at each.

It is instructive to consider this interaction for a single electron moving *slowly* in an external potential $A_\mu(x)$. The interaction energy is given by the matrix element of $H'$ between one electron states:

$$\begin{aligned} E_{\text{int}} &= \langle \mathbf{p}_2 | H' | \mathbf{p}_1 \rangle \\ &= -ie\langle \mathbf{p}_2 | \int A_\mu(x)(:\bar{\psi}(x)\gamma_\mu\psi(x):)\, d^3\mathbf{x} | \mathbf{p}_1 \rangle \\ &= -[ie/(2\pi)^3] \int d^3\mathbf{x} A_\mu(x) e^{i(p_1-p_2)x}\bar{u}(p_2)\gamma_\mu u(p_1), \end{aligned}$$

with $\mathbf{p}_1$, $\mathbf{p}_2 \to 0$. Using the representation introduced, in Chapter 2, Section 2.2 [Eqs. (2.19)], and Eq. (2.32),

$$\bar{u}(p_2)\gamma_4 u(p_1) \to \chi_2^{\dagger}\chi_1,$$

while

$$\bar{u}(p_2)\gamma u(p_1) \to -(i/2m)\chi_2^{\dagger}(\boldsymbol{\sigma}\ \boldsymbol{\sigma}\cdot\mathbf{p}_1 + \boldsymbol{\sigma}\cdot\mathbf{p}_2\ \boldsymbol{\sigma})\chi_1.$$

Therefore, using

$$\boldsymbol{\sigma}\cdot\mathbf{a}\ \boldsymbol{\sigma}\cdot\mathbf{b} = \mathbf{a}\cdot\mathbf{b} + i\boldsymbol{\sigma}\cdot\mathbf{a}\mathrm{x}\mathbf{b},$$

$$\mathbf{A}\cdot[u(p_2)\gamma u(p_1)] \to (i/2m)\chi_2^{\dagger}[\mathbf{A}\cdot(\mathbf{p}_1+\mathbf{p}_2) + i\mathbf{A}\mathrm{x}(\mathbf{p}_1-\mathbf{p}_2)\cdot\boldsymbol{\sigma}]\chi_1.$$

Thus

$$E_{\text{int}} \simeq (2\pi)^3 \int d^3\mathbf{x} e^{i(p_1-p_2)x}$$

$$\times\chi_2^{\dagger}\left[e\phi(x) - e\frac{(\mathbf{p}_1+\mathbf{p}_2)}{2m}\cdot\mathbf{A}(x) - \left(\frac{e}{2m}\right)\boldsymbol{\sigma}\cdot\nabla\mathrm{x}\mathbf{A}(x)\right]\chi_1.$$

The first two terms on the right-hand side of this expression are the same as we would find using the Schrödinger equation for the electron; the first expresses the interaction of the charge density with the scalar potential $\phi(x)$, the second the interaction of the convective current with the vector potential. The third term represents the interaction of the electron's *magnetic moment*

$$\boldsymbol{\mu} = (e/2m)\boldsymbol{\sigma},$$

with the *magnetic field* $\mathbf{B} = \nabla\mathrm{x}\mathbf{A}$.

Thus we automatically have the correct *g-factor* for the electron spin:

$$\boldsymbol{\mu} = g(e/2m)\mathbf{s},$$

and $\mathbf{s} = \frac{1}{2}\boldsymbol{\sigma}$, so that $g = 2$. [Small $(O(e^2))$ corrections to $g$ will be discussed later.]

## Problems

**4.1.** Verify that $\phi(x)$ given by Eq. (4.5) satisfies the Klein–Gordon equation.

**4.2.** Show that Eqs. (4.7)–(4.9) together imply the Klein-Gordon equation for $\phi(x)$.

**4.3.** Derive Eq. (4.10).

**4.4.** Use Eqs. (4.7) to evaluate $\dot{\pi}\,(x)$ and derive the Klein–Gordon equation for $\phi(x)$ from the Hamiltonian density for the charged Klein–Gordon field, Eq. (4.15).

**4.5.** Derive Eqs. (4.16) and (4.17).

**4.6.** Using the expansion for $\phi(x)$, Eq. (4.13), show that the charge Klein–Gordon field operators satisfy *equal-time commutation relations*

$$\begin{aligned}[\phi(\mathbf{x}, t), \phi(\mathbf{y}, t)] &= [\pi(\mathbf{x}, t), \pi(\mathbf{y}, t)] \\ &= [\phi(\mathbf{x}, t), \pi^\dagger(\mathbf{y}, t)] = \cdots = 0, \\ [\phi(\mathbf{x}, t), \pi(\mathbf{y}, t)] &= [\phi^\dagger(\mathbf{x}, t), \pi^\dagger(\mathbf{y}, t)] = i\delta(\mathbf{x} - \mathbf{y}).\end{aligned}$$

**4.7.** Using the second of Eqs. (4.7) show that Hamiltonian, Eq. (4.20), and the choice $\pi = i\psi^\dagger$ yield the Dirac equation.

**4.8.** By inserting the expansions, Eqs. (4.18) and (4.19), into the Dirac Hamiltonian, Eq. (4.20), find the expression for $H$ in terms of the number operators, using (a) commutation relations, (b) anticommutation relations, for the creation and destruction operators.

**4.9.** Consider fermions and neutral scalar bosons with an interaction

$$\mathcal{H}' = g : \phi\bar{\psi}\psi : .$$

Give the diagrams that contribute, in lowest nontrivial order, to (a) fermion–fermion scattering, (b) fermion–antifermion scattering, (c) boson–fermion scattering, (d) boson–boson scattering.

**4.10.** Show that

$$\Delta_F(x - y) = \langle 0 | T\{\phi(x)\phi(y)\} | 0 \rangle.$$

**4.11.** Show that

$$(\Box_x - m^2)\Delta_F(x - y) = i\delta^4(x - y),$$

and hence that $\Delta_F$ is a *Green's function* of the Klein–Gordon equation.

**4.12.** Show that

$$A_\mu(x) = i \int d^4y \; D_F(x - y)\mathfrak{j}_\mu(y)$$

satisfies the field equation with source (see Problem 2.6):

$$\Box A_\mu = -\mathfrak{j}_\mu .$$

## Bibliography

J. D. BJORKEN and S. D. DRELL. *Relativistic Quantum Fields.* McGraw-Hill, New York 1965.

R. P. FEYNMAN. *Theory of Fundamental Processes.* Benjamin, New York, 1962.

F. MANDL. *Introduction to Quantum Field Theory.* Wiley (Interscience), New York, 1959.

A. MARCH. *Quantum Mechanics of Particles and Wave Fields.* Wiley, 1951.

H. MUIRHEAD. *Physics of Elementary Particles.* Pergamon, Oxford, 1965.

J. J. SAKURAI. *Advanced Quantum Theory.* Addison-Wesley, Reading, Massachusetts, 1967.

CHAPTER FIVE

# Quantum Electrodynamics

The ideas introduced in the preceding chapters—specifically (1) interaction representation and perturbation theory, (2) second quantization, and (3) minimal electromagnetic coupling—enable. one to calculate purely electrodynamic processes, i.e., those involving only electrons (or muons) and photons, to *arbitrarily high precision*. There are, in fact, difficulties with this particular way of doing quantum electrodynamics, but we will not discuss these. Here we merely want to indicate how the theory works, first in lowest order of perturbation, then in the higher order corrections.

## 5.1. Lowest Order Terms

The first nonzero contribution to a scattering amplitude involving electrons and photons comes from the $n = 2$ term of the perturbation expansion. Using the interaction equation (4.51) and indicating the normal ordering explicitly, the first contributing term is

$$S_2 = (-1)[(-ie)^2/2] \int d^4x \int d^4y T[A_\mu(x) : \bar{\psi}(x)\gamma_\mu \psi(x) : A_\nu(y) : \bar{\psi}(y)\gamma_\nu \psi(y) :]. \tag{5.1}$$

### *a. Compton Scattering*

Let us first consider *Compton scattering*: $e + \gamma \to e + \gamma$. Only terms of the form $A_+\bar{\psi}_+\psi_-A_-$ will contribute to the Compton amplitude, so we want the terms that come from reduction from time-to-normal-ordering of Eq.

(5.1) with a $\bar{\psi}$, $\psi$ contraction. Thus an electron propagator appears. The contributing Feynman diagrams are shown in Fig. 5.1. Because there are only two ways to attach the photon lines to the electron line, we have only two diagrams; these correspond to the two terms in the $S_2$ sum that contribute to the matrix element.

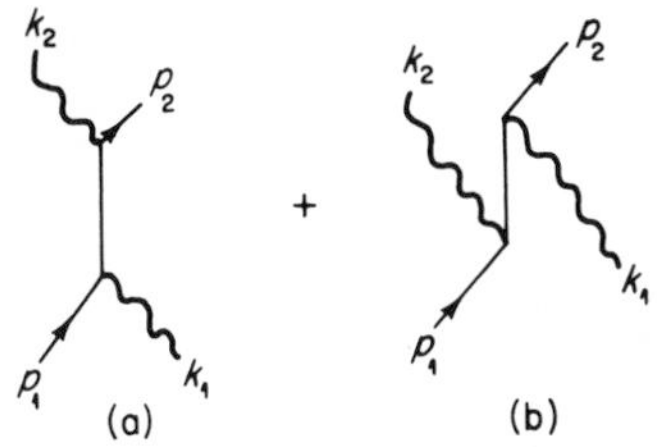


**Fig. 5.1.** Lowest order contributions to Compton scattering.

A look at the expansions for $\psi$, $\bar{\psi}$, and $A$ in terms of the creation and destruction operators and the momentum space electron propagator $S_F(p)$ enables us to write for the matrix element ($\delta$-function normalization)

$$\langle p_2 k_2 \varepsilon_2 | S_2 | p_1 k_1 \varepsilon_1 \rangle = e^2 \frac{(2\pi)^4 \delta(p_2 + k_2 - p_1 - k_1)}{(2\pi)^6} \left( \frac{1}{2\omega_1} \frac{1}{2\omega_2} \frac{m}{E_1} \frac{m}{E_2} \right)^{1/2}$$

$$\times \bar{u}(p_2) \left\{ \not{\varepsilon}_2 \frac{i}{i(\not{p}_1 + \not{k}_1) + m} \not{\varepsilon}_1 \right.$$

$$\left. + \not{\varepsilon}_1 \frac{i}{i(\not{p}_1 - \not{k}_2) + m} \not{\varepsilon}_2 \right\} u(p_1). \qquad (5.2)$$

The propagators in { } can be rewritten as

$$i \frac{1}{i(\not{p}_1 + \not{k}_1) + m} = -i \frac{i(\not{p}_1 + \not{k}_1) - m}{(p_1 + k_1)^2 + m^2} = i \frac{i(\not{p}_1 + \not{k}_1) - m}{s - m^2} = i \frac{i(\not{p}_1 + \not{k}_1) - m}{2m\omega_1}$$

$$i \frac{1}{i(\not{p}_1 - \not{k}_2) + m} = -i \frac{i(\not{p}_1 - \not{k}_2) - m}{(p_1 - k_2)^2 + m} = i \frac{i(\not{p}_1 - \not{k}_2) - m}{u - m^2} = -i \frac{i(\not{p}_1 - \not{k}_2) - m}{2m\omega_2}$$

where $\omega_1$ and $\omega_2$ are the photon energies in the lab system. Now

$$\not{A}\not{B} + \not{B}\not{A} = 2A \cdot B,$$

since

$$\gamma_\mu \gamma_\nu + \gamma_\nu \gamma_\mu = 2\delta_{\mu\nu}. \qquad (2.13)$$

Then

$$(\not{p}_1 + \not{k}_1)\not{\varepsilon}_1 = -\not{\varepsilon}_1(\not{p}_1 + \not{k}_1) + 2(p_1 + k_1) \cdot \varepsilon_1$$
$$= -\not{\varepsilon}_1(\not{p}_1 + \not{k}_1)$$

because $k_1 \cdot \varepsilon_1 = 0$, and we can choose $\varepsilon_1$ so that $p_1 \cdot \varepsilon_1 = 0$.[1] Furthermore, each spinor satisfies the appropriate Dirac equation:

$$(i\not{p}_1 + m)u(p_1) = 0 \qquad \text{and} \qquad \bar{u}(p_2)(i\not{p}_2 + m) = 0.$$

These relations enable us to write the interesting part of Eq. (5.2) in the slightly simpler form

$$\bar{u}(p_2)\{\quad\}u(p_1) = -\bar{u}(p_2)\left\{\frac{\not{\varepsilon}_2 \not{k}_1 \not{\varepsilon}_1}{s - m^2} - \frac{\not{\varepsilon}_1 \not{k}_2 \not{\varepsilon}_1}{u - m^2}\right\} u(p_1). \tag{5.3}$$

At this point, one can proceed in either of two ways: (1) By making definite choices for the spins and polarizations, one can evaluate Eq. (5.3) explicitly and use it to compute the Compton scattering cross section via Eq. (3.72), or (2) one can form the absolute square of Eq. (5.3) and sum over initial and final electron spins. The relation

$$\sum_{\sigma=1,2} u(p, \sigma)\bar{u}(p, \sigma) = \frac{m - i\not{p}}{2m}$$

(which is easy to see in the rest system and holds generally by covariance) can then be used to eliminate the spinors and emerge with

$$\sum_{\sigma,\sigma'} |\bar{u}(p_2)\{\quad\}u(p_1)|^2 = \frac{1}{4m^2}\,\mathrm{tr}[\{\quad\}(m - i\not{p}_1)\{\quad\}^{\dagger}(m - i\not{p}_2)].$$

The trace (tr) can be evaluated in a straightforward though tedious manner. (Evaluating these traces is especially tedious for higher order processes; fortunately the algebra can now be done by digital computer.) For details of the calculation, the student is referred to, for example, Feynman, *Quantum Electrodynamics.* The result, for plane polarized photons, is

$$\frac{d\sigma}{d\Omega} = \frac{\alpha^2}{m^2} \cdot \frac{1}{4} \cdot \left[\frac{\omega_2}{\omega_1}\right]^2 \left[\frac{\omega_2}{\omega_1} + \frac{\omega_1}{\omega_2} - 2 + 4(\boldsymbol{\varepsilon}_1 \cdot \boldsymbol{\varepsilon}_2)^2\right], \tag{5.4}$$

and is known as the *Klein–Nishina formula.* Here

$$\alpha \equiv e^2/4\pi \simeq 1/137$$

is the *fine structure constant.*[2] In the low energy limit, this agrees with the classical *Rayleigh–Thomson* result

$$d\sigma/d\Omega \simeq (\alpha^2/m^2)(\boldsymbol{\varepsilon}_1 \cdot \boldsymbol{\varepsilon}_2)^2.$$

---

[1] We choose the *Coulomb gauge* in the lab system; in that system, $\varepsilon_{1_\mu} = (\boldsymbol{\varepsilon}_1, 0)$, $\varepsilon_{2_\mu} = (\boldsymbol{\varepsilon}_2, 0)$, while $p_1 = (\mathbf{0}, im)$, so that $\varepsilon_1 \cdot p_1 = \varepsilon_2 \cdot p_1 = 0$.

[2] The $(4\pi)^{-1}$ appears with $e^2$ because of our choice of electrical units; in these units the potential from a point charge $q$ is $q/4\pi r$.

Averaged over polarization and integrated over angle, this gives

$$\sigma \simeq (8/3)\pi(\alpha^2/m^2) \simeq 6.6 \times 10^{-25} \quad \text{cm}^2.$$

Note that

$$\alpha/m = r_0 \simeq 2.8F = 2.8 \times 10^{-13} \quad \text{cm}$$

is the "classical electron radius." We might have guessed that

$$\sigma \sim \alpha^2/m^2 \equiv r_0{}^2,$$

since two factors of $e$ in the matrix element imply

$$\sigma \propto \alpha^2,$$

and the only mass available to give the length factors is the electron mass $m$.

### *b. Electron–Positron Annihilation*

Let us now examine the amplitude for the annihilation process $e^+e^- \to 2\gamma$, again in lowest order of perturbation. The part of $S_2$ that now contributes has a factor $A_+A_+\psi_-\psi_-$, and the contributing diagrams are shown in Fig. 5.2. The matrix element is

$$\langle \mathbf{k}_1\boldsymbol{\varepsilon}_1\mathbf{k}_2\,\boldsymbol{\varepsilon}_2 | S | \mathbf{p}_1\mathbf{p}_2\rangle = e^2\,\frac{(2\pi)^4\delta(k_1 + k_2 - p_1 - p_2)}{(2\pi)^6}\left(\frac{1}{2\omega_1}\frac{1}{2\omega_2}\frac{m}{E_1}\frac{m}{E_2}\right)^{1/2}$$

$$\times\, \bar{v}(p_2)\left\{\not{\varepsilon}_2\,\frac{i}{i(\not{p}_1 - \not{k}_1) + m}\,\not{\varepsilon}_1 + \not{\varepsilon}_1\,\frac{i}{i(\not{p}_1 - \not{k}_2) + m}\,\not{\varepsilon}_2\right\}u(p_1). \tag{5.5}$$

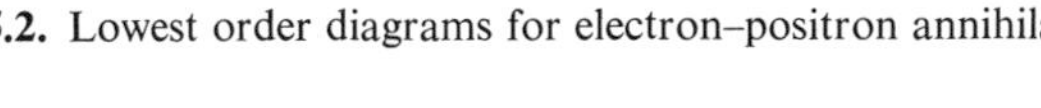


**Fig. 5.2.** Lowest order diagrams for electron–positron annihilation.

We see that the *same* expression appears between the spinors as for the $e\gamma \to e\gamma$ case, except that $k_1 \to -k_1$. If we adopt the convention that all momenta are to be considered *incoming* (Fig. 5.3), then we have

$$\{\ \} = \left\{\not{\varepsilon}_2\,\frac{i}{i(\not{p}_1 + \not{k}_1)}\,\not{\varepsilon}_1 + \not{\varepsilon}_1\,\frac{i}{i(\not{p}_1 + \not{k}_2)}\,\not{\varepsilon}_1\right\}$$

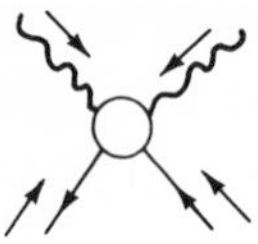

**Fig. 5.3.** All momenta are defined as ingoing, so that $p_1 + p_2 + k_1 + k_2 = 0$.

for *both* processes, and for $\gamma\gamma \to e^+e^-$ and $e^+\gamma \to e^+\gamma$ as well, the only difference being that some of the momentum four-vectors change the sign of their energy components. This property of the scattering amplitudes is known as *crossing symmetry* and clearly persists for higher order contributions as well.

### c. Møller and Bhabha Scattering

Another set of processes is contained in $S_2$: $e^\pm e^\pm \to e^\pm e^\pm$. Here it is the A's that are contracted, so that the photon propagator $D_F$ appears. The matrix element for $e^-e^- \to e^-e^-$ (*Møller scattering*) in second order has the diagrams of Fig. 5.4 and can readily be written down as

$$\langle p_3 p_4 | S_2 | p_1 p_2 \rangle = e^2 \frac{(2\pi)^4 \delta(p_3 + p_4 - p_1 - p_2)}{(2\pi)^6} \left( \frac{m^4}{E_1 E_2 E_3 E_4} \right)^{1/2}$$
$$\times \{ \bar{u}(p_4)\gamma_\mu u(p_2)\bar{u}(p_3)\gamma_\mu u(p_1) \cdot (p_1 - p_3)^{-2}$$
$$- \bar{u}(p_4)\gamma_\mu u(p_1)\bar{u}(p_3)\gamma_\mu u(p_2) \cdot (p_1 - p_4)^{-2} \}. \tag{5.6}$$

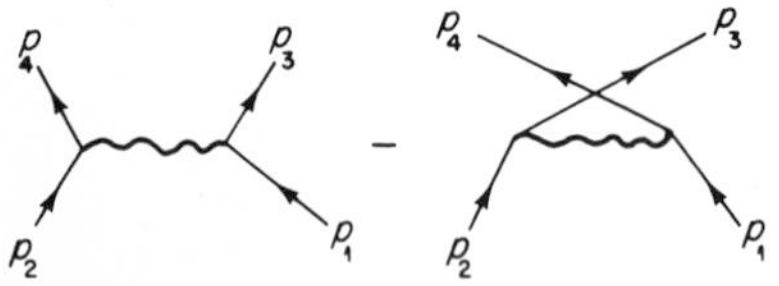


**Fig. 5.4.** Diagrams for Møller scattering ($e^-e^- \to e^-e^-$) in lowest order.

The minus sign appears because the amplitude must be antisymmetric under interchange of the identical fermions. Similarly, $e^-e^+ \to e^-e^+$ (*Bhabha scattering*) is given by the diagrams of Fig. 5.5, which can be obtained by *crossing*, i.e., twisting the 1 and 3 legs of the Møller scattering diagrams, Fig. 5.4. Finally the positron scattering ($e^+e^+ \to e^+e^+$) diagrams are obtained from Fig. 5.5 by crossing the 2 and 4 legs. The spinors do change under

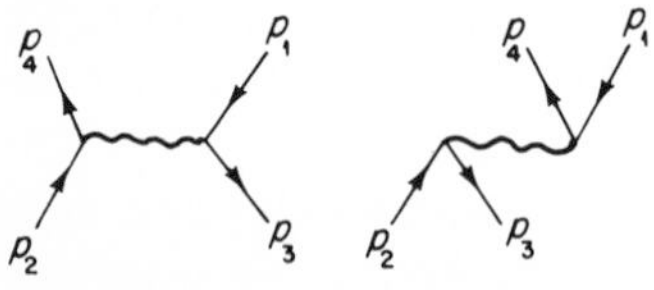


**Fig. 5.5.** Bhabha scattering ($e^-e^+ \to e^-e^+$) diagrams in lowest order.

crossing, $u \leftrightarrows v$, but if we agree to take the momentum in the direction of the arrow, then

$$\sum_{i=1,2} v(p)\bar{v}(p) = (m - i\not{p})/2m,$$

just like $\sum u\bar{u}$, and therefore $\sum |T_{\text{if}}|^2 = \text{tr}[\ \ ]$ is the same expression for each process. Note that the photon propagators for the two diagrams are $t^{-1}$ and $u^{-1}$, respectively.

## 5.2. Higher Order Corrections

We now want to describe what happens when one includes corrections of higher order of perturbation theory. Our discussion will be very sketchy; for the details the reader is referred to any of the standard texts on quantum field theory. Since they came from $S_2$, the matrix elements we have discussed so far are proportional to $\alpha$. If we desire a more accurate answer, we must include in our amplitude terms proportional to $\alpha^2$, which comes from $S_4$. (There are no correction terms $\propto e^3$, since each $e$ comes from attaching a photon line, and such a line must have both ends inside the Feynman diagram if it is to contribute to the matrix elements we have been discussing.) Let us put down all the possible $e^4$ corrections to one of the lowest order electron scattering diagrams, Fig. 5.4, for example. Figure 5.6 shows all the different types of diagrams we can think of.

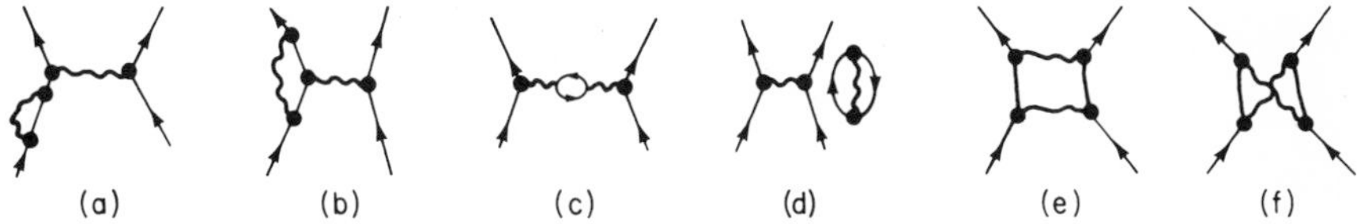


**Fig. 5.6.** Fourth-order corrections to Møller scattering (Fig. 5.4).

Each of the diagrams of Fig. 5.6 contains a *closed loop*. Now, as we have seen, in any Feynman graph each vertex introduces an energy–momentum $\delta$-function, while each propagator brings a $\int d^4k$. In any subdivision of a graph, however, the $\delta$-functions can always be rearranged so that one of them expresses overall energy–momentum conservation. Therefore, a closed loop, which has as many propagators as vertices, will always have one $\int d^4k$ left over when all the $\delta$-functions (except the overall one) are used up. But these integrals turn out to be divergent! These divergences are of two sorts: the *infrared divergence* coming from the $k \to 0$ part of the integral, and the *ultraviolet divergence* coming from the $k \to \infty$ part.[3]

The infrared divergence is really a result of our formalism, and goes away when the problem is done more carefully. In any process involving electron scattering, there is always the possibility of emitting a photon which

[3] For the infrared divergence, we have a term $\int dk/k$ for small $k$; for the ultraviolet divergence, we have a term $\int k^n\,(dk/k)$ for large $k$, with $n = 0$, 1, or 2.

goes undetected. For instance if the energy resolution of the detectors is $\Delta E$, then an extra photon of energy less than $\Delta E$ will be unobserved. Therefore, to get the cross section that has to do with experiment, we must add the probability for "soft" ($<\Delta E$) photon emission to the probability for the process without this photon. When this is done, it turns out that the infrared divergence cancels (in all orders of perturbation theory)!

This infrared divergence appears for graphs (b), (e), and (f) of Fig. 5.6, and once they are treated as mentioned above, graphs (e) and (f) are finite. The remaining graphs (a), (b), (c), and (d) have the much more troublesome ultraviolet divergence. These infinities, in fact, prevented progress in quantum electrodynamics for two decades, until Feynman and Schwinger showed how to circumvent them. The method, known as *renormalization*, will be discussed briefly in connection with these remaining diagrams of Fig. 5.6.

Figure 5.6 (d) contains what is known as a *disconnected vacuum–vacuum* graph. This is no problem at all, for the following reason: There appears in the expansion for $S$ an infinite number of diagrams which connect the vacuum state with the vacuum state; these are shown in Fig. 5.7. Suppose

**Fig. 5.7.** Vacuum–vacuum diagrams.

their sum is denoted by $C \equiv \langle 0|S|0\rangle$. Then since the $S$-matrix can only connect $|0\rangle$ with itself (because of energy–momentum conservation), $S|0\rangle = C|0\rangle$, so that the unitarity of $S$ implies $|C|^2 = 1$. Thus $C$ is a phase factor. But since the same set of diagrams appear as corrections to *any* process, the *same* phase factor multiplies *all* transition matrix elements, and hence $C$ can always be ignored. Therefore, we can always ignore disconnected vacuum-vacuum additions to any diagram.

Figure 5.6 (a) contains the *electron self-energy* term. This corrections should be included in every electron line, and, in fact, all of the terms of Fig. 5.8 are easily included. This results in a mass shift; that is, the observed,

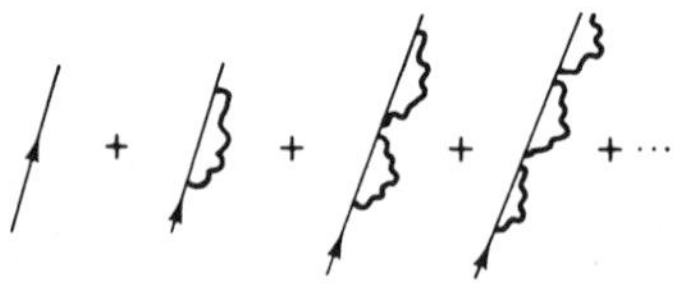


**Fig. 5.8.** The simplest self-energy corrections, which are easily summed.

*physical* mass is not just the mass that appears in the free electron Hamiltonian, now call $m_0$ (*bare* mass), but has an (infinite) correction $\delta m$ due to the electromagnetic interaction:

$$m = m_0 + \delta m.$$

It is then convenient to write the Hamiltonian in terms of $m$ and $\delta m$, and to put the $\delta m$ term in with the interaction Hamiltonian to form a new $\mathcal{H}'$:

$$\begin{aligned}\mathcal{H} &= \bar{\psi}(\gamma \cdot \nabla + m_0)\,\psi - ieA_\mu\,\bar{\psi}\gamma_\mu\psi \\ &= \bar{\psi}(\gamma \cdot \nabla + m)\,\psi - [ieA_\mu\bar{\psi}\gamma_\mu\psi + \delta m\bar{\psi}\psi],\end{aligned}$$

and thus

$$\mathcal{H}' = -ie\bar{\psi}A\!\!\!/\psi - \delta m\bar{\psi}\psi.$$

Then when one does perturbation theory with this $\mathcal{H}'$, there are more Feynman diagrams, but the total correction now gives *zero* mass shift. This technique is known as *mass renormalization*.

Figure 5.9 shows the *vertex correction* and *vacuum polarization* graphs which appear in Fig. 5.6(b) and (c). Like the electron self-energy term, these corrections diverge. It turns out, however, that the major effect of the three

**Fig. 5.9.** (a) Vertex correction and (b) vacuum polarization graphs.

divergent integrals, in addition to the electron mass shift just discussed, is to multiply the *bare charge* $e_0$ by a (infinite) constant to give the observed *renormalized charge* $e_R$. Once the physical amplitudes are expressed in terms of the quantities $m$ and $e_R$, no further fiddling is necessary, and the amplitudes are finite in *all orders of perturbation theory*! (A theory which can be made convergent to all orders by a finite number of these adjustments is deemed *renormalizable*.) It is these renormalized quantities that are identified with the physically measured mass and charge.

There are some finite effects of diagrams (a), (b), and (c) of Fig. 5.6 that remain after renormalization is carried out. For example, the vertex correction adds an *anomalous magnetic moment* to the Dirac moment $\mu_0 = e/2m$,

$$\delta\mu = (\alpha/2\pi)\mu_0,$$

which is in very good agreement with experiment. Another prediction is the very small splitting, known as the *Lamb shift*, between the $2s_{1/2}$ and $2p_{1/2}$ energy levels of hydrogen, which are exactly degenerate in the Dirac theory. All three diagrams contribute, but by far the largest contribution comes from the electron self-energy. The recent theoretical value of 1057.6 mHz (which includes even higher order corrections) agrees very well with the experimental result of 1058.05 $\pm$ 0.10 mHz ($\simeq 4.4 \times 10^{-6}$ eV).

One occasionally reads of experiments which show a "breakdown of quantum electrodynamics"—usually disagreement with the higher order corrections to $\mu$ or the Lamb shift—but so far these have all faded away after a while. We do expect some breakdown for interactions at small enough distances (because eventually particles other than photons and electrons contribute), but so far there have been no unambiguous experimental conflicts with the predictions of quantum electrodynamics. Considering the fantastic accuracy of some of the experimental work, this is quite an accomplishment.

## Problems

**5.1.** Show that $\sum_{\text{sp}} u\bar{u} = (m - i\not{p})/2m$.

**5.2.*** (a) Write the matrix element $T_{\text{fi}}$ for electron proton scattering using the fiction of purely minimal electromagnetic coupling for the proton.

(b) Form $|T_{\text{fi}}|^2$, sum over spins, and use the identity proved in Problem 5.1 to write the result as a product tr{ } · tr{ }.

(c) Using the identities

$$\text{tr}\{\gamma_\mu \gamma_\nu\} = 4\delta_{\mu\nu} \qquad \text{and} \qquad \text{tr}\{\gamma_\mu \gamma_\nu \gamma_\alpha \gamma_\beta\} = 4(\delta_{\mu\nu}\delta_{\alpha\beta} - \delta_{\mu\alpha}\delta_{\nu\beta} + \delta_{\mu\beta}\delta_{\nu\alpha}),$$

and the fact that the trace of an *odd* number of $\gamma$-matrices is zero, evaluate the traces in part (b).

(d) Proceed to write the lab differential cross section for high-energy ep scattering, using the approximation of neglecting the electron's mass $(m_e/E \simeq m_e/M_p \simeq 0)$.

### Bibliography


R. P. Feynman. *Quantum Electrodynamics*. Benjamin, New York, 1961.

R. P. Feynman. *Theory of Fundamental Processes*. Benjamin, New York, 1962.

D. Lurié. *Particles and Fields*. Wiley, New York, 1968.

H. Muirhead. *Physics of Elementary Particles*. Pergamon, Oxford, 1965.

J. J. Sakurai. *Advanced Quantum Mechanics*. Addison-Wesley, Reading, Massachusetts, 1967.

CHAPTER SIX

# Pions

The pion was postulated by Yukawa in 1935 to explain nuclear forces, but succeeded in evading detection by the experimenters until 1947, when Lattes, Occhialini, and Powell discovered positive and negative $\pi$'s in cosmic rays at high altitudes. Both charged and neutral pions are produced copiously in high-energy collisions of nucleons (e.g., proton beam + target $\to$ nucleons + pions), are subject to strong interactions, and thus are classified as hadrons. The pion masses are intermediate between the electron and the nucleon masses[1] ($m_\pi \simeq \frac{1}{7} m_N$), which is the origin of the term *meson*. Like almost all particles, the pions are unstable, with $\pi^\pm$ decaying rather slowly ($\tau \sim 10^{-8}$ sec) via the weak interaction while $\pi^0$ decays electromagnetically with a much shorter lifetime ($\tau \sim 10^{-16}$ sec). We now want to discuss some additional properties of the pions, introducing some powerful theoretical tools along the way.

---

[1] We find $m_{\pi^\pm} \simeq 140$ MeV, $m_{\pi^0} \simeq 135$ MeV, while $m_e \simeq 0.51$ MeV, and $m_N \simeq 940$ MeV. For precise values, see Appendix.

## 6.1. Time Reversal. Pion Spins

### *a.* $\pi^{\pm}$ *Spin Determination. Detailed Balance*

It was suggested by Marshak in 1951 that the spin of the $\pi^+$ could most easily be determined running the reaction

$$p + p \rightleftarrows d + \pi^+$$

both ways and comparing the cross sections. By employing a weak form of the *principle of detailed balance*,[2] namely,

$$\sum_{sp} |S_{fi}|^2 = \sum_{sp} |S_{if}|^2 \tag{6.1}$$

(the momenta are the *same* for the respective particles on both sides of the equation), we can get information about the remaining factors that make up the cross section. The (CM) differential cross section for the reaction $ab \to cd$, summed and averaged over spins, is (see Chapter 3, Section 3.4)

$$\frac{d\sigma(\theta)}{d\Omega} = \frac{1}{64\pi^2 s} \frac{|\mathbf{p}_f|}{|\mathbf{p}_i|} \frac{1}{(2s_a + 1)(2s_b + 1)} \sum_{sp} |T_{fi}|^2. \tag{3.96b}$$

We see that Eq. (6.1) implies

$$\frac{(d\sigma(\theta)/d\Omega)_{ab\to cd}}{(d\sigma(\theta)/d\Omega)_{cd\to ab}} = \frac{|\mathbf{p}_{cd}|^2}{|\mathbf{p}_{ab}|^2} \frac{(2s_c + 1)(2s_d + 1)}{(2s_a + 1)(2s_b + 1)}. \tag{6.2}$$

For the reaction in question, the spin factor is $\frac{3}{4}(2s_\pi + 1)$, and since $s_\pi$ must be integral, that can be $\frac{3}{4} \cdot 1, \frac{3}{4} \cdot 3, \ldots$. Thus even a relatively imprecise determination of the cross sections suffices to determine the $\pi^+$ spin. The experiments indicate spin zero. This is also the $\pi^-$ spin, since particle and antiparticle must have the same spin (they belong to the same field operator).

The form of detailed balance used to get this result comes most generally from *time-reversal invariance*, plus a little more.

### *b. Time-Reversal Invariance*

In classical physics, invariance under time reversal means that if we made a movie of some simple enough system, and then ran the film backward so that the time evolution and the velocities of the particles that make up the

---

[2] Equation (6.1) is sometimes called *semi-detailed balance*, while *detailed balance* is the stronger statement

$$S_{fi} = S_{if}$$

(same momenta and spins).

system are reversed, the "backward" motion is just as possible as the original. That is, it would be impossible for a viewer who was not tipped off to tell that the film was being run the wrong way.

For macroscopic systems, of course, we can usually tell which is the right way; when the wine rises up from the glass and goes into the bottle, we all agree that this is wrong. This motion is perfectly possible, but we do not believe it. That is because in order for the wine to rise out of the glass, the wine molecules all had to be moving in a very special way to begin with. Since the number of these special initial states is far less than the number of "ordinary" initial states of molecules in random motion, we say that the *probability* of the wine rising is extremely small.

Quantitatively, time reversal is accomplished by making the transformation $t \to -t$ everywhere in the equations of motion. Thus

$$\partial/\partial t \to -(\partial/\partial t),$$
$$q \to q, \qquad \dot{q} \to -\dot{q}, \qquad p \to -p, \qquad \dot{p} \to +\dot{p},$$

so that for *Hamilton's equations*,

$$\dot{q} = \partial H(p,q)/\partial p, \qquad \dot{p} = -[\partial H(p,q)]/\partial q,$$

to remain valid, we need $H \to H$; but this is what we expect since the kinetic energy is generally a quadratic function of the momenta.

In quantum mechanics we have $[q, p] = i$. Since we must have $q \to q$, $p \to -p$ here too, when we introduce an operator $T$ to do the transforming, we must have

$$TqT^{-1} = q, \qquad TpT^{-1} = -p, \qquad \text{and also} \qquad TiT^{-1} = -i, \tag{6.3}$$

in order to preserve the commutation relation. Thus $T$ must be not a linear operator but *antilinear*; i.e., instead of

$$T(aA + bB) = aTA + bTB,$$

we need

$$T(ab + bB) = a^*TA + b^*TB. \tag{6.4}$$

(Here $A$ and $B$ are operators *or* state vectors; $a$ and $b$ are complex numbers.)

The equation of motion in the Schrödinger representation,

$$i\,(\partial/\partial t)\,|\psi_S\rangle = H|\psi_S\rangle, \tag{3.30}$$

then transforms under time reversal to

$$T[i(\partial/\partial t)]T^{-1}\,|\psi_S'(t)\rangle = H'\,|\psi_S'(t)\rangle,$$

where

$$|\psi_S'(t)\rangle \equiv T|\psi_S(t)\rangle \qquad \text{and} \qquad H' \equiv THT^{-1}.$$

Now

$$T\left(i\frac{\partial}{\partial t}\right)T^{-1} = TiT^{-1}\frac{\partial}{\partial t} = -i\frac{\partial}{\partial t} = i\frac{\partial}{\partial(-t)}$$

(the parameter $t$ is unaffected by the $T$ operator), so that we have

$$i[\partial/\partial(-t)]|\psi_S'(t)\rangle = H'|\psi_S'(t)\rangle,$$

which can be rewritten by setting $t' \equiv -t$:

$$i(\partial/\partial t')|\psi_S'(-t')\rangle = H'|\psi_S'(-t')\rangle. \tag{6.5}$$

Then if $H$ is invariant under $T$, i.e., if

$$H' = THT^{-1} = H, \tag{6.6}$$

then $|\psi_S'(-t)\rangle$ obeys exactly the same equation as $|\psi_S(t)\rangle$. We say that the time-reversed state evolves exactly as the original one, but in the opposite time direction. This is just like running the film backward and just what we want for time-reversal invariance.

Let us look further at the properties of $T$. Matrix elements (in any representation) transform like

$$\langle\phi|\psi\rangle \xrightarrow{T} \langle T\phi|T\psi\rangle.$$

Then the complex conjugation property and the desired invariance of the probability $|\langle\phi|\psi\rangle|^2$ lead us to require

$$\langle\phi|\psi\rangle \xrightarrow{T} \langle\phi|\psi\rangle^*.$$

Since

$$\langle\phi|\psi\rangle^* = \langle\psi|\phi\rangle,$$

we conclude that

$$\langle T\phi|T\psi\rangle = \langle\psi|\phi\rangle. \tag{6.7}$$

This is in contrast to the relation

$$\langle U\phi|U\psi\rangle = \langle\phi|\psi\rangle,$$

which holds for unitary operators $U$. Because of this, $T$ is called *antiunitary* rather than unitary. A theory that allows a $T$ with all the required properties (see Table 6.1) is said to be *time-reversal invariant.*[3]

---

[3] This formulation is known as *Wigner time reversal*; an alternative formulation is known as *Schwinger time reversal*.

**Table 6.1**

*Behavior under Time Reversal for Invariance*

| | | |
|---|---|---|
| $\mathbf{r} \to \mathbf{r}$ | $\mathbf{j} \to -\mathbf{j}$ | $\mathbf{E} \to \mathbf{E}$ |
| $\mathbf{p} \to -\mathbf{p}$ | $\rho \to \rho$ | $\mathbf{B} \to -\mathbf{B}$ |
| $\mathbf{L} \to -\mathbf{L}$ | $\mathbf{A} \to -\mathbf{A}$ | $H \to H$ |
| $\boldsymbol{\sigma} \to -\boldsymbol{\sigma}$ | $A_4 \to -A_4$ | $\langle\phi\mid\psi\rangle \to \langle\phi\mid\psi\rangle^*$ |
| $\mathbf{J} \to -\mathbf{J}$ | | |

What does $T$-invariance imply about $S$-matrix elements? Here it is simplest to think in terms of the Schrödinger representation. We use $|\psi_0{}^+\rangle$ to denote the state which before collision ($t \to -\infty$) is purely plane wave with quantum numbers (momenta, spins, kinds of particles) specified by $c$; after collision ($t \to +\infty$), this state will have outgoing spherical waves, different kinds of particles, etc. Furthermore $|\psi_c{}^-\rangle$ denotes a state that evolves the opposite way: pure plane waves at $t \to +\infty$ which developed from *incoming* spherical waves at $t \to -\infty$. Now if we want the transition amplitude from one kind of plane wave ($a$) in the past to another kind of plane wave ($b$) in the future, this means the overlap $\langle\psi_b{}^-|\psi_a{}^+\rangle$. Thus we say that

$$S_{ba} = \langle\psi_b{}^-|\psi_a{}^+\rangle. \tag{6.8}$$

Now because $T$ precisely reverses the time evolution,

$$T|\psi_c{}^\pm\rangle = |\psi_{\tilde{c}}{}^\mp\rangle, \tag{6.9}$$

where $\sim$ means that the momenta and spins are reversed.[4] Using the anti-unitary property of $T$ [Eq. (6.7)],

$$\langle\psi_b{}^-|\psi_a{}^+\rangle = \langle T\psi_a{}^+|T\psi_b{}^-\rangle = \langle\psi_{\tilde{a}}{}^-|\psi_{\tilde{b}}{}^+\rangle, \tag{6.10}$$

and hence

$$S_{ba} = S_{\tilde{a}\tilde{b}}\,. \tag{6.11}$$

This expression of $T$-invariance is often called the *reciprocity relation.*

To proceed to Eq. (6.1), form

$$\sum_{\text{sp}} |S_{ab}|^2 = \sum_{\text{sp}} |S_{\tilde{b}\tilde{a}}|^2. \tag{6.12}$$

The $S$-matrix elements must be rotation invariant, and since, for two-body scattering, a rotation of 180° around the normal to the scattering plane reverses the momenta, we have that

$$\sum |S_{\tilde{b}\tilde{a}}|^2 = \sum |S_{ba}|^2. \tag{6.13}$$

---

[4] In Eq. (6.9), an arbitrary phase factor has been taken as unity.

Hence

$$\sum |S_{ab}|^2 = \sum |S_{ba}|^2, \tag{6.1}$$

the desired result.

Time reversal is usually assumed to hold in strong, electromagnetic, and most weak interactions. It is in fact rather difficult to test; the only violation that has been demonstrated at present is a small one in the (weak) decay of one particular kind of meson, the $K^0$.

### c. Spin of the $\pi^0$

The $\pi^0$ decays very rapidly into two photons. This is sufficient to establish that the $\pi^0$ spin is not one. The argument illustrates the power and simplicity of using the *tensor* properties of *S*-matrix elements.

Since the final state contains two photons, the decay amplitude

$$\langle \mathbf{k}_1 \boldsymbol{\varepsilon}_1 \mathbf{k}_2 \boldsymbol{\varepsilon}_2 | S | \mathbf{p} S_\pi \rangle$$

must have as factors the polarization vectors $\boldsymbol{\varepsilon}_1$ and $\boldsymbol{\varepsilon}_2$. If the $\pi^0$ had spin one, its spin state could also be described by a polarization vector which we call $\boldsymbol{\pi}$, and the amplitude must also have a factor of $\boldsymbol{\pi}$. But the amplitude must be rotationally invariant, i.e., scalar. In the CM system ($\mathbf{p} = 0$), the only vectors available are $\boldsymbol{\pi}$, $\boldsymbol{\varepsilon}_1$, $\boldsymbol{\varepsilon}_2$, and $\mathbf{k}$, where

$$\mathbf{k} \equiv \mathbf{k}_1 = -\mathbf{k}_2 = \tfrac{1}{2}(\mathbf{k}_1 - \mathbf{k}_2).$$

Since

$$\boldsymbol{\varepsilon}_1 \cdot \mathbf{k} = \boldsymbol{\varepsilon}_2 \cdot \mathbf{k} = 0,$$

the only scalars possible are

$$\boldsymbol{\pi} \cdot (\boldsymbol{\varepsilon}_1 \times \boldsymbol{\varepsilon}_2) \qquad \text{and} \qquad (\boldsymbol{\pi} \cdot \mathbf{k})(\boldsymbol{\varepsilon}_1 \cdot \boldsymbol{\varepsilon}_2).$$

Both of these change sign when we exchange the coordinates of the two photons

$$\boldsymbol{\varepsilon}_1 \leftrightarrows \boldsymbol{\varepsilon}_2, \qquad \mathbf{k} \leftrightarrows -\mathbf{k},$$

but the amplitude must be *symmetric* under interchange of identical bosons. Hence there is no acceptable amplitude; a spin one object *cannot* decay into two photons.

This, of course, does not rule out spin two or higher for the $\pi^0$, but there is no evidence that contradicts the assignment

$$s_{\pi^0} = s_{\pi^\pm} = 0.$$

## 6.2. Space Inversion. Pion Parities

Before going on to the *intrinsic parity* of the pions, we want to discuss space inversion symmetry in general. Present evidence indicates that this symmetry holds for the strong and electromagnetic interactions, but not for the weak interaction.

### *a. Space-Inversion Invariance*

The idea behind *space-inversion invariance*, or *parity* for short, is that nature has no inherent preference as to "handedness," i.e., right-handed coordinate systems and right-handed screws are just as good as left-handed ones. The fact that biological systems seem to favor molecules of a certain handedness is viewed as an historical accident.

The change from a right- to left-handed system is brought about by space inversion: $\mathbf{x} \to -\mathbf{x}$, while $t \to t$. Thus $\mathbf{p} \to -\mathbf{p}$, but $\mathbf{L} = \mathbf{x} \times \mathbf{p} \to +\mathbf{L}$, so that we demand that $\mathbf{S} \to \mathbf{S}$, $\mathbf{J} \to \mathbf{J}$. In addition, $\mathbf{E} \to -\mathbf{E}$, $\mathbf{A} \to -\mathbf{A}$, and therefore $\boldsymbol{\varepsilon} \to -\boldsymbol{\varepsilon}$, but since $\boldsymbol{\nabla} \to -\boldsymbol{\nabla}$, $\mathbf{B} = \boldsymbol{\nabla} \times \mathbf{A} \to \mathbf{B}$. Under space inversion $\mathbf{L}$, $\mathbf{S}$, $\mathbf{J}$, and $\mathbf{B}$ are *pseudovector* (or *axial vector*), while those changing sign are simply *vector*.

In classical mechanics, the equations

$$\dot{x} = \partial H/\partial p, \qquad \dot{p} = -\partial H/\partial x$$

remain unchanged under space inversion if $H \to H$. Thus terms like $a\mathbf{p}^2$, $b\mathbf{x}^2$, and even $j_\mu A_\mu$ can appear in the Hamiltonian, but not $\mathbf{x} \cdot \mathbf{B}$, etc.

In ordinary quantum mechanics, space inversion transforms the Schrödinger equation to

$$i(\partial/\partial t)\psi(-\mathbf{x}, t) = H'\psi(-\mathbf{x}, t). \tag{6.14}$$

Invariance requires that $\psi(-\mathbf{x}, t)$ satisfies the *same* equation as $\psi(\mathbf{x}, t)$, and therefore that $H' = H$; for this to hold the potential must, of course, be invariant: $V(-\mathbf{x}) = V(\mathbf{x})$. We introduce the unitary space inversion operator $P$, with the properties

$$P\psi(\mathbf{x}, t) = \psi(-\mathbf{x}, t), \tag{6.15}$$

and

$$P^2 = 1.$$

Then

$$H' = PHP^{-1} = H,$$

which implies that

$$d/dt\langle P \rangle = 0. \tag{6.16}$$

The behavior of various quantities under space inversion is shown in Table 6.2.

**Table 6.2**

*Behavior under Space Inversion for Invariance*

| | | |
|---|---|---|
| $\mathbf{r} \to -\mathbf{r}$ | $\mathbf{j} \to -\mathbf{j}$ | $\mathbf{B} \to \mathbf{B}$ |
| $\mathbf{p} \to -\mathbf{p}$ | $\rho \to \rho$ | $\mathbf{E} \cdot \mathbf{B} \to -\mathbf{E} \cdot \mathbf{B}$ |
| $\mathbf{L} \to \mathbf{L}$ | $\mathbf{A} \to -\mathbf{A}$ | $\mathbf{E} \times \mathbf{B} \to -\mathbf{E} \times \mathbf{B}$ |
| $\boldsymbol{\sigma} \to \boldsymbol{\sigma}$ | $A_4 \to A_4$ | $H \to H$ |
| $\mathbf{J} \to \mathbf{J}$ | $\mathbf{E} \to -\mathbf{E}$ | $\langle\phi\vert\psi\rangle \to \langle\phi\vert\psi\rangle$ |

Energy eigenfunctions of such a Hamiltonian are either degenerate (e.g., continum states) or have a definite *parity*, i.e.,

$$P\psi(\mathbf{x}) = \pm\,\psi(\mathbf{x}). \tag{6.17}$$

This is because

$$H\psi(\mathbf{x}) = E\psi(\mathbf{x}) \qquad \text{and} \qquad HP\psi = PH\psi = EP\psi;$$

so $P\psi$ is a eigenfunction with the same $E$. No degeneracy then implies that

$$P\psi(\mathbf{x}) = a\psi(\mathbf{x}),$$

with $a^2 = 1$. Therefore $a = \pm 1$. Wave functions of definite orbital angular momentum have parity $(-1)^l$, since, in terms of the spherical harmonics $Y_{lm}$,

$$PY_{lm}(\theta, \phi) = Y_{lm}(\pi - \theta, \phi + \pi) = (-1)^l Y_{lm}(\theta, \phi). \tag{6.18}$$

When we include the possibility of creating or transmuting particles, as in relativistic quantum mechanics, we can have something new. Specifically, for a single-particle plane-wave state, for example, we can have

$$P|\mathbf{p}\rangle = \eta|-\mathbf{p}\rangle, \qquad \eta = \pm 1. \tag{6.19}$$

This extra factor $\eta$, called the *intrinsic parity*, makes no difference if the particles retain their identity as in ordinary quantum mechanics, or the reaction $a + b \to a + b$. For $a + b \to c + d$, on the other hand,

$$P|\mathbf{p}_a, \mathbf{p}_b\rangle = \eta_a\eta_b\,|-\mathbf{p}_a, -\mathbf{p}_b\rangle,$$

while

$$P|\mathbf{p}_c, \mathbf{p}_b\rangle = \eta_c\eta_d|-\mathbf{p}_c, -\mathbf{p}_d\rangle.$$

We can have

$$\eta_a\eta_b \neq \eta_c\eta_d,$$

which, as we shall see, is important. An example is a one-photon state $|\mathbf{k}, \boldsymbol{\varepsilon}\rangle$; since the polarization vector transforms like $\mathbf{A}$, $\boldsymbol{\varepsilon} \to -\boldsymbol{\varepsilon}$, while the spin of the photon must remain unchanged. Furthermore, the photon amplitude is linear in $\boldsymbol{\varepsilon}$; hence

$$P|\mathbf{k}, \boldsymbol{\varepsilon}\rangle = |-\mathbf{k}, -\boldsymbol{\varepsilon}\rangle = -|-\mathbf{k}, \boldsymbol{\varepsilon}\rangle,$$

and therefore, $\eta = -1$, i.e. the photon has negative intrinsic parity.

In addition to the Hamiltonian being invariant, $P$-invariance implies that the $S$-operator is invariant:

$$PSP^{-1} = S. \tag{6.20}$$

Thus

$$\begin{aligned}\langle \mathrm{f}|S|\mathrm{i}\rangle &= \langle \mathrm{f}|(P^{-1}P)S(P^{-1}P)|\mathrm{i}\rangle \\ &= \langle P\mathrm{f}|PSP^{-1}|P\mathrm{i}\rangle \\ &= \langle P\mathrm{f}|S|P\mathrm{i}\rangle.\end{aligned}$$

In terms of momentum eigenstates (plane waves), this reads

$$\begin{aligned}\langle \mathbf{p}_c s_c, \mathbf{p}_d s_d|S|\mathbf{p}_a s_a, \mathbf{p}_b, s_b\rangle = \eta_a\eta_b\eta_c\eta_d \\ \times\langle(-\mathbf{p}_c)s_c, (-\mathbf{p}_d)s_d|S|(-\mathbf{p}_a)s_a, (-\mathbf{p}_b)s_b\rangle.\end{aligned}$$

(Here $s_a$, etc., specify the spin states.) But $P$-invariance becomes especially useful when we work with states of definite orbital angular momentum, because these are also eigenstates of $P$:

$$P|l, m_l, s_a, s_b\rangle = (-1)^l\eta_a\eta_b|l, m_l, s_a, s_b\rangle, \tag{6.21}$$

using Eq. (6.18). Then Eq. (6.20) implies that

$$\langle l's_c s_d|S|ls_a s_b\rangle = (-1)^{l'+l}\eta_c\eta_d\eta_a\eta_b\langle l's_c s_d|S|ls_a s_b\rangle,$$

so that $S_{\mathrm{fi}} = 0$, unless

$$(-1)^{l'+l}\eta_c\eta_d\eta_a\eta_b = 1. \tag{6.22}$$

Since the *overall parity* is given by

$$\eta = (-1)^l\eta_a\eta_b, \tag{6.23}$$

Eq. (6.22) expresses *parity conservation*: $\eta_{\mathrm{f}} = \eta_{\mathrm{i}}$; this also follows directly from Eq. (6.16).

### b. $\pi^{\pm}$ Parity

The $\pi^-$ parity is determined by allowing $\pi^-$ capture in deuterium. This is done by slowing down the $\pi^-$ beam enough so that the $\pi^-$'s stop in a liquid deuterium target. Atomic processes then insure that a *mesic atom* is formed (the $\pi^-$ replaces an electron) and that the $\pi^-$ reacts with the deuteron while in an $S(L=0)$-orbit. The observed reactions are

$$\begin{aligned}\pi^- \mathrm{d} &\to \mathrm{nn} \\ &\to \mathrm{nn}\gamma\end{aligned}$$

but *not* $\to \mathrm{nn}\pi^0$. For mesic atom orbital angular momentum $L$ [Eq. (6.23)],

$$P|\pi^-\mathrm{d}\rangle = (-1)^L \eta_{\pi^-} \eta_d |\pi^-\mathrm{d}\rangle,$$

while the intrinsic deuteron parity is given by

$$\eta_\mathrm{d} = (-1)^l \eta_\mathrm{p} \eta_\mathrm{n} = \eta_\mathrm{p} \eta_\mathrm{n},$$

since the deuteron is a superposition of $S$ and $D$ ($l = 0$ and 2) proton plus neutron states. For the nn state,

$$P|\mathrm{nn}\rangle = (-1)^{l'} \eta_\mathrm{n}^2 |\mathrm{nn}\rangle.$$

Parity conservation then implies

$$(-1)^L \eta_\mathrm{p} \eta_\mathrm{n} \eta_{\pi^-} = (-1)^{l'} \eta_\mathrm{n}^2. \tag{6.24}$$

To discover $l'$, we must consider what nn states are possible, considering the requirement of overall antisymmetry of identical fermions. First note that when two spin-one-half particles couple to give total spin $s = 1$, the spin wave function is symmetric under interchange, while $s = 0$ means antisym-

**Table 6.3**

*Spin Wave Functions for Two Spin-One-Half Objects*

| | $s = 1$, symmetric | $s = 0$, antisymmetric |
|---|---|---|
| $M_s = 1$ | $\uparrow\uparrow$ | |
| 0 | $1/\sqrt{2}(\uparrow\downarrow + \downarrow\uparrow)$ | $1/\sqrt{2}(\uparrow\downarrow - \downarrow\uparrow)$ |
| $-1$ | $\downarrow\downarrow$ | |

metric spin wave function (Table 6.3); then since, under interchange, the space part of the wave function obeys

$$\psi(r) = R(r)Y_{l'm}(\theta, \phi) \to R(r)Y_{l'm}(\pi - \theta, \phi + \pi) = (-1)^{l'} R(r)Y_{l'm}(\theta, \phi),$$

we have all together

$$\psi \to (-1)^{s+1+l'} \psi.$$

Overall antisymmetry then requires

$$(-1)^{s+1+l'} = -1,$$

or $s + l'$ *even*. Furthermore, angular momentum conservation requires that the total angular momentum $J$ of the two neutrons be one, because the deuteron has spin one and the $\pi^-$ is captured from an S-state. These two requirements, overall antisymmetry and $J = 1$, eliminate all of the possible nn states except $^3P_1$ (see Table 6.4).

**Table 6.4**

*Two Neutron States*[a]

| State | Complaint |
|---|---|
| $^1S_0$ | $J = 0$ |
| $^3S_0$ | $s + l'$ odd |
| $^1P_1$ | $s + l'$ odd |
| $^3P_0$ | $J = 0$ |
| $^3P_1$ | OK |
| $^3P_2$ | $J = 2$ |
| $^1D_2$ | $J = 2$ |
| $^3D_{1,2,3}$ | $s + l'$ odd |
| F and higher | $J > 1$ |

[a] Spectroscopic notation is used here: $^{(2S+1)}L_J$, with S, P, D, F, ..., for $L = 0, 1, 2, 3, \ldots$.

Putting $L = 0$ and $l' = 1$ into Eq. (6.24) yields

$$\eta_p \eta_{\pi^-} = -\eta_n . \tag{6.25}$$

At this point, we must recognize that two of these parities can be chosen arbitrarily. That is, *baryon conservation* tells us that we will always have the same number of $\eta_B$'s on both sides of every equation like Eq. (6.25), and *charge conservation* that we will always have a $\pi^\pm$, and hence an $\eta_{\pi^\pm}$, whenever we change from p to n or vice versa. (The weak decay,

$$n \to p + \text{leptons},$$

cannot be used to fix the relative parity because the weak interactions do *not* conserve parity.) We say from Eq. (6.25) that *the relative parity* of $\pi^-$, p, and n is *odd*. The usual *convention* is to take

$$\eta_p = \eta_n = +1.$$

Then Eq. (6.25) implies that

$$\eta_{\pi^-} = -1. \tag{6.26}$$

With this convention, then, the $\pi^-$ is *pseudoscalar*. Thus we demand that the pion field operator obey

$$\phi(\mathbf{x}, t) \xrightarrow{P} P\phi(\mathbf{x}, t)P^{-1} = -\phi(-\mathbf{x}, t) \tag{6.27}$$

under space inversion, and because it annihilates $\pi^+$'s as well as creating $\pi^-$'s, $\pi^+$ also has negative parity.[5] Since the strong interactions (as far as we know) conserve parity and therefore must be invariant under space inversion, interactions terms like

$$\phi\bar{\psi}_p\gamma_5\psi_n \qquad \text{and} \qquad \partial_\mu\phi\cdot\bar{\psi}_p\gamma_5\gamma_\mu\psi_n$$

are allowed, while terms like

$$\phi\bar{\psi}_p\psi_n \qquad \text{and} \qquad \partial_\mu\phi\cdot\bar{\psi}_p\gamma_\mu\psi_n$$

are not allowed (see Table 2.1).

### c. $\pi^0$ *Parity*

In order to explain why $\pi^-d \nleftrightarrow nn\pi^0$, we first observe that the $Q$ (kinetic energy released) for this reaction is very small. Now the amplitude for $\pi^0$ production must contain as a factor the wave function of the outgoing $\pi^0$. But the radial wave equation contains the angular momentum ($l$) or *centrifugal barrier* term coming from the Laplacian:

$$l(l+1)/r^2.$$

This acts like a repulsive potential. As a result one finds that for small $r$ the wave function is proportional to $(kr)^l$. Then since the *range* of nuclear forces is $\sim m_\pi^{-1}$ (compare Section 6.6), this leads to a factor of

$$(k/m_\pi)^{2l} \sim (Q/m_\pi)^l$$

in the probability. Since $Q \ll m_\pi$, we expect this centrifugal barrier factor to inhibit anything but $S$-wave $\pi^0$'s. In this case, the nn state must still be $^3P_1$, and hence the final state parity would be

$$(-1)\eta_n{}^2\eta_{\pi^0} = \eta_p\eta_n\eta_{\pi^-}.$$

Since we have already established that

$$\eta_p\eta_n\eta_{\pi^-} = -1,$$

this would say that

$$\eta_{\pi^0} = +1.$$

---

[5] Fermion and antifermion operators are contained in the same operator $\psi$. This results in the fermion and antifermion having *opposite* intrinsic parities! This behavior is due to the spinor nature of $\psi$.

Therefore,

$$\eta_{\pi^0} = -1$$

implies that *no* $S$-wave $\pi^0$ is possible and hence explains why $nn\pi^0$ is not seen.

"Cleaner" evidence for the pseudoscalar nature of the $\pi_0$ is provided by the electromagnetic decay $\pi^0 \to 2\gamma$, which *does* conserve parity. Once we have settled on spin zero for the $\pi^0$, the possible forms for the decay amplitude (and hence the $2\gamma$ wave function) are (1) $A\boldsymbol{\varepsilon}_1 \cdot \boldsymbol{\varepsilon}_2$ and (2) $B\mathbf{k} \cdot \boldsymbol{\varepsilon}_1 \times \boldsymbol{\varepsilon}_2$, where as before (Section 6.1c) $\boldsymbol{\varepsilon}_1$ and $\boldsymbol{\varepsilon}_2$ are the polarization vectors of the photons and $\mathbf{k}$ is the momentum of one of them in the $\pi^0$ rest system. Now (1) is unchanged under space inversion, while (2) changes sign. This means that the overall parity of the $2\gamma$ state is $+1$ for the first and $-1$ for the second form.

The two possibilities are in principle distinguishable, since (1) means that the photon polarizations are predominately parallel and (2) that they are perpendicular. However, the polarization of these high-energy ($\sim$70 MeV) photons is difficult to determine directly. Instead, one relies on the so-called *Dalitz pairs*. That is, in a fraction of about $\alpha$ of the $\pi^0$ decays, one of the photons is replaced by an electron–positron pair, and in about $\alpha^2 \sim 10^{-4}$ of the decays, two pairs are produced (Fig. 6.1).

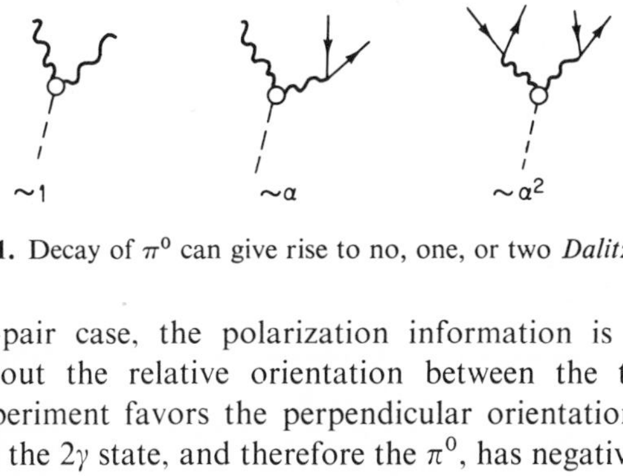


**Fig. 6.1.** Decay of $\pi^0$ can give rise to no, one, or two *Dalitz pairs.*

In the two-pair case, the polarization information is translated into information about the relative orientation between the two production planes. The experiment favors the perpendicular orientation and therefore form (2). Hence the $2\gamma$ state, and therefore the $\pi^0$, has negative parity. This is in agreement with the conclusion we reached above. Notice that the $\pi^0$ parity does *not* depend upon a convention.

## 6.3. Charge Conjugation

We now want to turn to the behavior of pions under a different symmetry operation, that of *charge conjugation*. Charge-conjugation invariance, like space-inversion invariance, appears to hold for strong and electromagnetic interactions, but not weak interactions. The charge-conjugation operation changes all particles into antiparticles, changes the sign of the charge, etc. Thus

$$j_\mu \to -j_\mu,$$

and therefore

$$A_\mu \to -A_\mu, \qquad E \to -E, \qquad \text{and} \qquad B \to -B.$$

A theory is invariant under charge conjugation if there exists a unitary operator $C$ such that particles $\leftrightarrows$ antiparticles,

$$j_\mu \xrightarrow{C} -j_\mu,$$

etc., and the Hamiltonian is invariant:

$$CHC^{-1} = H.$$

From this follows the invariance of the $S$-operator:

$$CSC^{-1} = S.$$

This behavior is summarized in Table 6.5.

**Table 6.5**

*Behavior under Charge Conjugation for Invariance*

| | |
|---|---|
| $Q \to -Q$ | $B \to -B$ |
| $j \to -j_\mu$ | $H \to H$ |
| $A_\mu \to -A_\mu$ | $\langle\phi\|\psi\rangle \to \langle\phi\|\psi\rangle$ |
| $E \to -E$ | particle $\to$ antiparticle |

It is easy to see what $C$ should do to a field operator for a charged Klein–Gordon field. We take[6]

$$Ca_k C^{-1} = b_k \qquad \text{and} \qquad Cb_k C^{-1} = a_k$$

to interchange particle and antiparticle [cf. Eq. (4.13)]. Therefore,

$$C|\pi^\pm\rangle = |\pi^\mp\rangle,$$

$$C\phi C^{-1} = \phi^\dagger, \qquad C\phi^\dagger C^{-1} = \phi, \tag{6.28}$$

and, for $C$-invariance, $H$ must be invariant under $\phi \leftrightarrows \phi^\dagger$.

[6] We could equally well take the convention $C\phi C^{-1} = -\phi^\dagger$. Several conventions are, in fact, in use for the $\pi^\pm$ field.

There are many interesting eigenstates of $C$. We start with the vacuum state and the assignment $C|0\rangle = |0\rangle$, innocently enough. Since

$$A_\mu \to -A_\mu, \qquad \varepsilon \to -\varepsilon,$$

and therefore the one photon state obeys

$$C|\mathbf{k}, \boldsymbol{\varepsilon}\rangle = -|\mathbf{k}, \boldsymbol{\varepsilon}\rangle,$$

or more concisely,

$$C|\gamma\rangle = -|\gamma\rangle.$$

Similarly,

$$C|\mathrm{n}\gamma\rangle = (-1)^n|\mathrm{n}\gamma\rangle$$

for an $n$ photon state. An immediate consequence is the vanishing of any amplitude that connects a state of an even number of photons with a state of an odd number of photons (*Furry's theorem*). The proof is an immediate application of the invariance of $S$ under $C$:

$$\langle \mathrm{n}\gamma|S|\mathrm{m}\gamma\rangle = \langle \mathrm{n}\gamma|C^{-1}SC|\mathrm{m}\gamma\rangle = (-1)^{n+m}\langle \mathrm{n}\gamma|S|\mathrm{m}\gamma\rangle,$$

and therefore

$$\langle \mathrm{n}\gamma|S|\mathrm{m}\gamma\rangle = 0,$$

unless $n + m$ is even.

The decay $\pi^0 \to 2\gamma$ tells us that

$$C|\pi^0\rangle = +|\pi^0\rangle.$$

Furthermore we can conclude that the process $\pi^0 \to 3\gamma$ is not just depressed by a factor of $\alpha$ but is *strictly forbidden.*

A state must of course be neutral (indeed have *all* additive quantum numbers equal to zero) in order to be an eigenstate of $C$, but not all such states are $C$-eigenstates. For example, an arbitrary $\pi^+\pi^-$ state is not an eigenstate, but one with definite orbital angular momentum is. For such a state, $C$ in effect changes $\mathbf{r}$ to $-\mathbf{r}$ in the wave function (Fig. 6.2), which produces a $(-1)^l$; that is,

$$C|\pi^+\pi^- l\mathrm{m}\rangle = (-1)^l|\pi^+\pi^- l\mathrm{m}\rangle.$$

[The corresponding analysis for positronium (bound $e^+e^-$ pair) is more complicated because of the effect of $C$ on the Dirac spinors.]

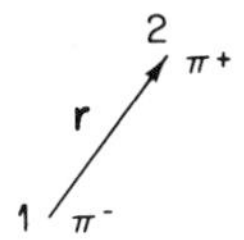


**Fig. 6.2.** The transformation $\mathbf{r} \to -\mathbf{r}$ has the same effect as $C$ for the $\pi^+\pi^-$ state.

## 6.4. *CPT* Theorem

We have seen how charge conjugation relates particle and antiparticle; we can use $C$-invariance to prove that particle and antiparticle have the same decay rate; for example,

$$\langle bc|S|a\rangle = \langle bc|C^{-1}SC|a\rangle = \langle \bar{b}\bar{c}|S|\bar{a}\rangle,$$

and therefore

$$\Gamma_{a\to bc} = \Gamma_{\bar{a}\to\bar{b}\bar{c}}.$$

But what if $C$-invariance does not hold, as we know is the case for weak interactions (e.g., $\pi^{\pm}$ decay); does the equality break down? The answer (for a *total* decay rate) is *no*, because this result is also a consequence of the *CPT theorem*, which holds much more generally than either $C$, $P$, or $T$.

The theorem states that any "sensible" theory is invariant under the *combined* operations of $C$, $P$, and $T$ (in any order). "Sensible" here means any local Lagrangian theory invariant under proper Lorentz transformations. There are a number of interesting results that follow from this theorem; for example:

1. Masses, lifetimes, and (except for sign) magnetic moments of particle and antiparticle are equal.
2. Invariance under $C$, $P$, or $T$ implies invariance under the product of the remaining two transformations.
3. Noninvariance under $C$, $P$, or $T$ implies lack of invariance under the product of the remaining two transformations.

The theorem is grounded on very general arguments, but experimental confirmation is still desirable. Perhaps the strongest piece of evidence is the equality of $K^0$ and $\overline{K}^0$ masses, which is established with very high precision —much higher than for any other particle—antiparticle pair. (Such evidence shows lack of violation, which, of course, is not the same as demonstrating the theorem.)

We will discuss pion decay and parity nonconservation later; the *CPT* result we cite here is the equality of $\pi^{\pm}$ masses and lifetimes.

## 6.5. Isospin

Isospin symmetry, which appears to hold exactly for the strong interactions, simplifies the table of hadrons (strongly interacting particles) by organizing them into multiplets.

### *a. Isospin Multiplets and Rotations. Invariance*

The three pions have nearly the same mass:

$$m_{\pi^\pm} - m_{\pi^0} \simeq 4.5 \quad \text{MeV},$$

which is small compared to the mass itself:

$$m_\pi \simeq 140 \quad \text{MeV}.$$

This leads us to believe that, if nature were so kind as to turn off the electromagnetic interaction, the three masses would be the same. The reader is probably aware that we now have many such *isospin multiplets* in particle physics, but there was only one when Heisenberg introduced the concept of *isotopic spin* (or *isospin*) in 1932. Heisenberg suggested that the neutron and proton be considered as two states of the same object, the *nucleon*. This is analogous to the two states of a spin-one-half object, so we represent the nucleon by a two-component spinor:

$$|\mathrm{p}\rangle \equiv |\mathrm{proton}\rangle \equiv \begin{pmatrix}1\\0\end{pmatrix}, \qquad |\mathrm{n}\rangle \equiv |\mathrm{neutron}\rangle \equiv \begin{pmatrix}0\\1\end{pmatrix},$$

i.e.,

$$T = \tfrac{1}{2}, \qquad T_3 = \pm\tfrac{1}{2}. \tag{6.29}$$

In the same way, the pion states are the three components of an isospin one object:

$$T = 1, \qquad T_3 = 1, 0, -1, \qquad \text{for} \quad \pi^+, \pi^0, \pi^-.$$

We can include both of these cases by writing for the charge $Q$,

$$Q = T_3 + B/2, \tag{6.30}$$

where $B$ is the *baryon number*. Just analogous to ordinary spin and angular momentum, the assumption of *isospin invariance* implies that the total isospin $T$ is conserved in the strong interactions. Of course, isospin conservation is violated by the electromagnetic interactions, but these are about 100 times weaker than the strong interactions.

We will illustrate how this is exploited shortly. Before doing so, it may be instructive to look more closely at what is meant by isospin invariance. We start with the statement that isospin invariance means that the nucleon–nucleon force is invariant under the replacement of each proton and neutron amplitude, $|\mathrm{p}\rangle$ and $|\mathrm{n}\rangle$, by some complex linear combination of $|\mathrm{p}\rangle$ and $|\mathrm{n}\rangle$;

$$\begin{aligned} |\mathrm{p}\rangle &\to |\mathrm{p}'\rangle = a|\mathrm{p}\rangle + b|\mathrm{n}\rangle, \\ |\mathrm{n}\rangle &\to |\mathrm{n}'\rangle = c|\mathrm{p}\rangle + d|\mathrm{n}\rangle. \end{aligned} \tag{6.31}$$

In terms of the isospinors

$$\begin{pmatrix} p' \\ n' \end{pmatrix} = \begin{pmatrix} a & b \\ c & d \end{pmatrix} \begin{pmatrix} p \\ n \end{pmatrix}, \tag{6.31'}$$

or

$$N' = UN.$$

We want

$$N^\dagger N = p^*p + n^*n$$

to remain unchanged, so $U$ must be unitary. One can show that the most general unitary $2 \times 2$ matrix is expressed by

$$U = e^{i\phi}\begin{pmatrix} a & b \\ -b^* & a^* \end{pmatrix}, \qquad a^*a + b^*b = 1. \tag{6.32}$$

The overall phase angle $\phi$ gives us nothing interesting here, so we restrict ourselves to $\phi = 0$; then $U$ is *unimodular* (det $U = 1$) as well as being unitary. The collection of all such $U$'s forms a (continuous) *group*[7] known as SU(2) (for *special unitary group in two dimensions*).

Any $U$ can be formed by multiplying together a large number of infinitesimal $U$'s (as discussed earlier); these have the form

$$U = 1 + i\varepsilon G = \begin{pmatrix} 1 + \alpha & \beta \\ -\beta^* & 1 + \alpha^* \end{pmatrix}, \tag{6.33}$$

with $\alpha$ and $\beta$ small. Then $\det(U) = 1$ implies

$$\alpha^* = -\alpha \qquad \text{or} \qquad \alpha = i\varepsilon_3,$$

while

$$\beta = \varepsilon_2 + i\varepsilon_1 \qquad (\varepsilon_i \text{ real}).$$

Thus

$$U = \begin{pmatrix} 1 & 0 \\ 0 & 1 \end{pmatrix} + i\varepsilon_3\begin{pmatrix} 1 & 0 \\ 0 & -1 \end{pmatrix} + i\varepsilon_2\begin{pmatrix} 0 & -i \\ i & 0 \end{pmatrix} + i\varepsilon_1\begin{pmatrix} 0 & 1 \\ 1 & 0 \end{pmatrix} = 1 + i\boldsymbol{\varepsilon} \cdot \boldsymbol{\tau}. \tag{6.34}$$

The $\tau_i$ are of course just the Pauli matrices, and since there are *three real parameters* $\varepsilon_i$, the transformation $U$ can be thought of as a *rotation* in an abstract three-dimensional *isospin space*. In fact, since the $\tau_i$ satsify the commutation relations

$$[\tau_i, \tau_j] = 2i\varepsilon_{ijk}\tau_k$$

---

[7] For a collection of transformations $\{U\}$ to form a group, we must have *closure*: for any two group elements, $U_1$, $U_2 \in \{U\}$, the products $(U_1U_2)$, $(U_2U_1) \in \{U\}$; and an *inverse* $U_1^{-1} \in \{U\}$ for each element $U_1 \in \{U\}$. The transformations considered above all form groups.

or

$$[\tfrac{1}{2}\tau_i, \tfrac{1}{2}\tau_j] = i\varepsilon_{ijk}\tfrac{1}{2}\tau_k \tag{6.35}$$

—the angular momentum commutation relations—the $U$'s have the multiplication properties appropriate to *ordinary rotations* in *three dimensions*. [Compare Eqs. (3.51); there is nothing analogous to orbital angular momentum here, however.]

To see explicitly that this is so, we can inquire about the behavior of a vector **A** in isospin space; the vector property means that the result must be the same as in Eq. (3.46). Thus for a rotation of $\theta$ around the 3 axis,

$$\mathbf{A}' = R\mathbf{A},$$

with $R = 1 + i\theta t_3$, and for a rotation around the direction $\hat{\mathbf{n}}$,

$$R = 1 + i\theta\hat{\mathbf{n}} \cdot \mathbf{t}. \tag{6.36}$$

[The $t$ are the same matrices as the $S$ in Eqs. (3.47).] Since it is the $\tau/2$ that satisfy the angular momentum commutation relations, we rewrite Eq. (6.34) as

$$U = 1 + i\theta\hat{\mathbf{n}} \cdot \boldsymbol{\tau}/2. \tag{6.37}$$

Then Eqs. (6.36), (6.37), and the fact that the $\tau_i/2$ and the $t_i$ satisfy the same commutation relations tell us that the multiplication table for the $R$'s is exactly the same as that for the $U$'s. We say that the $R$-matrices form a three-dimensional representation of the SU(2) group, and that the $t_i$ form a three-dimensional representation of the commutation relations, Eq. (6.35).

Finite transformations are compounded out of infinitesimal ones in the manner of Eq. (3.24). For a rotation of finite angle $\theta$ around the 3-axis, for example,

$$\begin{aligned} U &= e^{i\theta\tau_3/2}, \\ R &= e^{i\theta t_3}, \end{aligned} \tag{6.38}$$

and for representations of higher dimension,

$$D = e^{i\theta T_3}. \tag{6.39}$$

In addition to organizing the hadrons into multiplets, isospin symmetry implies conservation of $T$ and $T_3$, and relates matrix elements involving different members of multiplets. These different members, because they are related by isospin rotations, must have the same spin, parity, and mass (except for small electromagnetic corrections). Many examples of isospin symmetry will appear in the following, so we consider just one case here. Since both the deuteron and $He^4$ have $T = 0$ (neither having a similar state with $Q' = Q \pm 1$), the reaction

$$\begin{array}{ccccc} & d & + \; d \rightarrow & He^4 & + \; \pi^0 \\ (T = & 0 & 0 & 0 & 1) \end{array}$$

is forbidden; and, in fact, the experimental cross section for this process is less than 1% of what we would expect without isospin conservation.

### b. Isospin Invariant Coupling. Charge Independence

The combination $N^\dagger \boldsymbol{\tau} N$ is very important because it transforms like a a vector in isospin space. That is,

$$N'^\dagger \boldsymbol{\tau} N' = N^\dagger U^{-1} \boldsymbol{\tau} U N, \tag{6.40}$$

and for an infinitesimal rotation around the 3-axis again,

$$\tau_i' \equiv U^{-1} \tau_i U = (1 - i\theta\tau_3/2)\tau_i(1 + i\theta\tau_3/2) = \tau_i + i\theta[\tau_i, \tau_3]/2. \tag{6.41}$$

Therefore,

$$\begin{aligned} \tau_1' &= U^{-1}\tau_1 U = \tau_1 + \theta\tau_2, \\ \tau_2' &= U^{-1}\tau_2 U = \tau_2 - \theta\tau_1, \\ \tau_3' &= U^{-1}\tau_3 U = \tau_3, \end{aligned} \tag{6.42}$$

as is appropriate for a vector quantity.

So far we have been forming isospin multiplets out of states, but we can equally well talk about the corresponding field operators. Since the pion clearly has isospin one, we can form an isospin vector out of the pion field operators. Let $\pi^+(x)$ be the field operator that *creates* a $\pi^+$ (and destroys a $\pi^-$), similarly for operators $\pi^-(x)$ and $\pi^0(x)$. We introduce the Hermitian field operators

$$\pi_1 \equiv \frac{\pi^+ + \pi^-}{\sqrt{2}}, \qquad \pi_2 \equiv -i\,\frac{\pi^+ - \pi^-}{\sqrt{2}}, \qquad \pi_3 \equiv \pi_0; \tag{6.43}$$

that is,

$$\pi^+ = \frac{\pi_1 + i\pi_2}{\sqrt{2}}, \qquad \pi^- = \frac{\pi_1 - i\pi_2}{\sqrt{2}}.$$

Then

$$\boldsymbol{\pi}(x) \equiv \begin{pmatrix} \pi_1 \\ \pi_3 \\ \pi_2 \end{pmatrix} \tag{6.44}$$

is the desired isospin vector. Furthermore, we now let $p$ stand for $\psi_p(x)$, the field operator that destroys protons (and creates antiprotons), and $n$ stand for $\psi_n(x)$, which destroys neutrons, etc. Thus

$$N \equiv \begin{pmatrix} p \\ n \end{pmatrix} \equiv \begin{pmatrix} \psi_p \\ \psi_n \end{pmatrix} \equiv \psi_N(x) \tag{6.45}$$

destroys nucleons (and creates antinucleons) while $N^\dagger$ creates nucleons, etc. The $N^\dagger N$ combination will insure that the interaction conserves *baryon number*, i.e., number of (baryons) – number of (antibaryons); baryon conservation is very well founded experimentally.

We can now write a pion–nucleon interaction term which is invariant under isospin rotations (so that its inclusion in the Hamiltonian will not damage isospin conservation). We start with

$$\boldsymbol{\pi} \cdot \boldsymbol{\tau} \equiv \pi_1 \tau_1 + \pi_2 \tau_2 + \pi_3 \tau_3 = \sqrt{2}\pi^+ \tau^- + \sqrt{2}\pi^- \tau^+ + \pi_3 \tau_3, \tag{6.46}$$

where we have introduced the raising and lowering operators for the nucleon isospinors:

$$\tau^+ \equiv \tfrac{1}{2}(\tau_1 + i\tau_2), \qquad \tau^- \equiv \tfrac{1}{2}(\tau_1 - i\tau_2); \tag{6.47}$$

$$\tau^+ = \begin{pmatrix} 0 & 1 \\ 0 & 0 \end{pmatrix} \quad \text{and} \quad \tau^- = \begin{pmatrix} 0 & 0 \\ 1 & 0 \end{pmatrix} \tag{6.48}$$

in our representation. Then, since $N^\dagger \boldsymbol{\tau} N$ transforms like an isospin vector,

$$g\boldsymbol{\pi} \cdot N^\dagger \boldsymbol{\tau} N$$

is the invariant interaction; if we write the Dirac spinors explicitly, this is, for example,

$$ig\boldsymbol{\pi} \cdot \bar{\psi}_N \gamma_5 \boldsymbol{\tau} \psi_N \qquad \text{(pseudoscalar interaction)},$$

or

$$ig(\partial_\mu \boldsymbol{\pi}) \cdot \bar{\psi}_N \gamma_\mu \gamma_5 \boldsymbol{\tau} \psi_N \qquad \text{(pseudovector interaction)}.$$

It is interesting to examine this interaction in terms of the different charge states. Since

$$N^\dagger \tau^+ N = (p^*, n^*) \begin{pmatrix} 0 & 1 \\ 0 & 0 \end{pmatrix} \begin{pmatrix} p \\ n \end{pmatrix} = (p^*, n^*) \begin{pmatrix} n \\ 0 \end{pmatrix} = p^* n,$$

and

$$N^\dagger \tau^- N = (p^*, n^*) \begin{pmatrix} 0 & 0 \\ 1 & 0 \end{pmatrix} \begin{pmatrix} p \\ n \end{pmatrix} = n^* p,$$

$$g\boldsymbol{\pi} \cdot N^\dagger \boldsymbol{\tau} N = g\{\sqrt{2}\pi^+ n^* p + \sqrt{2}\pi^- p^* n + \pi^0 p^* p - \pi^0 n^* n\}, \tag{6.49}$$

using Eq. (6.46). Thus the individual coupling constants are

$$g_{\pi^+ \mathrm{np}} = g_{\pi^- \mathrm{pn}} = \sqrt{2}\, g \qquad \text{and} \qquad g_{\pi^0 \mathrm{pp}} = -g_{\pi^0 \mathrm{nn}} = g. \tag{6.50}$$

These relations allow us to compare the nn, pp, and np *forces* (in a given angular momentum and spin state). Using Yukawa's idea that these forces arise from *pion exchange*, we have the diagrams of Fig. 6.3. Thus

$$\mathrm{nn} : \mathrm{pp} : \mathrm{np} = (-g)^2 : g^2 : (2g^2 - g) = 1 : 1 : 1, \tag{6.51}$$

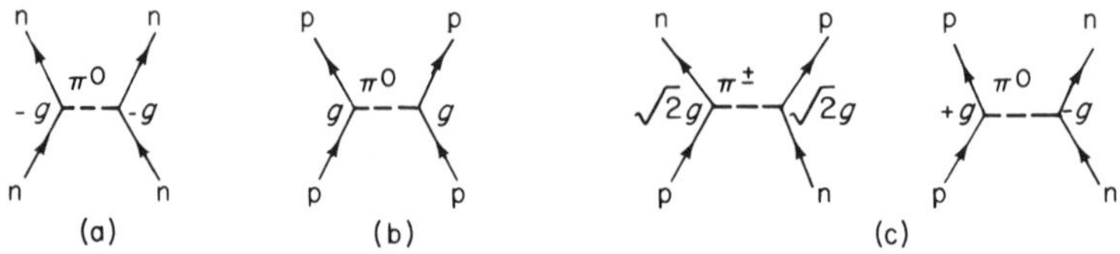


**Fig. 6.3.** The nucleon–nucleon force from pion exchange; (a) pp, (b) nn, (c) np.

which is just the statement of the *charge independence of nuclear forces*. In fact, Kemmer used this argument the other way around to predict the existence of neutral pions (in addition to the charged pions postulated by Yukawa), long before *any* pions were observed.

### c. Generalized Symmetry

We know that the total wave function for identical bosons must be *symmetric* under interchange of all (space and spin) coordinates, and *antisymmetric* for identical fermions; this can be generalized to include particles identical except for isospin state, simply by including the isospin (charge) coordinates among those that get interchanged. For example, a state of two $\pi^+$'s must be spatially symmetric. Since *all* the $2\pi$ $T = 2$ and $T = 0$ isospin wave functions are symmetric under interchange, *generalized symmetry* then requires that these states have symmetric spatial wave functions; similarly, since the $2\pi$ $T = 1$ isospin wave functions are antisymmetric, they must combine with antisymmetric spatial wave functions to have the correct overall symmetry. The two pion isospin wave functions and spatial symmetries are exhibited in Table 6.6. The same information for two nucleons is shown in Table 6.7.

Thus, for example, since the deuteron is isosinglet rather than triplet, even $L$ must combine with (ordinary) spin one in order to have overall antisymmetry.

**Table 6.6**

*Two Pions*[a,b]

| Particles | $\|T, T_3\rangle$ | Isospin wave function | Space wave function |
|---|---|---|---|
| $\pi^+\pi^+$ | $\|2, 2\rangle$ | $\|\pi^+, \pi^+\rangle$ | Symmetric |
| $\pi^+\pi^0$ | $\|2, 1\rangle$ | $1/\sqrt{2}\{\|\pi^+, \pi^0\rangle + \|\pi^0, \pi^+\rangle\}$ | Symmetric |
| $\pi^0\pi^0 + \pi^+\pi^-$ | $\|2, 0\rangle$ | $1/\sqrt{6}\{2\|\pi^0, \pi^0\rangle + \|\pi^+, \pi^-\rangle + \|\pi^-, \pi^+\rangle\}$ | Symmetric |
| $\pi^+\pi^0$ | $\|1, 1\rangle$ | $1/\sqrt{2}\{\|\pi^+, \pi^0\rangle - \|\pi^0, \pi^+\rangle\}$ | Antisymmetric |
| $\pi^+\pi^-$ | $\|1, 0\rangle$ | $1/\sqrt{2}\{\|\pi^+, \pi^-\rangle - \|\pi^-, \pi^+\rangle\}$ | Antisymmetric |
| $\pi^0\pi^0 + \pi^+\pi^-$ | $\|0, 0\rangle$ | $1/\sqrt{3}\{\|\pi^+, \pi^-\rangle + \|\pi^-, \pi^+\rangle - \|\pi^0, \pi^0\rangle\}$ | Symmetric |

[a] The wave functions are obtained by using the lowering operator $T_{\text{total}}^- = T_1^- + T_2^-$ on each $|T, T_3\rangle$ state and the relation

$$T^-|T, T_3\rangle = +[T(T+1) - T_3(T_3 - 1)]^{1/2}|T, T_3 - 1\rangle;$$

the $|1, 1\rangle$ wave function is constructed to be orthogonal to the $|2, 1\rangle$ wave function, etc. Note that applying $T^-$ does not affect the symmetry of the wave function.

[b] In this table, we have used

$$|\pi^+\rangle \equiv |1, 1\rangle,$$
$$|\pi^0\rangle \equiv |1, 0\rangle,$$
$$|\pi^-\rangle \equiv |1, -1\rangle$$

for simplicity. However, it is actually more convenient to use the phase convention $|\pi^+\rangle = -|1, 1\rangle$ with the others unchanged. This is because

$$e^{i\pi T_2}\,|1, 1\rangle = +|1, -1\rangle,$$

with the phase convention for the Clebsch–Gordan coefficients that has $T^-|T, T_3\rangle = +\sqrt{\ }\ |T, T_3 - 1\rangle$. Since we want

$$e^{i\pi T_2}\,|\pi^+\rangle = -|\pi^-\rangle$$

to agree with Eq. (6.52), we require the extra minus. To make Table 6.6. conform with this convention, we need merely insert a factor of $(-1)$ for each $\pi^+$. In general, we will omit these factors of $(-1)$.

**Table 6.7**

*Two Nucleons*

| Particles | $\|T, T^3\rangle$ | Isospin wave function | Spin–space wave function |
|---|---|---|---|
| pp | $\|1, 1\rangle$ | $\|\mathrm{p, p}\rangle$ | Antisymmetric |
| np | $\|1, 0\rangle$ | $\frac{1}{\sqrt{2}}\{\|\mathrm{p, n}\rangle + \|\mathrm{n, p}\rangle\}$ | Antisymmetric |
| nn | $\|1, -1\rangle$ | $\|\mathrm{n, n}\rangle$ | Antisymmetric |
| np | $\|0, 0\rangle$ | $\frac{1}{\sqrt{2}}\{\|\mathrm{p, n}\rangle - \|\mathrm{n, p}\rangle\}$ | Symmetric |

### d. G-Parity

By combining it with a rotation in isospin space, the $C$ operator can be made more useful. Consider the rotation

$$R \equiv e^{i\pi T_2}.$$

This must have the effect

$$\pi_1 \xrightarrow{R} -\pi_1, \qquad \pi_2 \xrightarrow{R} \pi_2, \qquad \pi_3 \xrightarrow{R} -\pi_3.$$

But under $C$,

$$\pi^0 \xrightarrow{C} \pi^0 \qquad \text{and} \qquad \pi^\pm \xrightarrow{C} \pi^\mp,$$

so that

$$\pi_1 \xrightarrow{C} \pi_1, \qquad \pi_2 \xrightarrow{C} -\pi_2, \qquad \text{and} \qquad \pi_3 \xrightarrow{C} \pi_3.$$

Therefore under the combined operation

$$G \equiv Ce^{i\pi T_2},$$
$$\boldsymbol{\pi} \xrightarrow{G} -\boldsymbol{\pi},$$

so that

$$G|\pi^{0\pm}\rangle = -|\pi^{0\pm}\rangle. \tag{6.52}$$

As a result, *any* state of $n$ pions is an eigenstate of $G$, with eigenvalue (*G-parity*) $(-1)^n$. This forbids any transition from an odd to an even number of pions, as long as $C$- and isospin invariance hold [note the analogy with Furry's theorem (Section 6.3)].

## 6.6. Pions and the Nucleon–Nucleon Force

Having done the isospin analysis in Section 6.5b, let us now go further and compute the lowest order pion exchange contribution to nucleon–nucleon scattering. First we do the calculation for *scalar* mesons and the scalar interaction

$$\mathcal{H}' = g\phi(x)\bar{\psi}(x)\psi(x).$$

From the Feynman rules (Table 4.2, using $\delta$-function normalization) and Fig. 6.4 we find

$$S_{2\mathrm{fi}} = (-ig)^2(2\pi)^{-6}\left(\frac{m_\mathrm{N}}{E_1}\frac{m_\mathrm{N}}{E_2}\frac{m_\mathrm{N}}{E_3}\frac{m_\mathrm{N}}{E_4}\right)^{1/2}(2\pi)^4\delta(P_\mathrm{i} - P_\mathrm{f})$$
$$\times\left\{(\bar{u}_3 u_1)(\bar{u}_4 u_2)\frac{-i}{(p_1 - p_3)^2 + m^2} - (\bar{u}_4 u_1)(\bar{u}_3 u_2)\frac{-i}{(p_1 - p_4)^2 + m^2}\right\}. \tag{6.53}$$

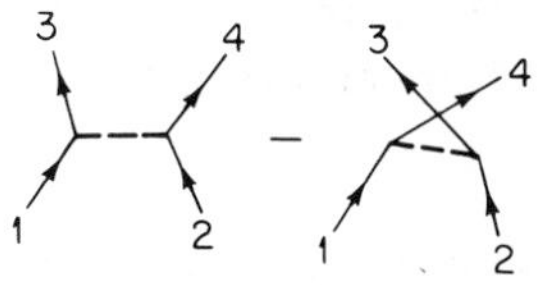


**Fig. 6.4.** Nucleon–nucleon scattering in lowest order.

In the CM system,

$$(p_1 - p_3)^2 \equiv -t = (\mathbf{p}_1 - \mathbf{p}_3)^2 = \mathbf{k}^2,$$

where

$$\mathbf{k} \equiv \mathbf{p}_1 - \mathbf{p}_3,$$

and

$$(p_1 - p_4)^2 \equiv -u = (\mathbf{p}_1 - \mathbf{p}_4)^2 = \mathbf{k}'^2,$$

where

$$\mathbf{k}' \equiv \mathbf{p}_1 - \mathbf{p}_4.$$

Furthermore, for *slow* nucleons,

$$E \simeq m_N$$

and

$$(\bar{u}_3 u_1) \simeq (\chi_3{}^\dagger \chi_1),$$

etc. Then factoring out

$$i(2\pi)^4 \delta(P_i - P_f)$$

to give $\overline{M}_{fi}$, Eq. (6.53) becomes simply

$$\overline{M}_{2fi} \simeq +g^2(2\pi)^{-6} \left\{ \frac{(\chi_3{}^\dagger\chi_1)(\chi_4{}^\dagger\chi_2)}{\mathbf{k}^2 + m^2} - \frac{(\chi_4{}^\dagger\chi_1)(\chi_3{}^\dagger\chi_2)}{\mathbf{k}'^2 + m^2} \right\}. \tag{6.54}$$

Now

$$(\mathbf{k}^2 + m^2)^{-1} = (4\pi)^{-1} \int e^{i\mathbf{k}\cdot\mathbf{r}} \frac{e^{-mr}}{r} \cdot d^3\mathbf{r}, \tag{6.55}$$

and comparison with Eq. (3.125) should convince the reader that the right-hand side of Eq. (6.54) is exactly what one would get from Born approximation for nucleon–nucleon scattering with interaction potential

$$V(\mathbf{r}) = -(g^2/4\pi)(e^{-mr}/r). \tag{6.56}$$

[The minus appears in front of the *exchange term* in Eq. (6.54) because we are now dealing with identical fermions rather than bosons.]

This form is known as the *Yukawa potential* and shows immediately how the *range* of the nuclear force is related to the mass of the exchanged meson: range $= m^{-1}$, and $m_\pi^{-1} \simeq 1.4$ fermi $= 1.4 \times 10^{-13}$ cm. This relation, in fact, led Yukawa to guess a value of about 200 $m_e$ for the pion mass (present value $\simeq 273\ m_e$ for $\pi^{\pm}$). We also see that exchange of scalar mesons gives rise to an *attractive* potential.

The calculation for *pseudoscalar* mesons (pions) is slightly more complicated. We use the pseudoscalar interaction

$$\mathcal{H}' = ig\pi(x)\bar{\psi}(x)\gamma_5\psi(x).$$

(The pseudovector form gives essentially the same results for slow nucleons. We can ignore isospin here because we have already done the isospin computation.) The resulting $S_{2fi}$ differs from Eq. (6.53) only in that

$$g \to ig,$$

and each

$$\bar{u}u \to \bar{u}\gamma_5 u.$$

Employing

$$u = \left(\frac{E+M}{2m}\right)^{1/2} \begin{pmatrix} \chi \\ \dfrac{\boldsymbol{\sigma}\cdot\mathbf{p}}{E+m}\chi \end{pmatrix}, \tag{2.32}$$

$$\gamma_4 = \begin{pmatrix} 1 & 0 \\ 0 & -1 \end{pmatrix}, \tag{2.19a}$$

and

$$\gamma_5 = -\begin{pmatrix} 0 & 1 \\ 1 & 0 \end{pmatrix},$$

we find for *slowly moving nucleons*

$$\begin{aligned} \bar{u}_3\gamma_5 u_1 &\simeq (1/2m_N)\chi_3{}^\dagger\boldsymbol{\sigma}\cdot\mathbf{k}\chi_1, \\ \bar{u}_4\gamma_5 u_2 &\simeq -(1/2m_N)\chi_4{}^\dagger\boldsymbol{\sigma}\cdot\mathbf{k}\chi_2, \\ \bar{u}_4\gamma_5 u_1 &\simeq (1/2m_N)\chi_4{}^\dagger\boldsymbol{\sigma}\cdot\mathbf{k}'\chi_1, \\ \bar{u}_3\gamma_5 u_2 &\simeq -(1/2m_N)\chi_3{}^\dagger\boldsymbol{\sigma}\cdot\mathbf{k}'\chi_2. \end{aligned}$$

Thus instead of Eq. (6.54), we now have

$$\overline{M}_{2fi} \simeq +\frac{g^2(2\pi)^{-6}}{(2m_N)^2}\left\{\frac{(\chi_3{}^\dagger\boldsymbol{\sigma}\cdot\mathbf{k}\chi_1)(\chi_4{}^\dagger\boldsymbol{\sigma}\cdot\mathbf{k}\chi_2)}{\mathbf{k}^2+m^2} - \frac{(\chi_4{}^\dagger\boldsymbol{\sigma}\cdot\mathbf{k}'\chi_1)(\chi_3{}^\dagger\boldsymbol{\sigma}\cdot\mathbf{k}'\chi_2)}{\mathbf{k}'^2+m^2}\right\}. \tag{6.57}$$

This corresponds to a nucleon–nucleon interaction potential

$$V(r) = -(g^2/4\pi)(2m_N)^{-2}(\boldsymbol{\sigma}_1\cdot\boldsymbol{\nabla}_1)(\boldsymbol{\sigma}\cdot\boldsymbol{\nabla}_2)(e^{-mr}/r), \tag{6.58}$$

which can be attractive or repulsive.

However, since the strong interaction coupling constant $g$ is not small, there is no *a priori* reason to expect the higher order contributions to be any smaller than the lowest order term we have just calculated. But then how can the above, or any, strong interaction perturbation theory calculation be any good? In general, such calculations are not much good; however, the preceeding *one-pion-exchange contribution* (*OPEC*) does successfully give the *long-range* part of the nucleon–nucleon force. This is because the longest range part of any force comes from the exchange of an object of the lowest possible mass and hence the diagrams of Fig. 6.4. (This property is exhibited by the Yukawa potential, and can be demonstrated more generally.) Now because the *high* $l$ N–N phase shifts (for which $l \geq kR$) are determined by this long-range part of the interaction, the OPEC idea can readily be tested and used to determine a value of $g^2$; this in turn can be compared with the value that emerges from (nonperturbative) analysis of $\pi$–N scattering and photopion production ($\gamma\text{N} \to \pi\text{N}$). The value $g^2/4\pi \simeq 14$ fits the three types of scattering.

## Problems

**6.1.** The time reversal formalism of Section 6.1b is consistent with the behavior (for spinless particles)

$$Ta^\dagger(\mathbf{k})T^{-1} = a^\dagger(-\mathbf{k}), \qquad Ta(\mathbf{k})T^{-1} = a(-\mathbf{k}),$$

under $T$. Show that, for a charged Klein–Gordon operator, the properties of $T$ consequently imply that

$$\phi(\mathbf{r}, t) \xrightarrow{T} \phi(\mathbf{r}, -t),$$

and, that for the accompanying current density operator [Eq. (2.7)],

$$\mathfrak{j}_\mu(\mathbf{r}, t) \to -\mathfrak{j}_\mu(\mathbf{r}, -t).$$

(Check this with Table 6.1.)

**6.2.** Investigate the behavior of the equation $\partial_\mu F_{\mu\nu} = -\mathfrak{j}_\nu$(cf. Problem 2.6) under time reversal, under space inversion, under charge conjugation.

**6.3.** By requiring complete antisymmetry under the exchange fermion ⇄ antifermion (accomplished by charge conjugation plus spin and space exchange), show that positronium has $C = (-1)^{l+s}$. (The antisymmetry requirement follows from the fact that the fermion and antifermion operators anticommute.)

**6.4.*** From the covariance of the Dirac equation, we found that $L^{-1}\gamma_\mu L = a_{\mu\nu}\gamma_\nu$. In the case where $a_{\mu\nu}$ represents *space inversion*, show that $L \equiv P = \eta\gamma_4$

has the required properties. Use this and our representation in the rest system to prove that fermion and antifermion have *opposite* intrinsic parity.

**6.5.** Show that the (rare) annihilation $\bar{p}p \to \pi^0\pi^0$ can only take place from $\bar{p}p$ states of *odd* orbital angular momentum, with total spin $S = 1$. (Hint: Use Bose symmetry, $P$- and $C$-invariance, and the result of Problem 6.3.)

**6.6.** Consider a bound $\pi^+\pi^-$ system. For what angular momentum states can the system decay into two photons? Three photons? What about competition from the $\pi^+\pi^- \to 2\pi^0$ channels?

**6.7.** Prove that the most general unitary $2 \times 2$ matrix can be put into the form given in Eq. (6.33).

**6.8.** Derive the isospin wave functions for $2\pi$'s shown in Table 6.6. (Hint: See the first footnote to this table.)

**6.9.** By using the convention $\pi^+ = -|1, 1\rangle$, $\pi^0 = |1, 0\rangle$, $\pi^- = |1, -1\rangle$ (cf. second footnote to Table 6.6), show that we must have

$$CT_3C^{-1} = -T_3, \qquad CT_+C^{-1} = -T_-;$$

hence

$$CT_1C^{-1} = -T_1, \qquad CT_2C^{-1} = +T_2,$$

and therefore $[\mathbf{T}, G] = 0$. Thus we can have simultaneous eigenstates of $\mathbf{T}^2$, $T_3$, and $G$. (Since $C$ reverses all additive quantum numbers we *cannot*, of course, have simultaneous eigenstates of $T_3$ and $C$.)

**6.10.** Consider a $\pi$–$N$ interaction of the type $\mathcal{H}' = g\phi\bar{\psi}\psi$.

(a) Show that this implies the field equation

$$(\Box - m^2)\phi(x) = g\bar{\psi}(x)\psi(x)$$

(so that the nucleon "current" $\bar{\psi}\psi$ is a source of the pion field, just as the electric current is the source of the electromagnetic field).

(b) Consider a static, classical nucleon localized at the origin:

$$\bar{\psi}(x)\psi(x) = \delta(\bar{r}).$$

Show that

$$\phi(x) = -(g/4\pi)(e^{-mr}/r)$$

satisfies the field equation.

(c) Use this (classical, static) pion field to derive the Yukawa potential,

$$V(r) = V(\mathbf{x} - \mathbf{y}) = -(g^2/4\pi)(e^{-mr}/r),$$

for the interaction between two (classical, static) nucleons.

## Bibliography


G. Källén. *Elementary Particle Physics.* Addison-Wesley, Reading, Massachusetts, 1964.

F. E. Low. *Symmetry and Elementary Particles.* Gordon & Breach, New York, 1967.

J. J. Sakurai. *Invariance Principles and Elementary Particles.* Princeton Univ. Press, Princeton, New Jersey, 1964.

E. Segrè. *Nuclei and Particles.* Chap. 14. Benjamin, New York, 1965.

CHAPTER SEVEN

# Pion–Nucleon Scattering

With the production of pion beams, it became possible to do $\pi$N scattering experiments; this, it was hoped, would lead us to an understanding of the strong interactions. We now know a lot about $\pi$N scattering and have discovered a vast number of $\pi$N *resonances*, but our understanding of the strong interactions is still very limited. Before discussing the experimental results we consider the implications of isospin invariance and (ordinary) rotation invariance.

## 7.1. Isospin Analysis

Isospin invariance[1] for the strong (*or hadronic*) interactions does *not* imply that the pion–nucleon forces are independent of the charge states, which would be the analog of the charge independence of the nucleon–nucleon force; the symmetry is more subtle than that. As mentioned before, rotational invariance in isospin space means that both total $T$ and $T_3$ are conserved.

[1] We are here ignoring Coulomb forces, which destroy the isospin invariance. Coulomb scattering is fortunately confined to small scattering angles and can be calculated precisely so that its effect can be removed and the scattering can be discussed as if the electromagnetic interaction were not present.

Since for pion–nucleon scattering we are combining $T = 1$ with $T = \frac{1}{2}$, we can have total $T = \frac{3}{2}$ or $\frac{1}{2}$. Invariance further requires that the $S$-matrix elements, or scattering amplitudes, depend only on $T$ and not on $T_3$. Therefore, a mere two amplitudes will describe all the different possibilities contained in the reaction $\pi N \to \pi N$.

In practice, beams of charged pions and targets of hydrogen or deuterium are employed; because of the complications of extracting pion–neutron data from pion–deuteron data, we will discuss only the following reactions:

(a) $\pi^+ p \to \pi^+ p$;
(b) $\pi^- p \to \pi^- p$;
(c) $\pi^- p \to \pi^0 n$ (charge exchange).

Reaction (a) is purely $T = \frac{3}{2}$ (since $T_3 = \frac{3}{2}$), while (b) and (c) involve both $T = \frac{3}{2}$ and $\frac{1}{2}$. The isospin wave functions, which are obtained by using the lowering operator on the $|T, T_3\rangle = |\frac{3}{2}, \frac{3}{2}\rangle$ state and choosing $|\frac{1}{2}, \frac{1}{2}\rangle$ to be orthogonal to $|\frac{3}{2}, \frac{1}{2}\rangle$, are shown in Table 7.1. The coefficients of these wave

**Table 7.1**

*Pion–Nucleon Isospin Wave Functions*[a]

| $T_3$ | $T = \frac{3}{2}$ | $T = \frac{1}{2}$ |
|---|---|---|
| $\frac{3}{2}$ | $\pi^+ p$ | |
| $\frac{1}{2}$ | $(1/\sqrt{3})\{\sqrt{2}\,\pi^0 p + \pi^+ n\}$ | $(1/\sqrt{3})\{\sqrt{2}\,\pi^+ n - \pi^0 p\}$ |
| $-\frac{1}{2}$ | $(1/\sqrt{3})\{\sqrt{2}\,\pi^0 n + \pi^- p\}$ | $(1/\sqrt{3})\{\sqrt{2}\,\pi^- p - \pi^0 n\}$ |
| $-\frac{3}{2}$ | $\pi^- n$ | |

[a] Compare footnote to Table 6.6 for $\pi^+$ phase convention.

functions (the *Clebsch–Gordan coefficients*) are all we need to express the *physical amplitudes* (a), (b), and (c) in terms of the two *isospin amplitudes* $S_{3/2}$ and $S_{1/2}$.

To see how this goes, all we need is the completeness relation for isospin wave functions

$$\sum_{T,T_3} |T, T_3\rangle\langle T, T_3| = 1. \tag{7.1}$$

Here 1 is the unit operator in isospin space. Thus

$$\langle \pi^+ p|S|\pi^+ p\rangle = \sum_{T, T_3} \sum_{T', T_3'} \langle \pi^+ p|T, T_3\rangle\langle T, T_3|S|T', T_3'\rangle\langle T', T_3'|\pi^+ p\rangle. \tag{7.2}$$

The invariance of $S$ implies that, unless $T = T'$ and $T_3 = T_3'$,

$$\langle T, T_3 | S | T', T_3' \rangle = 0,$$

and that $\langle T, T_3 | S | T, T_3 \rangle$ is independent of $T_3$; these properties can be written as

$$\langle T, T_3 | S | T', T_3' \rangle = \delta_{TT'} \delta_{T_3 T_3'} S_T .$$

Thus Eq. (7.2) becomes

$$\langle \pi^+ p | S | \pi^+ p \rangle = \sum_T \langle \pi^+ p | T, T_3 \rangle S_T \langle T, T_3 | \pi^+ p \rangle = S_{3/2}, \qquad (7.3a)$$

using Table 7.1. Similarly,

$$\begin{aligned} \langle \pi^- p | S | \pi^- p \rangle &= \sum_T \langle \pi^- p | T, T_3 \rangle S_T \langle T, T_3 | \pi^- p \rangle \\ &= |\langle \pi^- p | \tfrac{3}{2}, -\tfrac{1}{2} \rangle|^2 S_{3/2} + |\langle \pi^- p | \tfrac{1}{2}, -\tfrac{1}{2} \rangle|^2 S_{1/2} \\ &= \tfrac{1}{3} S_{3/2} + \tfrac{2}{3} S_{1/2}, \end{aligned} \qquad (7.3b)$$

and

$$\begin{aligned} \langle \pi^- p | S | \pi^0 n \rangle &= \sum_T \langle \pi^- p | T, T_3 \rangle S_T \langle T, T_3 | \pi^0 n \rangle \\ &= \langle \pi^- p | \tfrac{3}{2}, -\tfrac{1}{2} \rangle \langle \tfrac{3}{2}, -\tfrac{1}{2} | \pi^0 n \rangle S_{3/2} \\ &\quad + \langle \pi^- p | \tfrac{1}{2}, -\tfrac{1}{2} \rangle \langle \tfrac{1}{2}, -\tfrac{1}{2} | \pi^0 n \rangle S_{1/2} \\ &= \sqrt{2}/3 (S_{3/2} - S_{1/2}). \end{aligned}$$

The cross sections implied by these relations (7.3) are then given by

$$\sigma_{\pi^+ p \to \pi^+ p} \propto |S_{3/2}|^2, \qquad (7.4a)$$

$$\sigma_{\pi^- p \to \pi^- p} \propto \tfrac{1}{9} |S_{3/2} + 2 S_{1/2}|^2, \qquad (7.4b)$$

$$\sigma_{\pi^- p \to \pi^0 n} \propto \tfrac{2}{9} |S_{3/2} - S_{1/2}|^2, \qquad (7.4c)$$

with the same proportionality constant for each. One can add the elastic and charge–exchange scattering for $\pi^- p$:

$$\begin{aligned} \sigma_{\pi^- p} &\propto \sum_{T, T_3} \langle \pi^- p | S^\dagger | T, T_3 \rangle \langle T, T_3 | S | \pi^- p \rangle \\ &= \sum_T |\langle \pi^- p | T, T_3 \rangle|^2 |S_T|^2, \end{aligned}$$

or, in terms of the cross sections for pure isospin states,

$$\sigma_{\pi^- p} = \tfrac{1}{3} \sigma_{3/2} + \tfrac{2}{3} \sigma_{1/2}, \qquad (7.5a)$$

while, of course,

$$\sigma_{\pi^+ p} = \sigma_{3/2} . \qquad (7.5b)$$

These equations also hold for energies for which inelastic scattering (e.g., $\pi N \to \pi\pi N$) becomes important. In that case, $\sigma_{\pi^- p}$, $\sigma_{3/2}$, and $\sigma_{1/2}$ are *total* cross sections, not simply the $\pi N \to \pi N$ cross sections.

## 7.2. Angular Dependence

We now want to consider the angular dependence of the $\pi N \to \pi N$ scattering. Here the introduction of partial waves and phase shifts facilitates the discussion, but of course the presence of the nucleon spin complicates the situation.

### *a. Elastic Scattering Amplitudes*

We first attack the nucleon spin complications. The asymptotic wave function will now be expected to have the form [compare Eq. (3.54)]

$$\psi e \simeq {}^{ikz}\chi_\uparrow + (e^{ikr}/r)[f_1(\theta, \phi)\chi_\uparrow + f_2(\theta, \phi)\chi_\downarrow]. \tag{7.6}$$

Here $e^{ikz}\chi_\uparrow$ represents the incident plane wave; the $\chi_\uparrow$ is the spinor for nucleon spin "up" in some direction. The scattered wave has *both* spin up, $f_1(\theta, \phi)\chi_\uparrow$, and spin down, $f_2(\theta, \phi)\chi_\downarrow$, parts.

It is convenient to take up as meaning the $+z$-direction.[2] Then in the standard representation where $\sigma_z$ is diagonal.

$$\chi_\uparrow = \begin{pmatrix}1\\0\end{pmatrix}, \qquad \chi_\downarrow = \begin{pmatrix}0\\1\end{pmatrix}.$$

Furthermore, the scattered wave can be written in terms of a $2\times 2$ scattering operator $F(\theta, \phi)$:

$$e^{ikr}/r \cdot F(\theta, \phi)\chi_\uparrow .$$

Since we want $F(\theta, \phi)$ to be invariant under rotation, it can only contain terms proportional to 1 and $\sigma \cdot$ *vector*, and the coefficients of those terms can only be functions of *scalars*. The only available vectors are $\mathbf{k}_i$, $\mathbf{k}_f$, and $\mathbf{k}_i \times \mathbf{k}_f$, but $\boldsymbol{\sigma} \cdot \mathbf{k}_j$ and $\boldsymbol{\sigma} \cdot \mathbf{k}_f$, since they change sign under space inversion, are ruled out by the requirement of $P$-invariance. Furthermore, the only scalar available is $\mathbf{k}_i \cdot \mathbf{k}_f = k^2 \cos\theta$, so the angular dependence of the coefficients is only on $\theta$, not on $\phi$ ($k \equiv |\mathbf{k}_i| = |\mathbf{k}_f|$ in the CM system). We write

$$F(\theta, \phi) \equiv 1 \cdot F_N(\theta) + \frac{i\boldsymbol{\sigma} \cdot \mathbf{k}_i \times \mathbf{k}_f}{k^2} F_S(\theta)$$

$$= 1 \cdot F_N(\theta) + i\boldsymbol{\sigma} \cdot \hat{\mathbf{n}} \sin\theta \cdot F_S(\theta). \tag{7.7}$$

---

[2] A different method of prescription of the spin state that has come into wide use is the *helicity representation*; here one uses eigenstates of the *helicity* operator $\boldsymbol{\sigma} \cdot \mathbf{p}/|\mathbf{p}|$ instead of $\sigma_z$ (with a fixed $z$-direction). The helicity operator is rotationally invariant, so a helicity eigenstate remains "pure" under rotation, in contrast to a $\sigma_z$ eigenstate.

Here

$$\hat{\mathbf{n}} \equiv \frac{\mathbf{k}_i \times \mathbf{k}_f}{|\mathbf{k}_i \times \mathbf{k}_f|}$$

is the normal to the scattering plane; from Fig. 7.1 we can see that

$$\hat{\mathbf{n}} = (-\sin\phi, \cos\phi, 0).$$

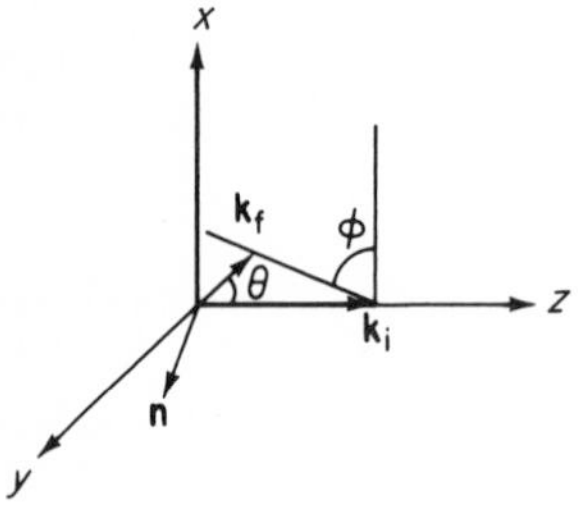


**Fig. 7.1.** Momenta and angles for $\pi N$ scattering.

Then using

$$\sigma_x = \begin{pmatrix} 0 & 1 \\ 1 & 0 \end{pmatrix} \quad \text{and} \quad \sigma_y = \begin{pmatrix} 0 & -i \\ i & 0 \end{pmatrix},$$

so that

$$i\boldsymbol{\sigma} \cdot \hat{\mathbf{n}} = \begin{pmatrix} 0 & e^{-i\phi} \\ -e^{i\phi} & 0 \end{pmatrix}, \tag{7.8}$$

we find

$$F\begin{pmatrix} 1 \\ 0 \end{pmatrix} = F_N \begin{pmatrix} 1 \\ 0 \end{pmatrix} - \sin\theta \cdot e^{i\phi} F_S \begin{pmatrix} 0 \\ 1 \end{pmatrix}, \tag{7.9}$$

which is just the form of the scattered wave in Eq. (7.6), with the $\phi$-dependence made explicit. $F_S(\theta)$ is called the *spin-flip* amplitude, $F_N(\theta)$ the *nonspin-flip* amplitude.

Now because $F(\theta, \phi)$ is rotationally invariant, an incident wave with arbitrary spinor $\chi_i$ produces scattered wave $F(\theta, \phi)\chi_i$. The amplitude for finding final state spinor $\chi_f$ is then

$$\chi_f^\dagger F \chi_i \equiv \langle f|F|i\rangle.$$

The scattering probability, i.e., differential cross section averaged over initial and summed over final spin, is therefore

$$\begin{aligned} d\sigma(\theta)/d\Omega &= \tfrac{1}{2} \sum_f \sum_i |\langle f|F|i\rangle|^2 \\ &= \tfrac{1}{2} \operatorname{tr}(FF^\dagger) \\ &= \tfrac{1}{2}\{|F_N|^2 + \sin^2\theta |F_S|^2\}. \end{aligned} \tag{7.10}$$

### b. Partial Wave Expansion

Of course, working with $F(\theta, \phi)$ is just one way of discussing the elastic scattering $S$-matrix elements; it is essentially the plane wave representation:

$$F_{\mathrm{N}}(\theta) \propto \langle \mathbf{k}_{\mathrm{f}}, \uparrow | S | \mathbf{k}_{\mathrm{i}}, \uparrow \rangle \tag{7.11a}$$

$$F_{\mathrm{S}}(\theta) \propto \langle \mathbf{k}_{\mathrm{f}}, \downarrow | S | \mathbf{k}_{\mathrm{i}}, \uparrow \rangle \tag{7.11b}$$

The angular momentum representation is another possibility; rotation invariance implies that $S$ is diagonal in angular momentum, which is a great advantage. However, there are two ways of forming an angular momentum $J$ in the presence of a single spin one-half:

$$J = l + \tfrac{1}{2}$$

and

$$J = l' - \tfrac{1}{2} \qquad \text{with} \quad l' = l + 1.$$

But the states of orbital angular momentum $l$ and $l'$ have opposite parity; so parity conservation implies that (elastic part of) $S$ is diagonal in $l$ and $l'$, too. The quantum numbers $J$, $l$, $m_J$ completely determine the states (for elastic scattering at a given energy). The diagonal property combined with conservation of probability (i.e., unitarity) then allows us to write [compare the spin-zero case, Eq. (3.62)]

$$S_{l_\pm} = \exp(2i\delta_{l_\pm}), \tag{7.12}$$

with the phase shifts $\delta_{l_+}$ real below the inelastic thresholds. Here $l_\pm$ or $J_\pm$ means $J = l \pm \frac{1}{2}$.

It is clear that $F(\theta, \phi)$ is completely determined by the phase shifts. Defining partial wave amplitudes

$$f_{l_\pm} \equiv \exp(i\delta_{l_\pm}) \sin \delta_{l_\pm}, \tag{7.13}$$

an analysis similar to Eqs. (3.56)–(3.63) gives the results

$$F_{\mathrm{N}}(\theta) = \frac{1}{k} \sum \{(l+1) f_{l_+} + l f_{l_-}\} P_l(\cos\theta), \tag{7.14a}$$

$$F_{\mathrm{S}}(\theta) = \frac{1}{k} \sum \{f_{l_+} - f_{l_-}\} \frac{dP_l(\cos\theta)}{d(\cos\theta)}, \tag{7.14b}$$

and

$$\sigma = \frac{2\pi}{k^2} \sum_J (2J+1)\{\sin^2 \delta_{J_+} + \sin^2 \delta_{J_-}\}. \tag{7.15}$$

We are now in a position to discuss the results of $\pi$N scattering experiments.

## 7.3. The (3, 3) Resonance

The total cross sections are determined by measuring the attenuation of $\pi^+$ and $\pi^-$ beams by liquid hydrogen targets and are plotted as functions of energy in Fig. 7.2. The striking peaks at about 200 MeV lab kinetic energy are evidence for the first resonance discovered in particle physics, now known

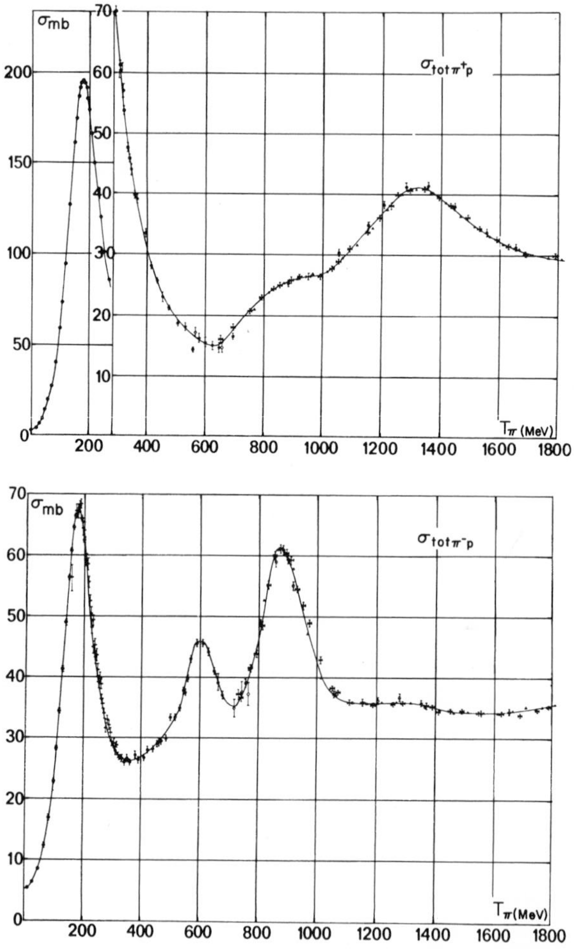


**Fig. 7.2.** Total cross sections for $\pi^+$p and $\pi^-$p scattering. [From P. Bareyre, C. Bricman, and G. Villet, Phase-shift analysis of pion–nucleon elastic scattering below 1.6 GeV. *Phys. Rev.* **165**, 1730–1749 (1968).]

as the N*(1236) or Δ(1236)[3] but originally called the (3, 3) *resonance*. Now in order to constitute a resonance, the scattering particles (if described in terms of wave packets) must experience a *time delay* between entering and leaving the range of interaction; the time delay varies with the energy and has its maximum at the resonance energy. This maximum time delay is the lifetime of the resonance. Furthermore, the associated peak in the cross section must correspond to unique values of spin, parity, and isospin. We discuss the evidence for this being so for the Δ(1236).

Considering isospin first, the peak in the $\pi^+$p cross section requires isospin $\frac{3}{2}$. If the scattering at the peak energy is pure isospin $\frac{3}{2}$, then from Eq. (7.5)

$$\sigma_{\pi^- p} : \sigma_{\pi^+ p} = 1 : 3, \tag{7.16}$$

which agrees with the experimental values. Thus the *background*, in this case $T = \frac{1}{2}$ scattering, must be very small. Another check comes from measuring $\pi^-$p elastic and charge exchange scattering separately. With the neglect of the $T = \frac{1}{2}$ contribution, Eqs. (7.4) require

$$\sigma_{\pi^+ p} : \sigma_{\pi^- p \to \pi^- p} : \sigma_{\pi^- p \to \pi^0 n} = 9 : 1 : 2, \tag{7.17}$$

which also agrees with experiment.

There are a number of measurements that can determine the spin of the resonance. To have a definite value for the spin and parity, the resonance arises from a single partial wave amplitude $f_{J_\pm}$ taking on its maximum value because of the phase shift *increasing* through 90° (or any odd multiple of $\pi/2$).[4]

Neglecting background, the cross section for pure isospin state at the peak is given by

$$\sigma_{\text{peak}} = 2\pi(2J + 1)/k^2. \tag{7.18}$$

The choice $J = \frac{3}{2}$ agrees with the observed 200 mb at the peak.

This choice is verified by examining the angular distributions. At the peak, the distribution, shown in Fig. 7.3, exhibits the $1 + 3\cos^2\theta$ behavior that is characteristic of $J = \frac{3}{2}$. Interestingly enough, the angular distribution cannot determine the orbital angular momentum (and hence the parity) of the resonance; in spite of the $l$ dependence of Eqs. (7. 14a, b), the differential cross section [Eq. (7.10)] depends only on $J$ and not on whether we have $J_+$ or $J_-$.

---

[3] The number inside the parenthesis is the energy in MeV.

[4] An interesting result of scattering theory is that the lifetime of a resonance (the time delay experienced by a wave packet) is given by $2(d\delta(E)/dE)$. This lifetime must be greater than 0; hence the phase shift must be increasing at resonance. See Goldberger and Watson, 1964, Chapter 8.

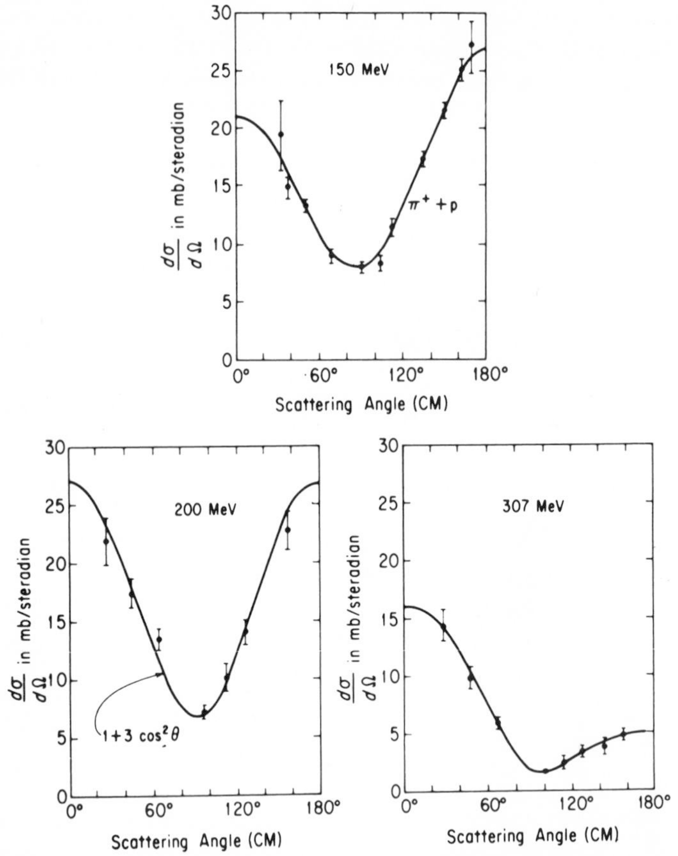


**Fig. 7.3.** Differential cross sections for $\pi^+$p scattering at several energies. At 200 MeV, the curve drawn is $(1 + 3\cos^2\theta)$. [From William R. Frazer, *Elementary Particles*, © 1966, Prentice-Hall, Inc., Englewood Cliffs, New Jersey. Originally from J. D. Jackson, *Physics of Elementary Particles* (copyright © 1958 by Princeton University Press).]

However, the fact that $kR \lesssim 1$ with the range $R \simeq (2m_\pi)^{-1}$ (corresponding to $2\pi$ exchange, since $G$-parity forbids single pion exchange) means $l > 1$ is unlikely. More convincing evidence has come from tracing the behavior of the individual phase shifts as the energy is increased from threshold; as $k \to 0$, the proportionality

$$\delta_{l_\pm} \propto k^{2l+1}$$

holds. This analysis, which of course requires data in the entire range of energies 0–200 MeV (lab) instead of just at 200 MeV, confirms the $P$-wave assignment.[5] Thus the $\Delta(1236)$ resonance is identified as having $T = \frac{3}{2}$ and $J^P = \frac{3}{2}^+$.

## 7.4. Results at Higher Energies

After a great deal of experimental work, the various phase shifts over an extensive energy range have been disentangled; some of these are shown as functions of energy in Fig. 7.4. For

$$K_\pi^{\mathrm{L}} \equiv E_\pi^{\mathrm{L}} - m_\pi \gtrsim 400 \quad \mathrm{MeV} \qquad (W \gtrsim 1400 \quad \mathrm{MeV}),$$

the inelastic channels become significant and we must introduce an inelasticity parameter $\eta_\pm$ for each $\delta_{l\pm}$ [as in Eq. (3.62′)].

Another way of exhibiting this $\delta$, $\eta$ behavior is by plotting each amplitude $A_{l_\pm}$ on an *Argand diagram.* Letting

$$A_{l_\pm} \equiv \frac{S_{l_\pm} - 1}{2i} = \frac{\eta_{l_\pm} \exp(2i\delta_{l_\pm}) - 1}{2i} = \frac{\eta_{l_\pm}}{2} \exp[2i\delta_{l_\pm} - i(\pi/2)] + \frac{i}{2}, \quad (7.19)$$

we plot points in the ($\mathrm{Re} A_{l_\pm}$, $\mathrm{Im}\, A_{l_\pm}$) plane for various values of the energy $E$. The quantities are shown in Fig. 7.5. For purely elastic scattering, the $A(E)$ point moves on the radius $\frac{1}{2}$ circle, and an elastic resonance corresponds to going through the point (0, 1) in the counterclockwise direction. This is shown for the (3, 3) resonance in Fig. 7.6. When there is appreciable inelasticity, the $A(E)$ point falls inside the circle and, in general, the behavior is much more complicated. Some experimental Argand diagrams for $\pi$N scattering are shown in Fig. 7.7; these correspond to the $\delta$'s and $\eta$'s of Fig. 7.4. Resonances correspond to counterclockwise loops in these diagrams.

The diagrams can be used to determine the positions and widths of the $\pi$N resonances in the energy range covered; these are shown in Table 7.2. This table also shows the values obtained by analyzing the energy dependence of each partial wave contribution to the total cross section; the two ways, in general, give somewhat different answers. The contributions of the higher energy resonances to the total cross sections (Fig. 7.2) are less striking than that of the (3, 3) resonance because (1) the resonances overlap; (2) any single partial wave forms only a small part of the total scattering once $kR \gg 1$; and (3) the inelasticity is large. The number of resonances that have been found is alarmingly large, and a continuing effort is being made to find more and determine their quantum numbers.

Needless to say, there has been a sustained effort on the part of theorists to "explain" this plethora of resonances. The difficulty of calculating hadronic processes has made progress very difficult, but there are definitely some bright spots too, which we discuss later. In particular, these resonances have helped to reveal some striking symmetries and regularities of the strong interaction.

---

[5] Both the cross sections and the polarization are also invariant under change of the *sign* of all the phase shifts. The resulting ambiguity is resolved by observing the interference with Coulomb scattering which occurs at small angles.

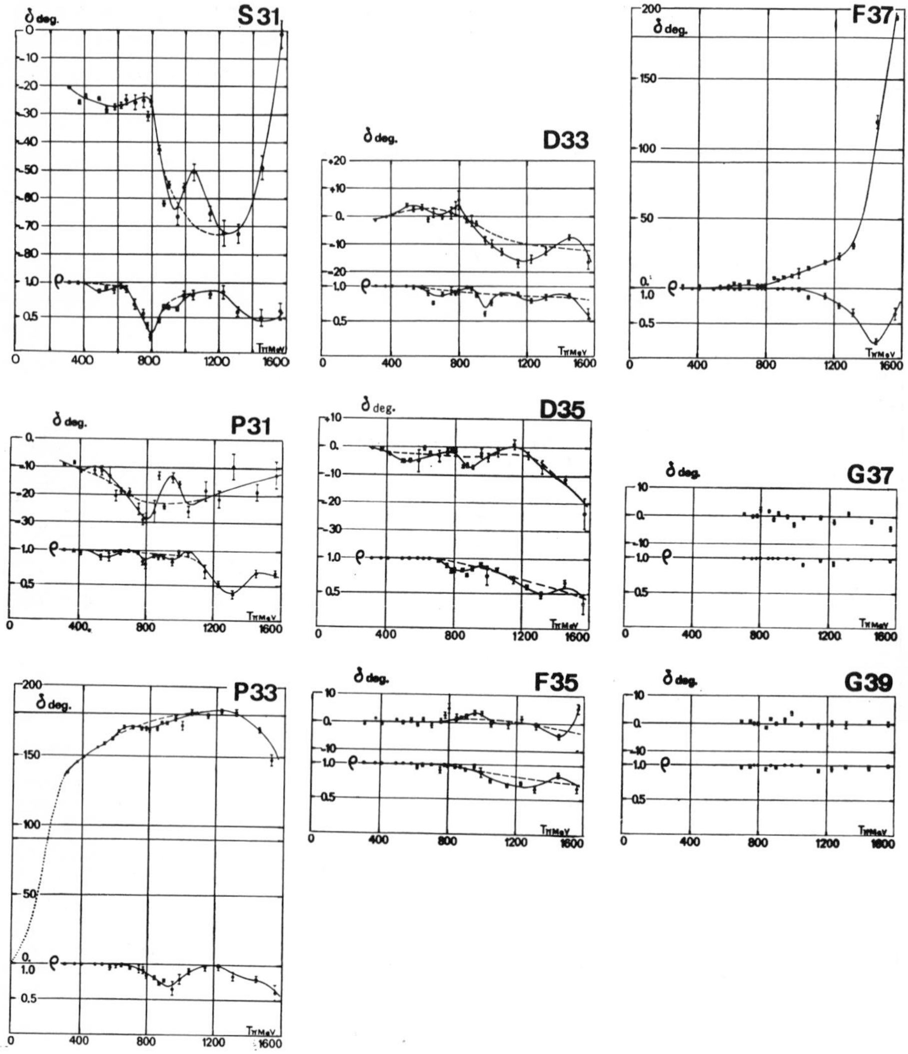


**Fig. 7.4(a).** Phase shifts and inelasticity parameters for $\pi$N scattering: $T = \frac{3}{2}$. The notation is $l_{2T,2J}$. [From P. Bareyre, C. Bricman, and G. Villet, Phase-shift analysis of pion–nucleon elastic scattering below 1.6 GeV. *Phys. Rev.* **165**, 1730–1749 (1968).]

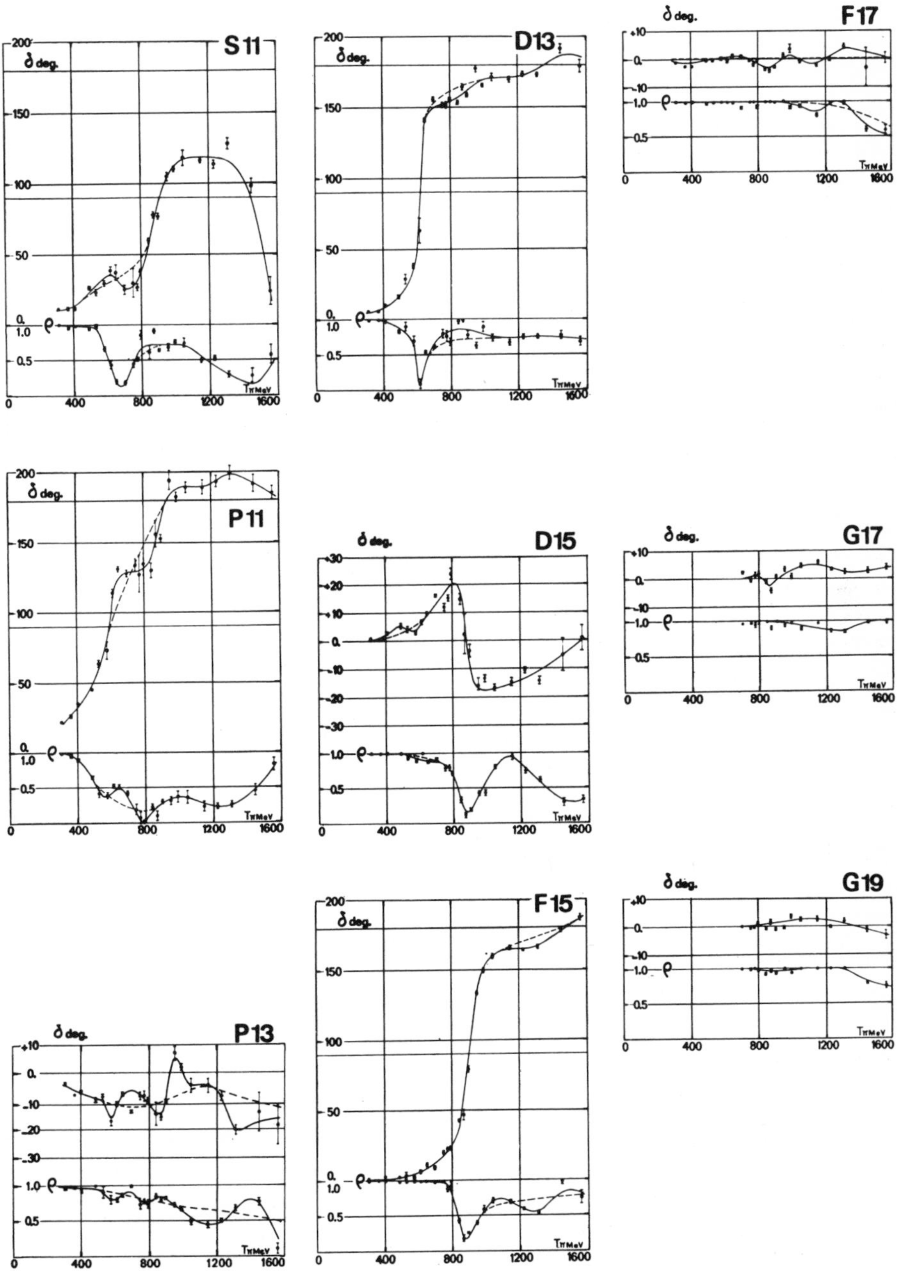


**Fig. 7.4(b).** Phase shifts and inelasticity parameters for $\pi N$ scattering: $T = \frac{1}{2}$. The notation is $l_{2T,2J}$. [From P. Bareyre, C. Bricman, and G. Villet, Phase-shift analysis of pion–nucleon elastic scattering below 1.6 GeV. *Phys. Rev.* **165**, 1730–1749 (1968).]

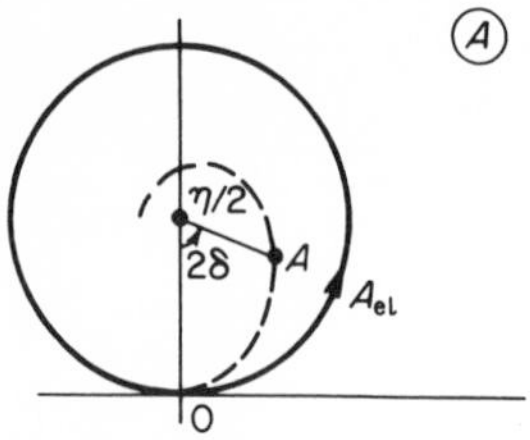


**Fig. 7.5.** Argand diagram for a partial wave scattering amplitude. Trajectories for elastic and inelastic scattering are shown.

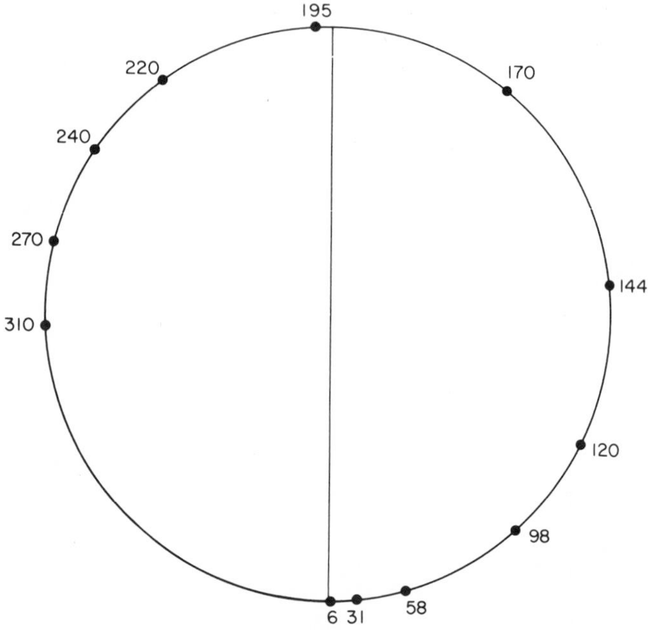


**Fig. 7.6.** Argand diagram for the $P_{33}$ partial wave of $\pi$N scattering. Numbers are $\pi$ lab energy in million electron volts. [Data from Roper *et al.*, *Phys. Rev.* **138**, B921 (1965).]

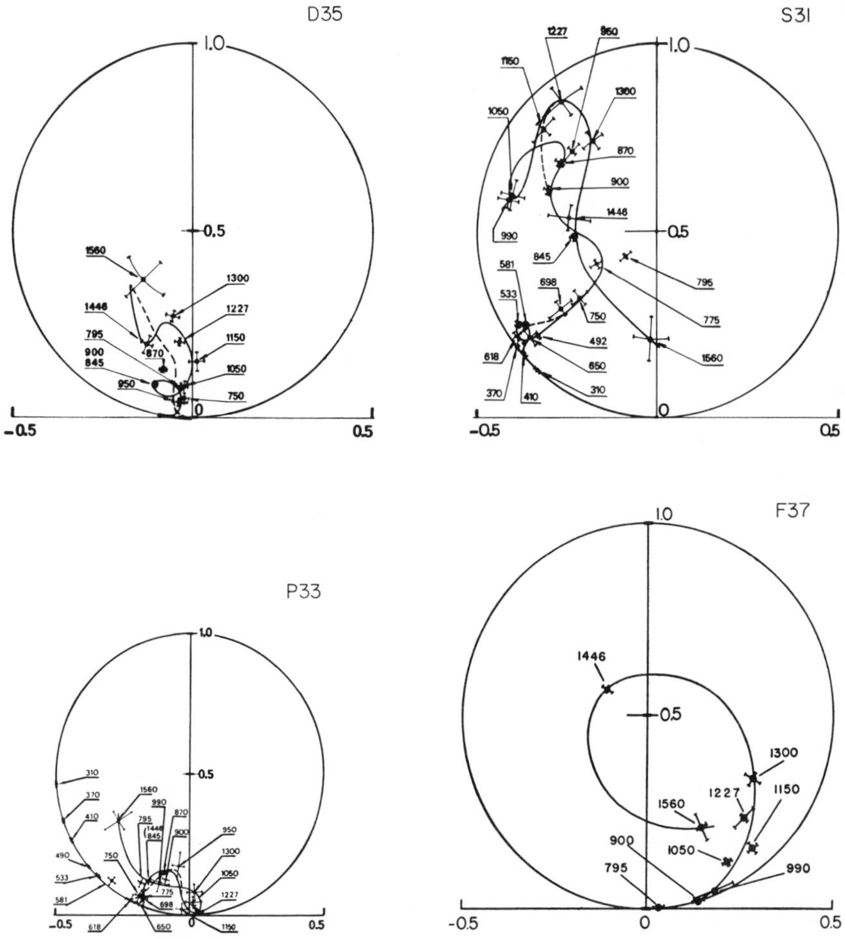


**Fig. 7.7(a).** Argand diagrams for $\pi$N scattering: $T = \frac{3}{2}$. [From P. Bareyre, C. Bricman, and G. Villet, Phase-shift analysis of pion–nucleon elastic scattering below 1.6 GeV. *Phys. Rev.* **165**, 1730–1749 (1968).]

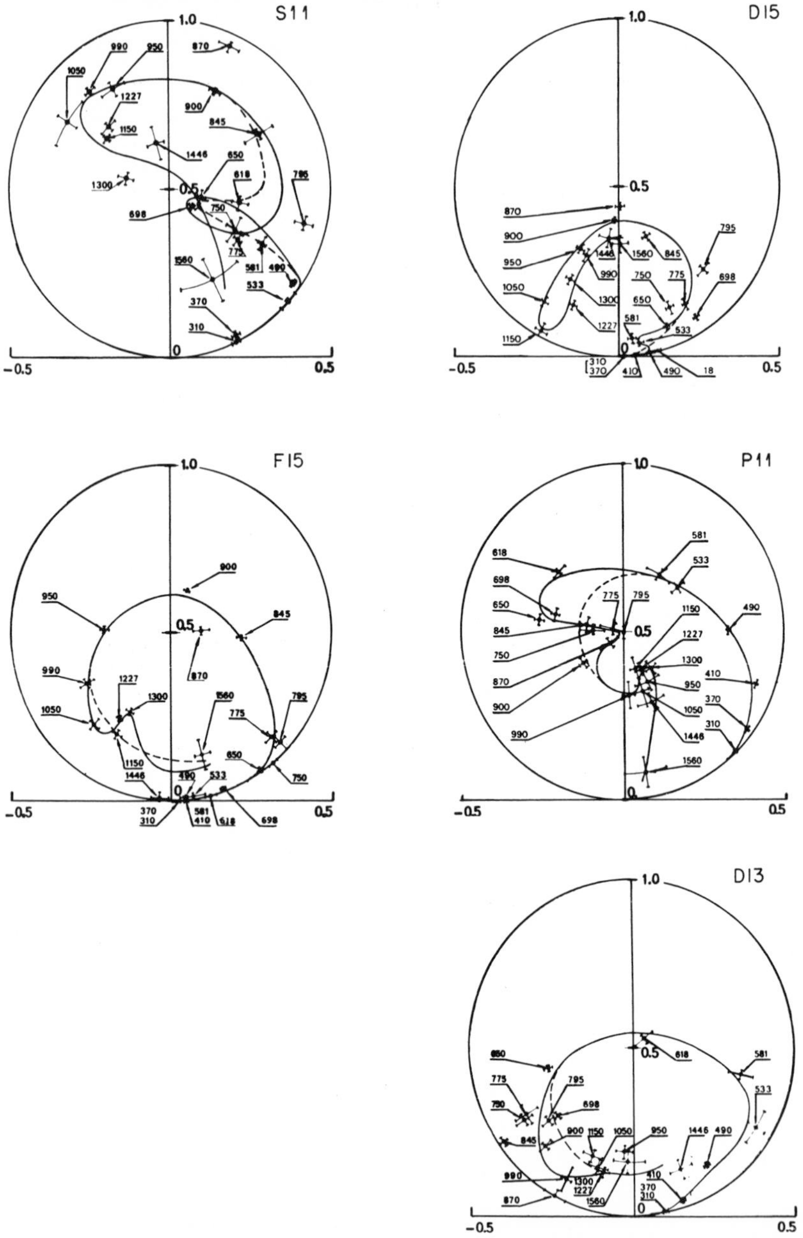


**Fig. 7.7(b).** Argand diagrams for $\pi N$ scattering: $T = \frac{1}{2}$. [From P. Bareyre, C. Bricman, and G. Villet, Phase-shift analysis of pion–nucleon elastic scattering below 1.6 GeV. *Phys. Rev.* **165**, 1730–1749 (1968).]

**Table 7.2**

*Mass and Width of the Resonant States*[a,b]

| State $l_{2T,2J}$ | From total cross section $M$ (MeV) | $\Gamma$ (MeV) | From velocity in complex plane $M$ (MeV) | $\Gamma$ (MeV) | Estimated elasticity $x = \Gamma_{el}/\Gamma_{tot}$ |
|---|---|---|---|---|---|
| $S_{31}$ | 1695 | 250 | 1650 | 130 | |
| $P_{33}$[c] | 1240 | 110 | | | 1 |
| $F_{37}$ | 1975 | 180 | 1980 | 140 | 0.57 |
| $S_{11}$ | 1535 | 155 | 1515 | 105 | |
| $S_{11}$ | 1710 | 260 | 1665 | 110 | |
| $P_{11}$ | 1470 | 255 | 1505 | 205 | 0.68 |
| $D_{13}$ | 1510 | 125 | 1515 | 110 | 0.54 |
| $D_{15}$ | 1680 | 135 | 1655 | 105 | 0.41 |
| $F_{15}$ | 1690 | 110 | 1680 | 105 | 0.64 |

[a] From P. Bareyre, C. Bricman, and G. Villet, Phase-shift analysis of pion–nucleon elastic scattering below 1.6 GeV. *Phys. Rev.* **165**, 1730–1749 (1968).]

[b] Estimated from the partial cross section and from the rate of variation with energy of the amplitude and approximate elasticity. The accuracy for mass and widths is not better than 20 MeV. Elasticities are taken at the maximum of $\sigma_{tot}$ with an accuracy of about 0.1.

[c] These values come from P. Auvil *et al.*, *Phys. Lett.* **12**, 76 (1964); A. Donnachie *et al.* in *Proceedings of the 13th Annual Conference on High-Energy Physics, Berkeley, 1966*, University of California Press, Berkeley, California, 1967.

## Problems

**7.1.** Using charge independence, show that

(a)
$$\frac{\sigma(\mathrm{pd} \to \mathrm{He}^3\pi^0)}{\sigma(\mathrm{pd} \to \mathrm{H}^3\pi^-)} = \frac{1}{2},$$

(b)
$$\frac{\sigma(\mathrm{pp} \to \mathrm{d}\pi^+)}{\sigma(\mathrm{np} \to \mathrm{d}\pi^0)} = 2.$$

(c) Find the branching ratio $\to\pi^+ n/\to\pi^0 p$, for $N^{*+}(1236)$ decay.

**7.2.** Derive the relations analogous to Eqs. (7.4) and (7.5) for $\pi^0 n$, $\pi^+ n$ scattering.

**7.3.*** The *polarization* (in direction $\hat{\mathbf{n}}$) of a collection of spin-one-half particles is given by

$$P(\hat{\mathbf{n}}) \equiv \frac{\#(\uparrow) - \#(\downarrow)}{\#(\uparrow) + \#(\downarrow)},$$

$\uparrow$ and $\downarrow$ being defined with respect to direction $\hat{\mathbf{n}}$. Show that, for $\pi N$ scattering with an initially unpolarized target (so that we can average over the initial spin), the polarization of the recoil protons is given by

$$P_{\text{i}} = \frac{\operatorname{tr}(F\sigma_{\text{i}}F^{\dagger})}{\operatorname{tr}(FF^{\dagger})}.$$

Express $\mathbf{P}$ in terms of $F_S$ and $F_{\text{N}}$.

**7.4.** A beam or target with spins can be described in terms of the *density matrix*:

$$\rho \equiv \sum a_{\text{i}}|\text{i}\rangle\langle\text{i}|;$$

$a_{\text{i}}$ represents the fraction of particles in the state i, so that

$$\sum a_{\text{i}} = \sum \langle\text{i}|\rho|\text{i}\rangle = \operatorname{tr}\rho = 1.$$

Show that the average value of any operator $A$ is given by

$$\bar{A} = \operatorname{tr}(A\rho).$$

If $F$ is the scattering operator, show that the scattered particle density matrix is

$$\rho' = F\rho F^{\dagger}.$$

For spin one-half

$$\rho = \tfrac{1}{2}(1 + \mathbf{P}\cdot\boldsymbol{\sigma})$$

is the most general $2 \times 2$ density matrix. Show that $\mathbf{P}$ is the polarization vector.

**7.5.*** Elastic scattering partial wave analysis with spin one-half. Given that

$$\mathcal{Y}^{1/2}_{l+1/2} = \left(\frac{l+1}{2l+1}\right)^{1/2} Y_l^{\,0}\chi_\uparrow + \left(\frac{l}{2l+1}\right)^{1/2} Y_l^{\,1}\chi_\downarrow,$$

$$\mathcal{Y}^{1/2}_{l-1/2} = -\left(\frac{l}{2l+1}\right)^{1/2} Y_l^{\,0}\chi_\uparrow + \left(\frac{l}{2l+1}\right)^{1/2} Y_l^{\,1}\chi_\downarrow$$

are the total angular momentum eigenfunctions, and hence that

$$Y_l^{\,0}\chi_\uparrow = \left(\frac{2l+1}{4\pi}\right)^{1/2} P_l^{\,0}\chi_\uparrow = \left(\frac{l+1}{2l+1}\right)^{1/2} \mathcal{Y}^{1/2}_{l+1/2} - \left(\frac{l}{2l+1}\right)^{1/2} \mathcal{Y}^{1/2}_{l-1/2},$$

and introducing the expansion for the scattered wave

$$\frac{e^{ikr}}{r} F(\theta)\chi_{\uparrow} = \frac{e^{ikr}}{2ikr} (4\pi)^{1/2} \sum_l [(l+1)^{1/2} a_{l+} \mathcal{Y}_{l+1/2}^{1/2} + l^{1/2} a_{l-} \mathcal{Y}_{l-1/2}^{1/2}],$$

show that the partial wave amplitudes must satisfy

$$a_{l\pm} = \exp 2i\delta_{l\pm} - 1.$$

Then using

$$Y_l^1 = -\left[\frac{2l+1}{4\pi} \frac{(l-1)!}{(l+1)!}\right]^{1/2} \sin\theta \frac{dP_l(\cos\theta)}{d(\cos\theta)},$$

derive Eqs. (7.14).

### Bibliography

W. R. Frazer. *Elementary Particles.* Chap. 3. Prentice-Hall, Englewood Cliffs, New Jersey, 1966.

M. L. Goldberger and K. M. Watson. *Collision Theory.* Wiley, New York, 1964.

G. Källén. *Elementary Particle Physics.* Chap. 4. Addison-Wesley, Reading, Massachusetts, 1964.

E. Segrè. *Nuclei and Particles.* Chap. 14. Benjamin, New York, 1965.

W. S. C. Williams. *An Introduction to Elementary Particles.* Academic Press, New York, 1971.

CHAPTER EIGHT

# Pion Resonances

When a final state consists of three or more particles, the distribution of energies among them is not fixed by the kinematics but is free instead to reveal dynamic effects—that is, effects due to the details of production and the forces between the particles. Thus many particle final states have been a rich source of interesting phenomena. Here we will discuss *pion resonances*; that is, short-lived bosons that decay into pions. The lifetimes involved are generally of the order of $10^{-23}$–$10^{-22}$ sec. Such short times ($\sim$ fermi/c) are characteristic of strong interactions and, of course, are not measured directly but only inferred from the uncertainty relation,

$$\Gamma\tau \simeq 1, \tag{8.1}$$

and the observed energy width $\Gamma$.

## 8.1. The ω Meson

### *a. Observation. Isospin*

A three pion resonance was discovered in 1961 by analyzing the pion spectrum produced by annihilation of antiprotons in a hydrogen bubble chamber. The reaction studied is

$$p\bar{p} \to 2\pi^+ 2\pi^- \pi^0.$$

The $\pi^0$ is not observed directly but deduced from a *missing mass* analysis. That is, the incoming **p** and the four visible outgoing momenta are determined by the curvatures (the chamber is in a magnetic field), the energies from the momenta (under the hypothesis that the four outgoing tracks are indeed pions), and the missing energy and momentum (if any) from the conservation laws; then if

$$(\Delta E)^2 - (\Delta \mathbf{p})^2 = m_{\pi^0}^2,$$

the missing particle is identified as a $\pi^0$.

The hypothesis that the reaction actually proceeds in two steps,

$$\bar{\mathrm{p}}\mathrm{p} \to X 2\pi, \quad X \to 3\pi,$$

can be investigated by looking at the distribution in *invariant mass* $M^*$ of any three of the pions:

$$M^{*2} \equiv (E_1 + E_2 + E_3)^2 - (\mathbf{p}_1 + \mathbf{p}_2 + \mathbf{p}_3)^2. \tag{8.2}$$

For each experimental event, there are ten ways to pick out the three pions, which divide into five possible charge states. The resulting distributions of experimental events are shown in Fig. 8.1. The solid lines indicate the distribution predicted by phase space alone, i.e., setting

$$|T_{\mathrm{fi}}|^2 = \text{const},$$

so that the distribution of events is proportional to $\partial I/\partial M^*$, where, from Eq. (3.72)

$$I = \int \delta^4(P_i - P_f) \prod_{k=1}^{5} (d^3\mathbf{p}_k/2E_k). \tag{8.3}$$

From Fig. 8.1 we see that only for neutral triplets is there any significant departure from the phase space curves. This object, the $\omega$, has a mass centered at about 780 MeV (later experiments give $782.7 \pm 0.5$) and a width of about 9 MeV ($9.3 \pm 1.7$). The lifetime associated with this width, $\tau \simeq 7 \times 10^{-23}$ sec, is long enough for us to expect the decay into $3\pi$'s to be independent of the production process; so now we turn the determination of the properties of $\omega$ from the observed decays.

Because it only appears in the neutral state, $\omega$ must have $T = 0$. Since the transition $\omega \to 3\pi$ proceeds by the strong interaction ($\tau$ is too short for electromagnetic decay), the $G$-parity of the $\omega$ is $-1$ and the $3\pi$ state has $T = 0$. That the $T = 0$ isospin wave function of $3\pi$'s is antisymmetric under interchange of any pair can be seen by forming the wave function, or more simply by

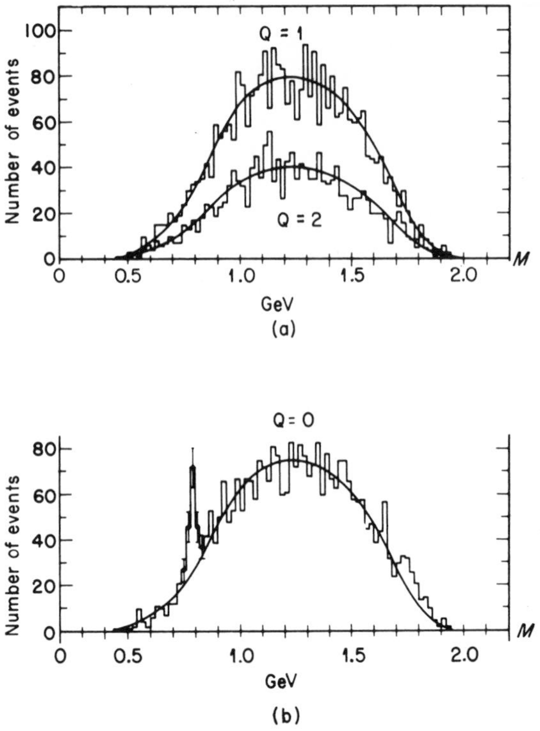


**Fig. 8.1.** Invariant mass spectrum for pion triplets in the reaction $p\bar{p} \to 2\pi^+ 2\pi^- \pi^0$. [From B. C. Maglic *et al.*, Evidence for a $T = 0$ three-pion resonance. *Phys. Rev. Lett.* **7**, 178–182 (1961).]

realizing that the only way to get an isoscalar (hence $T = 0$) object from three isovector ($T = 1$) objects is with the combination

$$\pi_1 \cdot \pi_2 \times \pi_3,$$

which is clearly completely antisymmetric. Generalized Bose symmetry then requires that the *spatial* amplitude (wave function) also be antisymmetric under any interchange. The experimental consequences of this fact is easiest to discuss in terms of the *Dalitz plot*, which has to do with the distribution of energy among the $3\pi$'s.

### *b. Dalitz Plot*

The decay rate of the $\omega$, in terms of $|T_{fi}|^2$, is given by Eq. (3.91). The required three-body phase space integral we have already worked out

(Chapter 3, Section 3.4c) to be

$$I_3 = \tfrac{1}{8}\int \delta(E)\, dE_1\, dE_2\, dE_3\, d\Omega_1\, d\phi_{12} \qquad (3.98)$$

$$= \tfrac{1}{8}\int dE_1\, dE_2\, d\Omega_1\, d\phi_{12}$$

in the rest system of the $\omega$. The variables $\Omega_1$ and $\phi_{12}$ have to do with the orientation of Fig. 8.2 and affect neither the magnitudes nor the relative

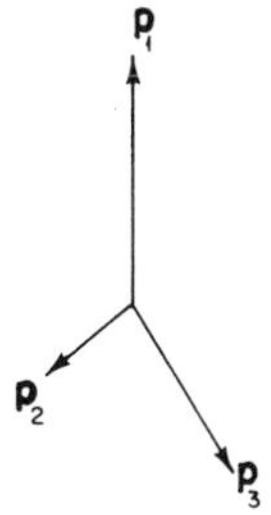


**Fig. 8.2.** The momenta from $\omega$ decay; the energies and relative angles are independent of the orientation of this figure.

directions of the three momenta. Only if the $\omega$ had a spin and was polarized could we get a dependence on $\Omega_1$ and $\phi_{12}$; in any case, we choose to ignore any such dependence here—i.e., we integrate over $\Omega_1$ and $\phi_{12}$ so that

$$|T_{\mathrm{fi}}|^2 \to \overline{|T_{\mathrm{fi}}|^2}.$$

The "interesting" quantities—energies and relative angles—are completely determined by $E_1$ and $E_2$, and

$$I_3 = \pi^2 \int dE_1\, dE_2\,. \qquad (8.4)$$

This expression tells us that, if $\overline{|T_{\mathrm{fi}}|^2}$ is constant, the distribution of experimental events in the available part of the $(E_1, E_2)$ plane is *uniform*. Thus looking at this distribution (the Dalitz plot) is a convenient way of (experimentally) investigating the variation of $\overline{|T_{\mathrm{fi}}|^2}$.

It is instructive to look at how choosing $E_1$ and $E_2$ fixes the relative momenta (in general, for $m_1 \neq m_2 \neq m_3$). We first choose $E_1$ ($>m_1$), which fixes $\mathbf{p}_1$ (its orientation is arbitrary); then taking some value of $E_2$ ($>m_2$) fixes both $|\mathbf{p}_2|$ and $|\mathbf{p}_3|$, since

$$E_1 + E_2 + E_3 = M^*. \qquad (8.5)$$

Because the three momentum vectors form a triangle, we draw one circle (radius $|\mathbf{p}_2|$) centered on one end of $\mathbf{p}_1$ and another (radius $|\mathbf{p}_3|$) centered on the other end of $\mathbf{p}_1$, as shown in Fig. 8.3. The intersections give the possible values of $\mathbf{p}_2$ and $\mathbf{p}_3$.

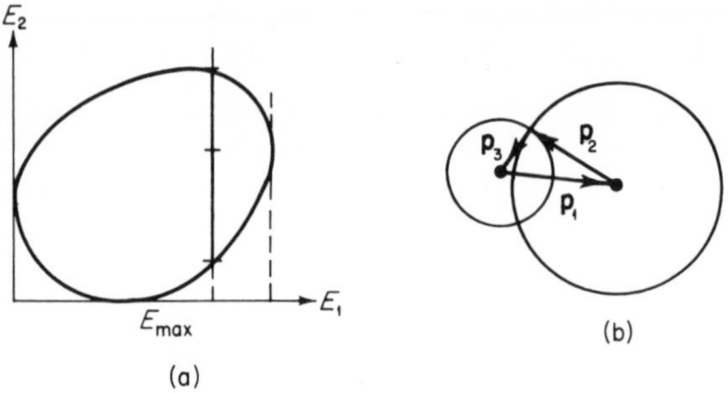


**Fig. 8.3.** (a) Boundary of the Dalitz plot; (b) construction used in finding (a).

It is clear from this diagram that the maximum allowed value of $E_2$ (fixed $E_1$) is obtained by enlarging $|\mathbf{p}_2|$ (and therefore shrinking $|\mathbf{p}_3|$) until the two circles intersect at only one point, i.e., until the triangle has collapsed and the three vectors are *collinear* [Fig. 8.4(a)]. Similarly, the minimum value of $E_2$ (fixed $E_1$) is obtained

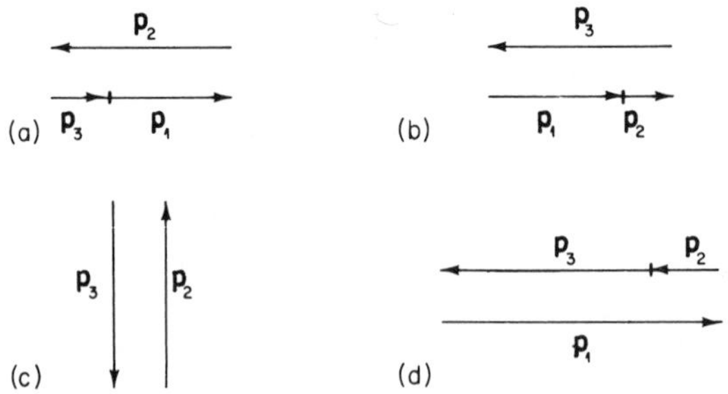


**Fig. 8.4.** Points on the boundary of the Dalitz plot correspond to *collinear* momenta. (a) Maximum $E_2$ (fixed $E_1$); (b) minimum $E_2$ (fixed $E_1$); (c) minimum $E_1$ (any $E_2$); (d) maximum $E_1$ (any $E_2$).

by shrinking $|\mathbf{p}_2|$ until the vectors are again collinear [Fig. 8.4(b)]. Finally the minimum value of $E_1$ is clearly $m_1$ [Fig. 8.4(c)], and the maximum has $\mathbf{p}_2 \parallel \mathbf{p}_3$ [Fig. 8.4(d)]. The equations defining the boundary in the $(E_1, E_2)$ plane [Fig. 8.3(a)] are thus

$$|\mathbf{p}_1| \pm |\mathbf{p}_2| = \pm |\mathbf{p}_3|, \tag{8.6}$$

which simply express the collinearity.

The boundary is especially simple if all three particles are *ultrarelativistic.* Then Eqs. (8.6) become

$$E_1 + E_2 = M^* - E_1 - E_2,$$

and

$$E_1 - E_2 = \pm(M^* \pm E_1 - E_2),$$

i.e.,

$$E_3 = M^*/2, \qquad E_1 = M^*/2, \qquad \text{and} \qquad E_2 = M^*/2.$$

The resulting allowed region is shown in Fig. 8.5(a). This way of presenting the Dalitz plot has the disadvantage of treating $E_3$ differently from $E_1$ and $E_2$. We remedy this by converting to the variables $E_2$ and $(E_1 - E_3)/\sqrt{3}$; since the transformation is linear, it does not disturb the uniformity of the Dalitz plot. Then $E_1$, $E_2$, and $E_3$ appear symmetrically [Fig. 8.5(b)].

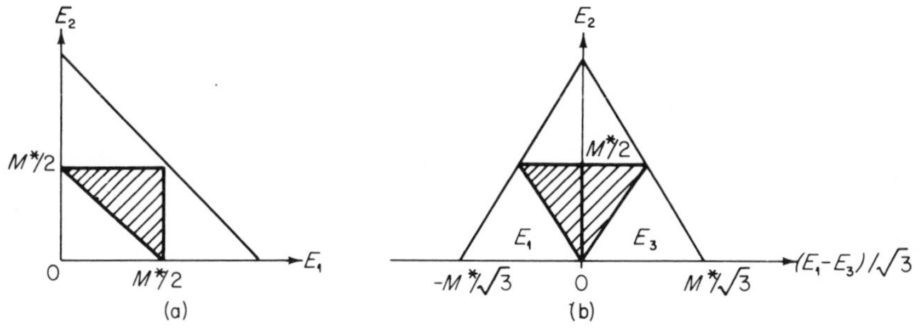


**Fig. 8.5.** Dalitz plot boundary and allowed region (shaded); (a) $(E_1, E_2)$ coordinates; (b) $[(E_1 - E_3)1/\sqrt{3}, E_2]$ coordinates.

At the other extreme, one can show that the boundary for three *nonrelativistic* particles with equal masses is a circle in the $[E_2, (E_1 - E_3)/\sqrt{3}]$ plane.

The decay of the $\omega$ falls between these two extremes; the boundary, which can be determined from Eqs. (8.6), is shown in Fig. 8.6, along with experimental points. We are now in a position to discuss how these experimental data can tell us the spin and parity of the $\omega$.

### c. *Spin and Parity of ω*

We consider several possibilities for $J^P$. First, $J = 0$. Suppose the relative orbital angular momentum of the $\pi^+$ and $\pi^-$ is $l_\pm$ (Fig. 8.7); the orbital angular momentum of the $\pi^0$ relative to the $\pi^+\pi^-$ CM, $l_0$, must be the same in order for them to add up to zero. Then the parity is given by

$$\eta = \eta_{\pi^+}\eta_{\pi^-}\eta_{\pi^0}(-1)^{l_\pm} \cdot (-1)^{l_0} = -1, \tag{8.7}$$

so we are considering $J^P = 0^-$. But as we noted above, we also have the requirement of spatial antisymmetry to impose. The $\omega$ decay matrix element $T_{\mathrm{fi}}$ is invariant and therefore a function only of invariants. Introduce the invariants

$$s_1 \equiv (P - p_1)^2, \qquad s_2 \equiv (P - p_2)^2, \qquad s_3 \equiv (P - p_3)^2 \tag{8.8}$$

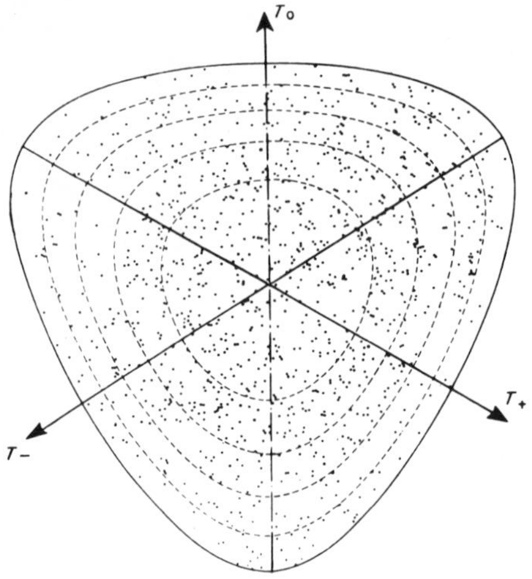


**Fig. 8.6.** The Dalitz plot for $\omega$ decay. [From G. Puppi, Interactions of pions, nucleons and antinucleons. In *1962 International Conference on High-Energy Physics at CERN*, J. Prentki, ed. CERN, Scientific Information Service, Geneva, Switzerland, 1962.]

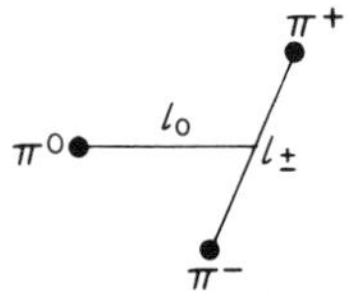


**Fig. 8.7.** Orbital angular momenta for three pions.

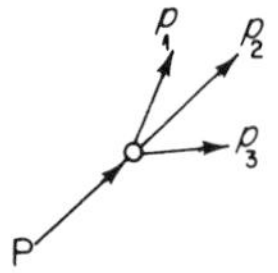


**Fig. 8.8.** The momenta of $\omega$ decay.

(Fig. 8.8); spatial antisymmetry requires that $T_{\mathrm{fi}}$ $(s_1, s_2, s_3)$ be antisymmetric under interchange of any pair. Now

$$T_{\mathrm{fi}} = (s_1 - s_2)(s_2 - s_3)(s_3 - s_1)f(s_1, s_2, s_3), \tag{8.9}$$

with $f(s_1, s_2, s_3)$ completely symmetric, satisfies the requirement, and it is at least plausible to assume that $f$ is a slowly varying function of its arguments. Then Eq. (8.9) tells us a lot about how the experimental Dalitz plot should look. Since in the $\omega$ rest system,

$$s_i = -P^2 - p_i^2 + 2Pp_i = M^{*2} + m_\pi^2 - 2M^*E_i,$$

we have

$$|T_{\mathrm{fi}}|^2 \propto (E_1 - E_2)^2(E_2 - E_3)^2(E_3 - E_1)^2 |f(E_1, E_2, E_3)|^2. \tag{8.10}$$

Thus the density of experimental points should be zero on the lines $E_1 = E_2$, etc. The probability distribution, neglecting any variation of $f$, is shown in Fig. 8.9(a). It is evident from Fig. 8.6(a) that the experimental distribution is nothing like this; therefore the $0^-$ assignment is ruled out.

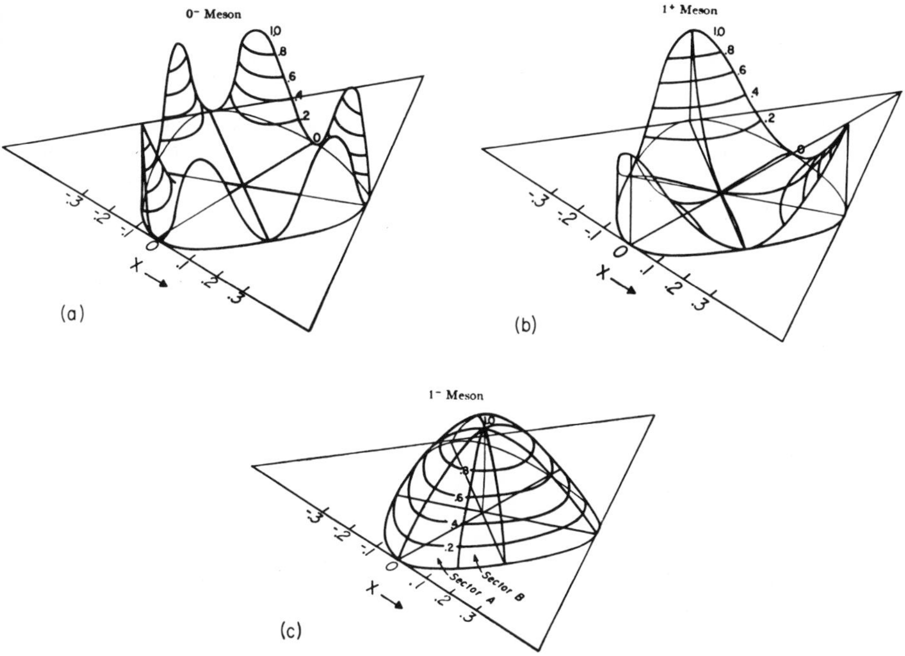


**Fig. 8.9.** The probability distribution over the Dalitz plot for $\omega$ decay, for $J^P$ values: (a) $0^-$; (b) $1^+$, (c) $1^-$. [From M. L. Stevenson *et al.*, Spin and parity of the $\omega$ meson. *Phys. Rev.* **125**, 687–690 (1962).]

We next consider the possibility $J = 1$. In that case the $\omega$ has a polarization vector $\boldsymbol{\varepsilon}$ which must appear linearly in the decay amplitude. (Recall the argument for $\pi^0$ decay, Chapter 6, Section 6.1c.) But the decay amplitude also has to be rotationally invariant and therefore proportional to $\boldsymbol{\varepsilon}\cdot\mathbf{K}$, $\mathbf{K}$ being some vector (or pseudovector). In the $\omega$ rest system,

$$\langle \mathbf{k}_1, \mathbf{k}_2, \mathbf{k}_3 | S | \boldsymbol{\varepsilon}\rangle = f\boldsymbol{\varepsilon}\cdot\mathbf{K}, \tag{8.11}$$

$f$ being an unknown scalar function (there are no pseudoscalars available for $f$ to depend upon). Now, under space inversion

$$\boldsymbol{\varepsilon} \to \boldsymbol{\varepsilon}$$

(since it specifies the spin),

$$\mathbf{k} \to -\mathbf{k}, \qquad \text{and} \qquad \mathbf{K} \to \mp\mathbf{K},$$

while the $S$-operator must be invariant. Thus

$$\begin{aligned}\langle \mathbf{k}_1, \mathbf{k}_2, \mathbf{k}_3 | S | \boldsymbol{\varepsilon}\rangle &= \langle \mathbf{k}_1, \mathbf{k}_2, \mathbf{k}_3 | P^{-1}PSP^{-1}P | \boldsymbol{\varepsilon}\rangle \\ &= \eta_\pi^{\,3}\eta_\omega\langle -\mathbf{k}_1, -\mathbf{k}_2, -\mathbf{k}_3 | S | \boldsymbol{\varepsilon}\rangle,\end{aligned}$$

so that

$$f\boldsymbol{\varepsilon}\cdot\mathbf{K} = \mp\eta_\pi{}^3\eta_\omega f\boldsymbol{\varepsilon}\cdot\mathbf{K}. \tag{8.12}$$

Therefore,

$$\mathbf{K}\text{ vector} \Rightarrow \eta_\pi{}^3\eta_\omega = -1;$$
$$\mathbf{K}\text{ pseudovector} \Rightarrow \eta_\pi{}^3\eta_\omega = +1.$$

Remembering the requirement of spatial antisymmetry, we can have

$$\text{vector:}\quad \mathbf{K} = \mathbf{k}_1(s_2 - s_3) + \mathbf{k}_2(s_3 - s_1) + \mathbf{k}_3(s_1 - s_2),$$
$$\text{pseudovector:}\quad \mathbf{K} = \mathbf{k}_1 \times \mathbf{k}_2 + \mathbf{k}_2 \times \mathbf{k}_3 + \mathbf{k}_3 \times \mathbf{k}_1 = 3\mathbf{k}_1 \times \mathbf{k}_2;$$

so

$$f \equiv f(s_1, s_2, s_3)$$

is again a symmetric function. Neglecting its variation and averaging $|T_{\text{fi}}|^2$ over initial spin states (and thus choices for $\boldsymbol{\varepsilon}$), we have

$$\overline{|T_{\text{fi}}|^2} \propto \mathbf{K}^2. \tag{8.13}$$

The two possibilities for $|T_{\text{fi}}|^2$ are plotted in Fig. 8.9(b) and (c). Comparison with the observed Dalitz plot (Fig. 8.6) rules out the vector choice for **K** [Fig. 8.9(b)] while giving good agreement (as one can show quantitatively) with the pseudovector choice. We conclude that $\eta_\omega = \eta_\pi{}^3 = -1$. The resulting assignment $J^{PG} = 1^{--}$ is also consistent with the analyses of other reactions in which $\omega$ appears.

## 8.2. The ρ Meson

This resonance was actually discovered shortly before the $\omega$, and, in fact, had been predicted on theoretical grounds having to do with the electromagnetic "shape" of the nucleon (cf. Chapter 12, Section 12.2). It shows up clearly in the reactions

$$\pi^-\text{p} \to \text{p}\pi^-\pi^0$$
$$\to \text{n}\pi^+\pi^-.$$

When the number of events was plotted as a function of invariant mass of the pion pair, the result deviated strikingly from the phase space distribution, as shown in Fig. 8.10. (Since we are dealing with three particles in the final state, the events can be exhibited on a Dalitz plot; Figure 8.10 is then the projection of such a Dalitz plot on one axis.) The $\rho$-meson peak is very broad (width about 150 MeV) and is centered at about 765 MeV (more recent values are 769 ± 3 MeV and 112 ± 4 MeV for the position and width). The peak is

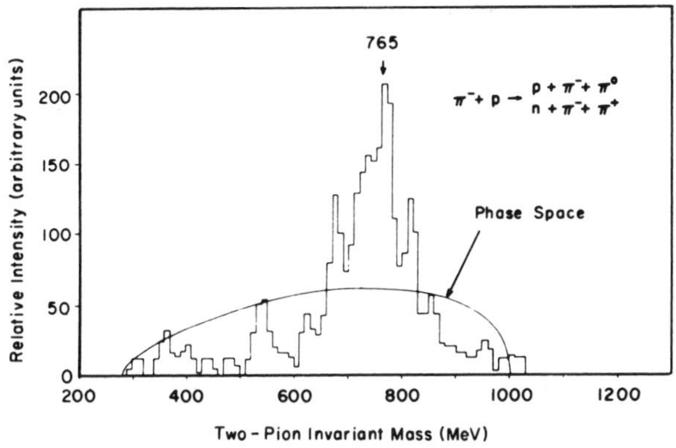


**Fig. 8.10.** Invariant mass spectrum for pion pairs in

$$\pi^- p \rightarrow p\pi^-\pi^0$$
$$\rightarrow n\pi^+\pi^-.$$

[From A. R. Erwin *et al.*, Evidence for a $\pi$–$\pi$ resonance in the $I = 1$, $J = 1$ state. *Phys. Rev. Lett.* **6**, 628–630 (1961).]

seen in both the neutral and negatively charged mode (the two are combined in Fig. 8.10); other experiments show the $\rho^+$ as well, but no doubly charged modes. Thus the $\rho$ has isospin one.

Now when we form isospin one out of two isospin one objects, the resulting isospin wave function must be antisymmetric under interchange (from Table 6.6, or the fact that the only isovector combination of two isovectors is the antisymmetric one, $\boldsymbol{\pi}_1 \times \boldsymbol{\pi}_2$). Generalized Bose symmetry then requires the space wave function of the pions to be antisymmetric, so that the orbital angular momentum must be odd. Thus $\rho$ is limited to $J = 1, 3, \ldots$; so far, all the data are consistent with the simplest possibility, $J = 1$. The $G$-parity is clearly $+1$, and the parity

$$\eta_\rho = (-1)^J \eta_\pi{}^2 = -1.$$

## 8.3. The η Meson

This curious object had been predicted by Gell-Mann on the basis of his *eightfold way* (see Chapter 10, Section 10.3) and was discovered (1961) in the reaction

$$\pi^+ d \rightarrow pp\pi^+\pi^-\pi^0,$$

by the same kind of analysis we have discussed in regard to the $\omega$. The distribution in invariant mass of the three pions is shown in Fig. 8.11. In addition

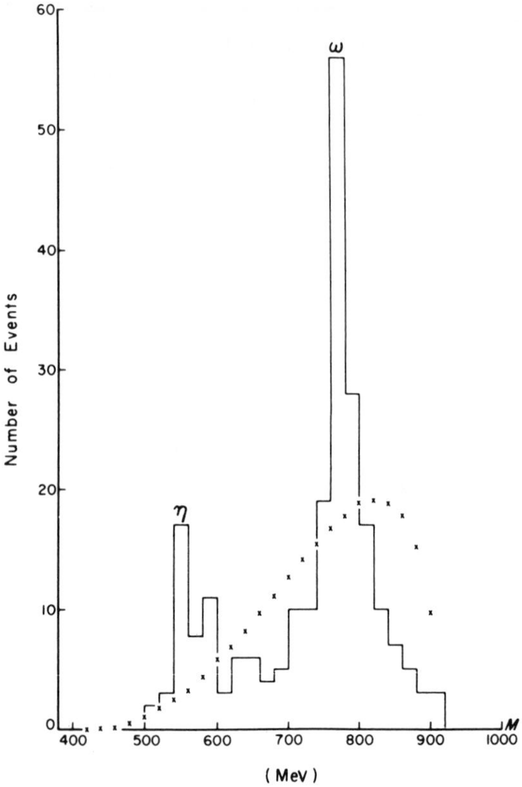


**Fig. 8.11.** Invariant mass spectrum for pion triplets in the reaction $\pi^+ d \to pp\pi^+\pi^-\pi^0$. [From A. Pevsner *et al.*, Evidence for a three-pion resonance near 550 MeV. *Phys. Rev. Lett.* 7, 420–423 (1961).]

to the prominent $\omega$ peak, we see the smaller and narrower $\eta$ peak with mass 550 MeV (more recent value, $548.8 \pm 0.5$ MeV) and width less than 10 MeV. For the events included under the $\eta$ peak,

$$\pi^+ d \to pp\eta$$
$$\hookrightarrow \pi^+\pi^-\pi^0.$$

Since the $\eta$ is only seen in the neutral mode, we conclude that $T = 0$. Then isospin conservation implies the $T = 0$ combination of pion isospins:

$$\boldsymbol{\pi}_1 \cdot \boldsymbol{\pi}_2 \times \boldsymbol{\pi}_3,$$

which is completely antisymmetric; therefore we cannot have decay into $3\pi^0$'s—only the $\pi^+\pi^-\pi^0$ combination can be antisymmetric. However, it is observed that, on the contrary, the branching ratio

$$\pi^+\pi^-\pi^0/\text{all neutrals} = 0.3 \pm 0.1.$$

Thus isospin *cannot* be conserved in $\eta$ decay! Since the electromagnetic interaction violates isospin conservation, we can blame it for the decay. Note that we have only an upper limit on the width and therefore on the decay rate; it might well be slow—e.g., $\sim 10^{-16}$ sec, like the $\pi^0$.

Another useful fact about the $\eta$ is the uniformity of the experimental events in the Dalitz plot; this leads us to believe that $J = 0$, because any nonzero orbital angular momentum of the daughter pions should produce a varying distribution (as we saw for $\omega$ decay). In addition, the decay $\eta \to 2\gamma$ has been observed, which proves (as in the $\pi^0$ case) that $J \neq 1$. Then orbital angular momentum zero implies that

$$\eta_\eta = (\eta_\pi)^3 = -1,$$

i.e., pseudoscalar. The $J^P = 0^-$ assignment accounts for the absence of the strong decay $\eta \to 2\pi$. What about the lack of *strong* three pion decay? We can account for this simply by assigning $\eta$ the $G$-parity $+1$.

Nothing discussed so far rules out $\eta \to 4\pi$ via strong interactions, but this mode is suppressed by the lack of kinetic energy. In fact, $\eta \to 2\pi^+2\pi^-$ cannot go at all, and the $2\pi^0\pi^+\pi^-$ and $4\pi^0$ modes have a $Q$-value of a mere several MeV and therefore not much phase space. This, combined with the fact that the pions must have some relative orbital angular momentum in order to provide the negative parity, so that the amplitude must have some centrifugal barrier factors (cf. Chapter 6, Section 6.2c), suffices to depress the rate for this strong decay below the electromagnetic three pion decay.

Many other pion resonances have been and presumably will continue to be discovered at higher energies. Since we are more interested in learning about the ideas that help us to cope with these things than in a complete catalog, we simply refer the reader to the current Rosenfeld *et al.* tables for up-to-date data on the pion resonances (cf. Appendix). As we will see later, these and other resonances can be fit into the regular patterns. These "spectra" of elementary particles are gradually leading us to an understanding of the strong interactions.

## Problems

**8.1.** For decay into three equal-mass *nonrelativistic* particles, show that the boundary of the Dalitz plot is a circle in the $[(E_1 - E_2)/\sqrt{3}, E_3]$-plane.

**8.2.*** (Frazer) For *any* decay into three equal-mass particles ($M \to 3m$), prove that the Dalitz plot boundary is given by [cf. Eqs. (8.8)]

$$s_1 s_2 s_3 = m^2(M^2 - m^2)^2.$$

(Hint: Using the $M$ rest system, show first that

$$I_\alpha \equiv \varepsilon_{\alpha\beta\gamma\delta} p_{1\beta} p_{2\gamma} p_{3\delta} = 0$$

on the boundary. Then express $I_\alpha I_\alpha$ in terms of the invariants.)

**8.3.** Discuss the behavior of $\omega$ and the $\omega$-decay amplitude under charge conjugation. Is the decay $\omega \to \pi^0\gamma$ possible?

**8.4.** What tensor forms are possible for the decay $\rho \to 2\pi$? Discuss the behavior of $\rho^0$ and its decay amplitude under $C$ and under $P$. Is the decay $\rho^0 \to \pi^0\gamma$ possible?

## Bibliography

W. R. Frazer. *Elementary Particles.* Chap. 5. Prentice-Hall, Englewood Cliffs, New Jersey, 1966.

G. Källén. *Elementary Particle Physics.* Addison-Wesley, Reading, Massachusetts, 1964.

W. S. C. Williams. *An Introduction to Elementary Particles.* Chap. 7. Academic Press, New York, 1971.

CHAPTER NINE

# Strange Particles

About the time of the experimental discovery of the pions (1947), some heavier particles were also found. Their existence was a complete surprise and their behavior puzzling. Two sorts of these *strange particles* were observed: K-*mesons*, which are bosons with mass about 500 MeV, and *hyperons* ($\Lambda$, $\Sigma$, $\Xi$), which are heavy baryons (masses about 1120, 1190, 1320 MeV, respectively; see Appendix).

An immediate problem was reconciling the copious production of these strange particles ($\sigma \sim m_\pi^{-2}$), which suggests strong interactions, with decays that are much too slow ($\tau \sim 10^{-8}$–$10^{-10}$ sec) to be proceeding via the strong interactions. For example, if the $\Lambda$ can interact strongly, why does not the (observed) decay

$$\Lambda \to \pi^- p$$

go via strong interactions with a lifetime of about $10^{-23}$ sec instead of the observed $10^{-10}$ sec? To explain this, Pais (in 1952) suggested the hypothesis of *associated production*: Strange particles can interact strongly, and therefore are produced, *only in pairs*; once separated, each member can decay into ordinary particles only through the weak interaction. This idea was afterward confirmed experimentally, typical production reactions being

$$\pi^- p \to \Lambda K^0 \qquad \text{and} \qquad pp \to \Lambda p K^+.$$

However, the idea of associated production does *not* explain why the reaction

$$nn \to \Lambda\Lambda,$$

which has a much lower energy threshold, does not occur. Gell-Mann and Nishijima (in 1953) answered this and predicted a lot more with their *strangeness scheme*.

## 9.1. Strangeness

The classification scheme of Gell-Mann and Nishijima rests on the assumption that isospin invariance applies to the strong interactions of strange particles as well as to pions and nucleons. This seems very natural now but was not at all obvious in 1953. The fact that there are no charged hyperons with mass approximately equal to that of the $\Lambda$, which is neutral, leads to assignment of $T = 0$ to the $\Lambda$. The reaction

$$\pi^- p \to \Lambda K^0,$$

and the assumption of isospin invariance implies that since the $\pi^- p$ system has $T_3 = -\frac{1}{2}$, and $T = \frac{3}{2}$ or $\frac{1}{2}$, the same must be true for the $K^0$. The lack of any doubly charged K's rules out $T = \frac{3}{2}$; so we are left with $T = \frac{1}{2}$. This means that there must be a $K^+$ to complete the isospin doublet, which indeed is observed. Now the observed reactions

$$\pi^- p \to \Sigma^- K^+ \qquad \text{and} \qquad \pi^+ p \to \Sigma^+ K^+$$

imply that the $\Sigma^\pm$ hyperons have $T_3 = \pm 1$, respectively. Since no doubly charged $\Sigma$'s are found, we conclude that $T = 1$. In this way, the existence of the $\Sigma^0$, which is the $T_3 = 0$ member of the isospin triplet, was predicted before its discovery. Note that isospin conservation means

$$nn \nrightarrow \Lambda\Lambda, \qquad \pi^- p \nrightarrow \Sigma^+ K^-,$$

etc., all in agreement with observation.

Clearly the relation between charge $Q$ and $T_3$ differs for different multiplets. We have already seen (Chapter 6, Section 6.5a) that for $\pi$ and N,

$$Q = T_3 + B/2, \tag{6.30}$$

where $B$ is the baryon number. For the strange particles, we can write

$$Q = T_3 + (B + S)/2, \tag{9.1}$$

defining the *strangeness* $S$. Note that

$$\begin{aligned} \text{conservation of charge} &\Rightarrow \Delta Q = 0, \\ \text{conservation of baryon number} &\Rightarrow \Delta B = 0, \\ \text{conservation of isospin} &\Rightarrow \Delta T_3 = 0 \end{aligned}$$

(the $\Delta$ refers to the change in a reaction), and all three together with Eq. (9.1) imply $\Delta S = 0$. Thus the assumption of isospin invariance implies the existence of a strangeness quantum number which is *conserved* in the strong interactions. Strangeness, and $T_3$, are also conserved in the electromagnetic interactions; experimentally, we observe $\Sigma^0 \to \Lambda\gamma$, but not $\Lambda \to n\gamma$. The weak interactions, being responsible for the decay of the strange particles, violate strangeness conservation.

The $S$ values implied by Eq. (9.1) are shown in Table 9.1. Using $S$ is slightly easier than using $T_3$ in deciding whether a reaction is permitted. Thus

$$\pi^- p \to \Sigma^- K^+$$
$$(S = 0, 0;\ -1,\ +1),$$

but

$$\pi^- p \not\to \Sigma^+ K^-$$
$$(S = 0, 0;\ -1,\ -1).$$

In Table 9.1 we have included the $\Xi$ hyperon, which is produced, e.g., in the reaction

$$\pi^- p \to K^0 K^+ \Xi^-.$$

The Gell-Mann–Nishijima scheme also predicted the $\Xi^0$ before its observation.

**Table 9.1**

*Gell-Mann–Nishijima Scheme*

| Particle | $B$ | $T$ | $S$ | $Y$ |
|---|---|---|---|---|
| $\pi^+, \pi^0$ | 0 | 1 | 0 | 0 |
| $K^+, K^0$ | 0 | $\frac{1}{2}$ | +1 | +1 |
| $\bar{K}^0, K^-$ | 0 | $\frac{1}{2}$ | −1 | −1 |
| p, n | 1 | $\frac{1}{2}$ | 0 | +1 |
| $\Lambda$ | 1 | 0 | −1 | 0 |
| $\Sigma^\pm, \Sigma^0$ | 1 | 1 | −1 | 0 |
| $\Xi^0, \Xi^-$ | 1 | $\frac{1}{2}$ | −2 | −1 |

It is important to recognize that $K^0$ and $\bar{K}^0$ are *different* particles; they belong to different isospin multiplets and have opposite strangeness. Because of this difference in strangeness, the interactions in matter of the K's are very different from those of the $\bar{K}$'s; the $\bar{K}$'s impinging on nucleons produce hyperons, but not the K's.

The *hypercharge* quantum number $Y$ has been introduced in Table 9.1. This alternative to $S$ is defined by

$$Y \equiv S + B. \tag{9.2}$$

As we shall see in Chapter 10, using $Y$ makes the assignments more symmetrical. The particles are displayed on $(T, Y)$ diagrams in Fig. 9.1.

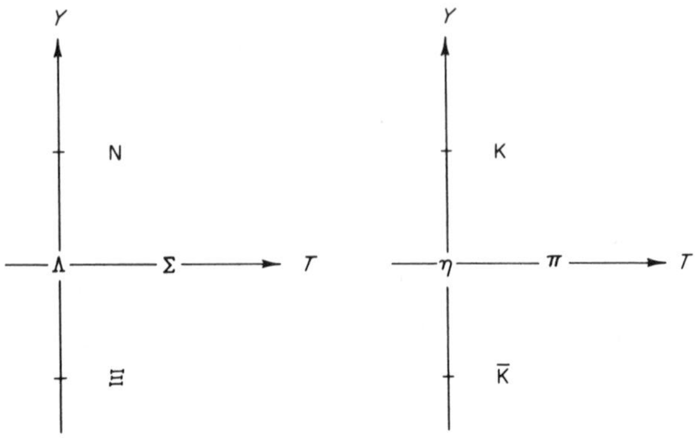


**Fig. 9.1.** Baryons and pseudoscalar mesons on a $T$ versus $Y$ plot.

## 9.2. Some Properties of the Strange Particles

### a. Spins

The observed weak decay mode

$$K^0 \to \pi^0\pi^0,$$

the fact that the $\pi^0$ has zero spin, and the requirement that the $2\pi^0$ wave function be symmetric under interchange imply that the $K^0$ has *even* angular momentum. Then lack of any decay asymmetry favors the $J = 0$ assignment. This leads to spin zero for the isotopic partner, $K^+$, and the antiparticles $\overline{K}^0$ and $K^-$.

Odd half-integral spin for the $\Lambda$, $\Sigma$, and $\Xi$ hyperons is already implied by the reactions given above. Spin one-half has been established for each of these particles.

### b. Parities

Now the strange particle decays are weak processes (except for the electromagnetic $\Sigma^0 \to \Lambda\gamma$ decay), and these, as we will discuss below, do not conserve parity. The electromagnetic interactions, like the strong interactions, do conserve parity but also conserve strangeness. Since only the weak interactions connect states of different strangeness, one relative parity between such states is arbitrary. A convention is needed to fix it and like the np case, we choose the ΛN relative parity to be *even*. Then experiment can determine the K, Σ, and Ξ parities.

Interesting evidence for negative KΛ parity comes from the *hypernucleus* ${}_{\Lambda}H^4$. A hypernucleus, denoted essentially by ${}_{\Lambda}Z^A$, is a nucleus with one of the nucleons replaced by a Λ. Many different hypernuclei have been studied, partly in order to learn about the ΛN force. These studies indicate that the spin–singlet force is the more attractive, which leads to spin zero for the ground state of ${}_{\Lambda}H^4$. But this hypernucleus is observed as a product of the capture (from an atomic orbit) reaction

$$K^-He^4 \to {}_{\Lambda}H^4\pi^0.$$

If all of the reactants have spin zero, the $K^-$ and the $\pi^0$ must have the same orbital angular momentum. Then parity conservation says that

$$(-1)^l \eta_p^{\ 2}\eta_n^{\ 2}\eta_{K^-} = (-1)^l \eta_{\pi^0}\eta_\Lambda \eta_p \eta_n^{\ 2},$$

or

$$\eta_p \eta_{K^-} = \eta_{\pi^0}\eta_\Lambda; \tag{9.3}$$

and with the convention

$$\eta_\Lambda = \eta_p = +1,$$

we have

$$\eta_{K^-} = \eta_{\pi^0} = -1. \tag{9.4}$$

This argument would fail if the ${}_{\Lambda}H^4$ were formed in an excited state of spin one, which then decays by an (unobserved) $\gamma$ emission to spin-zero ground state. There is no evidence for such a state and it seems very unlikely, but the possibility cannot be completely ruled out.

Conclusive evidence for the KΣ relative parity has come from the study of the reaction

$$K^-p \to \Sigma^+\pi^-$$

over a range of energies including an $S = -1$ resonance called $Y_0^*$ (1520) [or Λ′(1520)]. The $\Sigma^+$ polarization is very sensitive to the relative parity and implies

$$\eta_{K^-}\eta_{\Sigma^+} = -1. \tag{9.5}$$

Information on the $\Sigma^0\Lambda$ relative parity is contained in the decay

$$\Sigma^0 \to \Lambda\gamma,$$

but it is easier to extract when the $\gamma$ is replaced by a Dalitz pair:

$$\Sigma^0 \to \Lambda e^+ e^-.$$

Analysis of the invariant mass distribution of the pair indicates

$$\eta_\Sigma \text{-} \eta_\Lambda = +1, \tag{9.6}$$

which is consistent with Eqs. (9.4) and (9.5) (using the fact that different members of a isomultiplet must have the same parity).

The evidence also favors positive parity for $\Xi$, so all eight baryons $N, \Lambda, \Sigma, \Xi$ are $J^P = \frac{1}{2}^+$.

### 9.3. Strange Particle Resonances

The first strange particle resonance was discovered in 1960 in the reaction

$$K^- p \to \Lambda \pi^+ \pi^-.$$

The Dalitz plot is shown in Fig. 9.2. Note the concentration of events around the lines $T_\pi \simeq 285$ MeV; this corresponds to mass 1385 MeV for the $\Lambda\pi^\pm$ system. The assignment $J^P = \frac{3}{2}^+$ was later established, and the resonance is known as the $Y_1{}^*(1385)$ [or $\Sigma(1385)$]. The notation is $Y_T{}^*(M^*)$, with Y indicating $(\Sigma\pi)$ or $(\Lambda\pi)$ quantum numbers. Of course, the $\Lambda\pi$ combination must have $T = 1$.

Analysis of the reactions

$$\begin{aligned} K^- p &\to K^- p \quad \text{or} \quad K^0 n \\ &\to \Sigma^{0\pm} \pi^{0\mp} \\ &\to \Lambda \pi^0 \end{aligned}$$

predicted a $T = 0$ $S$-wave *bound state* of the KN system. This is analogous to predicting the existence of the deuteron from the np scattering. This KN bound state also shows up at the correct energy as a $(\Sigma^0\pi^0)$ or $(\Lambda\pi^0)$ resonance in the many-particle final state:

$$\begin{aligned} K^- p &\to (\Sigma^0 \pi^0) \pi^+ \pi \\ &\to (\Lambda \pi^0) \pi^+ \pi^-, \end{aligned}$$

but not in the charged $(\Sigma\pi)$, $(\Lambda\pi)$ states. This is the $Y_0{}^*(1405)$ [$\Lambda'(1405)$], with $J^P = \frac{1}{2}^-$. There are also $S = -2$ baryon resonances.

We also mention the earliest strange meson resonance to be found, the K*(890). This shows up very clearly in the reaction

$$K^+ p \to (K^+ \pi^-) \pi^+ p,$$

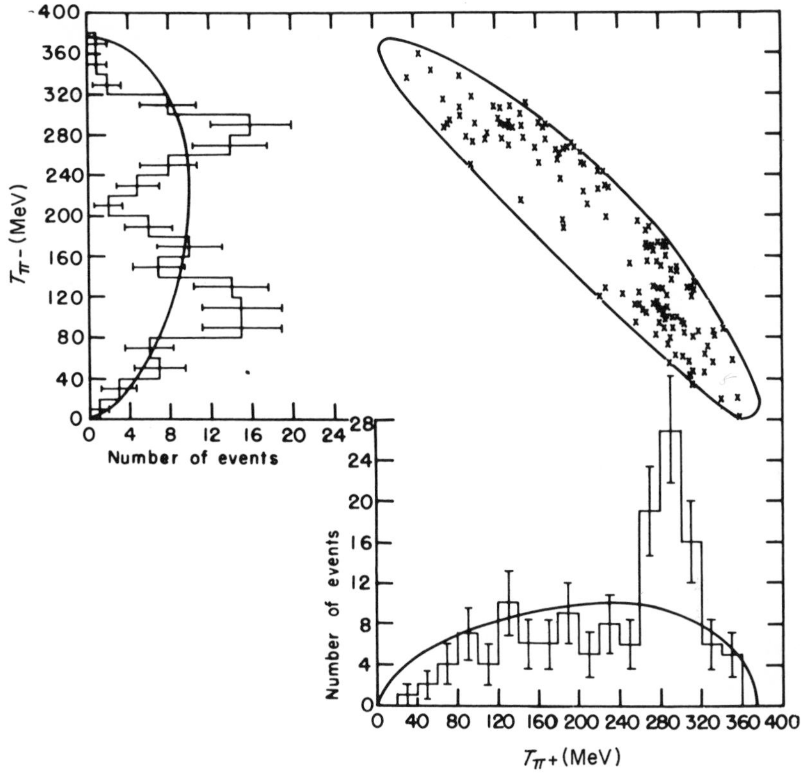


**Fig. 9.2.** Dalitz plot for $K^-p \to G^+\pi^-$, showing the concentration of events around $T_{\pi^-} = 285$ MeV. [From M. Alston *et al.*, Resonance in the $\Lambda\pi$ system. *Phys. Rev. Lett.* **5**, 520–524 (1960).]

as can be seen from Fig. 9.3. Clearly $Y = S = +1$, and furthermore $T = \frac{1}{2}$ and $J^P = 1^-$ have been established; the antiparticles of this doublet form the $Y = S = -1$ $\overline{K}^*(890)$ doublet.

Finally, we mention the $\phi$ [or $\phi(1019)$], a neutral, strangeness-zero meson that likes to decay into a $K\overline{K}$ pair. It was discovered in the reactions

$$\begin{aligned} K^-p &\to \Lambda K^+K^- \\ &\to \Lambda K^0\overline{K}^0 \end{aligned}$$

but also decays (about 20% of the time) into $\pi^+\pi^-\pi^0$. The mass is 1019 MeV, width $3.7 \pm 0.6$ MeV, but all the quantum numbers are the same as that of the $\omega$:

$$J^PT^G = 1^-0^-.$$

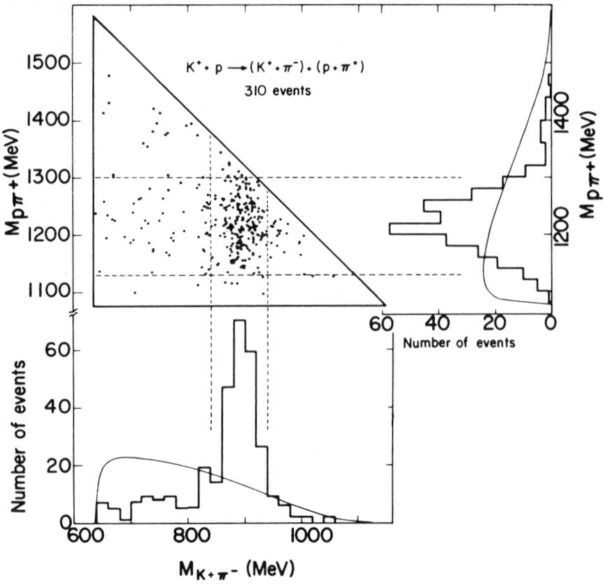


**Fig. 9.3.** Mass correlations for $K^+p \rightarrow K^+\pi^-\pi^+p$, which goes mostly via K*(890)N* (1238). [From W. Chinowsky *et al.*, On the spin of the K* resonance. *Phys. Rev. Lett.* **9**, 330–332 (1962).]

We have merely given a sampling of the hadron (i.e., strongly interacting particle) resonances; to get an idea of the vast number of resonances left out, the student is urged to investigate a current set of Rosenfeld *et al.* tables (see Appendix).

## Problems

**9.1.** Use isospin symmetry to discuss $K^{\pm}N$ elastic and charge exchange reactions, deriving relations between the various cross sections.

**9.2.** Consider $K^-$ absorption by $He^4$. What is the ratio of ${}_{\Lambda}He^4\,\pi^-$ to ${}_{\Lambda}H^4\pi^0$ produced, assuming isospin symmetry?

**9.3.** Consider the reactions

(1) $\pi^+p \rightarrow \Sigma^+K^+$,
(2) $\pi^-p \rightarrow \Sigma^0K^0$,
(3) $\pi^-p \rightarrow \Sigma^-K^+$.

Show that isospin symmetry implies that the associated amplitudes satisfy

$$f(1) - f(3) - \sqrt{2}\,f(2) = 0.$$

This implies that the amplitudes form the sides of a triangle in the complex plane, and hence that the cross sections satisfy "triangular inequalities," e.g., $[2\sigma(2)]^{1/2} + [\sigma(3)]^{1/2} \geq [\sigma(1)]^{1/2}$, etc.

## Bibliography

W. R. FRAZER. *Elementary Particles*. Chap. 4. Prentice-Hall, Englewood Cliffs, New Jersey, 1966.

J. D. JACKSON. *Physics of Elementary Particles*. Princeton Univ. Press, Princeton, New Jersey, 1958.

G. KÄLLÉN. *Elementary Particle Physics*. Addison-Wesley, Reading, Massachusetts, 1964.

E. SEGRÈ. *Nuclei and Particles*. Chap. 15. Benjamin, New York, 1965.

CHAPTER TEN

# SU(3)

SU(3) is a simple generalization of the isospin SU(2) group. One version of SU(3) symmetry, the *eightfold way*, proposed in 1961 by Gell-Mann and by Ne'eman, has been spectacularly successful in bringing some order to hardron physics. We will discuss SU(3) and its application to particle physics in some detail. To introduce the ideas, we review some aspects of SU(2).

## 10.1. SU(2) Group

We have discussed how the SU(2) (isospin) group arises when we considered transformations on the two component spinor $\binom{p}{n}$. These unitary, unimodular $U$'s form the SU(2) *Lie group*, while the commutation relations satisfied by the generators $\tau_i$ of the group form the SU(2) *Lie algebra*. In these commutation relations,

$$[\tfrac{1}{2}\tau_i, \tfrac{1}{2}\tau_j] = i\varepsilon_{ijk}\tfrac{1}{2}\tau_k, \tag{6.35}$$

the constants on the right-hand side (here $\varepsilon_{ijk}$) are known, in general, as *structure constants*. Through Eq. (6.35), these constants determine the commutation properties of the infinitesimal $U$'s and hence the structure of the group (i.e., the multiplication table).

### *a. Irreducible Representations*

We are interested in the *multiplets* implied by the SU(2) symmetry, and hence in the *irreducible representations* of SU(2). *Representation* means a set of $n \times n$ matrices that satisfy the commutation relations, Eq. (6.35), and therefore an $n \times n$ matrix for each group element:

$$[T_i, T_j] = i\varepsilon_{ijk} T_k$$

and

$$D(\theta) = e^{i\theta_j T_j} \equiv e^{i\boldsymbol{\theta}\cdot\mathbf{T}}. \tag{10.1}$$

These matrices operate on $n$-dimensional column vectors $\psi$:

$$\psi' = D(\theta)\psi. \tag{10.2}$$

Suppose we perform a unitary transformation $M$ on the $\psi$'s:

$$\tilde{\psi} = M\psi. \tag{10.3}$$

Then

$$\tilde{\psi}' = MD(\theta)M^{-1}\tilde{\psi} = \tilde{D}(\theta)\tilde{\psi}; \tag{10.4}$$

so

$$\tilde{D}(\theta) = MD(\theta)M^{-1}. \tag{10.5}$$

Suppose further, that for a given representation, there exists an $M$ such that $\tilde{D}(\theta)$ for all $\theta$ has the form

$$\tilde{D}(\theta) = \begin{pmatrix} D_1(\theta) & \cdots \\ \cdots & D_2(\theta) \end{pmatrix}. \tag{10.6}$$

(The $\cdots$ stand for zeros everywhere in that section; this form is called *block diagonal.*) Then $\psi$ has the form

$$\tilde{\psi} = \begin{pmatrix} \psi_1 \\ \psi_2 \end{pmatrix} \tag{10.7}$$

and

$$\tilde{\psi}' = \begin{pmatrix} \psi_1' \\ \psi_2' \end{pmatrix},$$

where

$$\psi_1' = D_1(\theta)\psi_1 \qquad \text{and} \qquad \psi_2' = D_2(\theta)\psi. \tag{10.8}$$

That is, the 1 and 2 parts of $\tilde{\psi}$ are not mixed together by the group transformations but remain separate. Such a representation $D$ is called *reducible*; a representation which cannot be reduced to the block diagonal form (no such $M$ exists) is *irreducible*.

### b. Example

All this is quite general, but an SU(2) example might be helpful. Form a four-dimensional representation by combining two two-dimensional ones (spinors) in a *direct product*:

$$\chi = \begin{pmatrix} \chi_1 \\ \chi_2 \end{pmatrix}, \qquad \eta = \begin{pmatrix} \eta_1 \\ \eta_2 \end{pmatrix}; \tag{10.9}$$

$$\psi_1 \equiv \chi_1\eta_1, \qquad \psi_2 \equiv \chi_1\eta_2, \qquad \psi_3 \equiv \chi_2\eta_2, \qquad \psi_4 \equiv \chi_2\eta_2. \tag{10.10}$$

Since we know how the $\chi$ and $\eta$ transform, we can deduce how $\psi$ transforms: Writing

$$D^{(2)} \equiv D^{(2)}(\theta),$$

for the two-dimensional irreducible representation,

$$\begin{aligned} \chi' &= D^{(2)}\chi, \\ \eta' &= D^{(2)}\eta, \\ \chi_i'\eta_j' &= (D^{(2)})_{ik}(D^{(2)})_{jl}\,\chi_k\eta_l; \end{aligned} \tag{10.11}$$

so

$$\psi' = D^{(2\times 2)}\psi \tag{10.12}$$

implies

$$(D^{(2\times 2)})_{11} = (D^{(2)})_{11}(D^{(2)})_{11}, \tag{10.13}$$

etc. This direct product is reducible; the reduction can be affected by the unitary transformation $U$ that transforms $\psi$ into

$$\tilde{\psi} = U\psi = \begin{pmatrix} \psi_1 \\ \frac{1}{\sqrt{2}}(\psi_2 + \psi_3) \\ \psi_4 \\ \frac{1}{\sqrt{2}}(\psi_2 - \psi_3) \end{pmatrix} = \begin{pmatrix} \chi_1\eta_1 \\ \frac{1}{\sqrt{2}}(\chi_1\eta_2 + \chi_2\eta_1) \\ \chi_2\eta_2 \\ \frac{1}{\sqrt{2}}(\chi_1\eta_2 - \chi_2\eta_1) \end{pmatrix}. \tag{10.14}$$

That is,

$$UD^{(2\times 2)}U^{-1} = \tilde{D}^{(2\times 2)} = \left(\begin{array}{c|c} D^{(3)} & \cdots \\ \hline \cdots & D^{(1)} \end{array}\right). \tag{10.15}$$

$D^{(3)}(\theta)$ is the three-dimensional (irreducible) representation, $D^{(1)}(\theta)$ $(\equiv 1)$ the one-dimensional one. This reduction of the direct product is denoted by

$$\mathbf{2} \otimes \mathbf{2} = \mathbf{1} \oplus \mathbf{3} \tag{10.16}$$

and, of course, amounts to combining two spin-one-half objects to form spin one or spin zero. This is the way multiplets are combined to form new multiplets.

### c. Weight Diagrams

A simple way to combine multiplets involves the use of *weight diagrams*; this is essentially the same as the *vector model* of introductory quantum mechanics. One of the generators (usually $T_3$) is chosen to be diagonal; the diagonal elements are eigenvalues, and each is used to label its eigenvector; the different vectors together compose the multiplet. Thus a plot exhibiting the different eigenvalues (the weight diagram) provides a picture of the entire multiplet. Now when multiplets are combined, the eigenvalues simply add, and the resulting weight diagram has all the possibilities. This is illustrated for two doublets again in Figs. 10.1(a) and (b). Clearly the result is not a single multiplet (since there are *two* ways of having $T_3 = 0$) but can be written as a sum of multiplets (singlet plus triplet) as in Fig. 10.1(c).

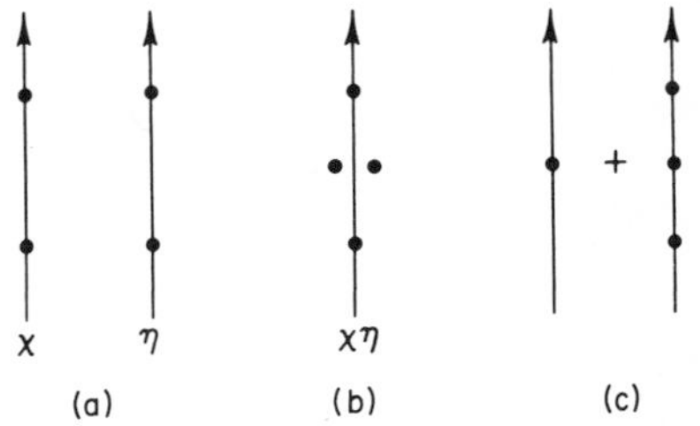


**Fig. 10.1.** Combining the SU(2) doublets: (a) Weight diagrams of the doublets; (b) the direct product weight diagram; (c) reduced to singlet plus triplet.

## 10.2. SU(3) Group

### a. Defining Representation

We now want to generalize the isospin group to include hypercharge. This is accomplished simply by considering the group of unitary unimodular transformations on *three* objects: SU(3). We name the three objects[1] p̂, n̂, and $\hat{\lambda}$; the corresponding linearly independent (three-) spinors we take as

$$\hat{\mathrm{p}} \equiv q(1) \equiv \begin{pmatrix} 1 \\ 0 \\ 0 \end{pmatrix}, \tag{10.17a}$$

$$\hat{\mathrm{n}} \equiv q(2) \equiv \begin{pmatrix} 0 \\ 1 \\ 0 \end{pmatrix}, \tag{10.17b}$$

[1] In the *Sakata model*, these were the actual p, n, and Λ; here they are just names.

$$\hat{\lambda} \equiv q(3) \equiv \begin{pmatrix} 0 \\ 0 \\ 1 \end{pmatrix}; \tag{10.17c}$$

and an arbitrary spinor can then be written as

$$q \equiv \begin{pmatrix} \mathrm{p} \\ \mathrm{n} \\ \lambda \end{pmatrix}. \tag{10.17'}$$

A transformation of such a $q$ is then

$$q' = Qq, \tag{10.18}$$

where $Q$ (the $3 \times 3$ matrix) can be expressed in terms of generators $F_i$:

$$Q = e^{i\alpha_j F_j} \equiv e^{i\boldsymbol{\alpha}\cdot\mathbf{F}}. \tag{10.19}$$

For an infinitesimal transformation (i.e., infinitesimal $\alpha_j$)

$$Q = 1 + i\alpha_j F_j \equiv 1 + i\boldsymbol{\alpha}\cdot\mathbf{F}. \tag{10.20}$$

Of course, $Q$ unitary implies $F_j$ Hermitian; $Q$ unimodular implies $F_j$ *traceless.* There are *eight* such generators that are linearly independent (a $3 \times 3$ Hermitian matrix has nine real parameters, and the zero trace condition reduces this to eight) and therefore $\boldsymbol{\alpha}$ and $\mathbf{F}$ each have eight components. In analogy with SU(2), the generators can be chosen as follows:

$$F_i = \tfrac{1}{2}\lambda_i; \tag{10.21}$$

$$\lambda_1 = \begin{pmatrix} 0 & 1 & 0 \\ 1 & 0 & 0 \\ 0 & 0 & 0 \end{pmatrix}, \quad \lambda_2 = \begin{pmatrix} 0 & -i & 0 \\ i & 0 & 0 \\ 0 & 0 & 0 \end{pmatrix}, \quad \lambda_3 = \begin{pmatrix} 1 & 0 & 0 \\ 0 & -1 & 0 \\ 0 & 0 & 0 \end{pmatrix},$$

$$\lambda_4 = \begin{pmatrix} 0 & 0 & 1 \\ 0 & 0 & 0 \\ 1 & 0 & 0 \end{pmatrix}, \quad \lambda_5 = \begin{pmatrix} 0 & 0 & -i \\ 0 & 0 & 0 \\ i & 0 & 0 \end{pmatrix}, \quad \lambda_6 = \begin{pmatrix} 0 & 0 & 0 \\ 0 & 0 & 1 \\ 0 & 1 & 0 \end{pmatrix},$$

$$\lambda_7 = \begin{pmatrix} 0 & 0 & 0 \\ 0 & 0 & -i \\ 0 & i & 0 \end{pmatrix}, \quad \lambda_8 = \frac{1}{\sqrt{3}} \begin{pmatrix} 1 & 0 & 0 \\ 0 & 1 & 0 \\ 0 & 0 & -2 \end{pmatrix}. \tag{10.22}$$

Notice that there exist *two* independent diagonal matrices $\lambda_3$ and $\lambda_8$; thus the SU(3) group is said to be of *rank* 2. This means that the spinors $q$ are characterized by *two* quantum numbers, as we desire. Equations (10.22) can now be used to calculate the structure constants $f_{ijk}$:

$$[F_i, F_j] = if_{ijk} F_k \tag{10.23}$$

or

$$[\lambda_i, \lambda_j] = 2if_{ijk}\lambda_k.$$

### b. Shift Operators

Instead of dealing directly with the $F$'s, it is convenient to introduce the *shift operators*:

$$T_\pm \equiv F_1 \pm iF_2, \tag{10.24a}$$

$$U_\pm \equiv F_6 \pm iF_7, \tag{10.24b}$$

$$V_\pm \equiv F_4 \pm iF_5, \tag{10.24c}$$

along with the diagonal operators

$$T_3 \equiv F_3, \tag{10.24d}$$

$$Y \equiv (2/\sqrt{3})F_8. \tag{10.24e}$$

We identify $T_3$, $T_\pm$ with isospin and $Y$ with hypercharge. Then one can show that

$$[T_3, T_\pm] = \pm T_\pm, \tag{10.25a}$$

$$[T_3, U_\pm] = \mp\tfrac{1}{2}U_\pm, \tag{10.25b}$$

$$[T_3, V_\pm] = \pm\tfrac{1}{2}V_\pm, \tag{10.25c}$$

$$[Y, T_\pm] = 0, \tag{10.25d}$$

$$[Y, U_\pm] = \pm U_\pm, \tag{10.25e}$$

$$[Y, V_\pm] = \pm V_\pm. \tag{10.25f}$$

These commutation relations suffice to establish the stepping properties. Specifically, suppose we have a certain irreducible respresentation of the commutation relations. The matrices corresponding to $T_\pm$, $U_\pm$, and $V_\pm$ take us from one substate (member of the multiplet) to another. For example, consider an eigenvector of $T_3$ and $Y$:

$$\begin{aligned} T_3|t_3, y\rangle &= t_3|t_3, y\rangle, \\ Y|t_3, y\rangle &= y|t_3, y\rangle. \end{aligned} \tag{10.26}$$

Then Eq. (10.25a) implies

$$T_3(T_+|t_3, y\rangle = T_+(T_3 + 1)|t_3, y\rangle = (t_3 + 1)T_+|t_3, y\rangle, \tag{10.27}$$

while Eq. (10.25d) implies

$$Y(T_+|t_3, y\rangle) = T_+(Y|t_3, y\rangle) = yT_+|t_3, y\rangle. \tag{10.28}$$

Thus on a $(t_3, y)$ weight diagram (Fig. 10.2), $T_+$ steps as shown. Similarly, Eqs. (10.25b) and (10.25c) imply that $U_+$ increases by $y$ by one while decreasing $t_3$ by one-half, and so forth. The effect of such stepping operations is shown in Fig. 10.2. The remaining commutation relations are easily written down; this is left as an exercise.

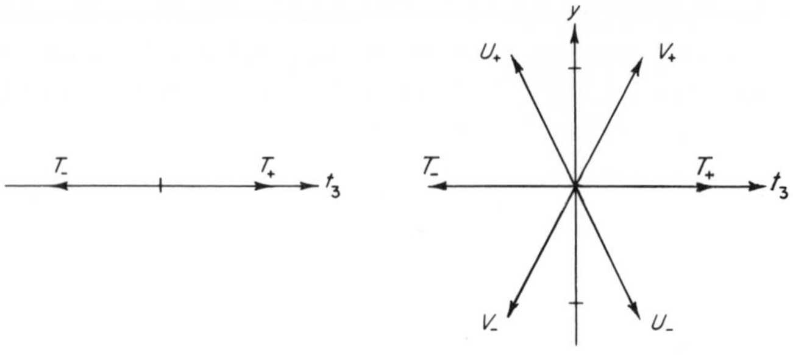


**Fig. 10.2.** The stepping operators for (a) SU(2); (b) SU(3).

## *c. SU(3) Multiplets*

The commutation relations can be used to prove some general properties of irreducible representations. Here we merely quote some of the results that will subsequently be most useful:

1. The multiplets in general possess hexagonal weight diagrams (Fig. 10.3), which (with proper choice of scale for the $y$-axis) are symmetric about the three axes shown.

2. The multiplets (irreducible representations) are completely specified by the two integers $p$ and $q$, which give the number of spaces in two adjacent sides of the outside boundary of the weight diagram (Fig. 10.3).

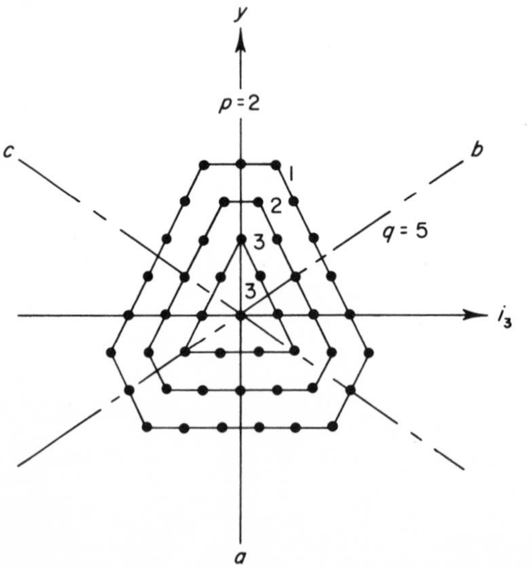


**Fig. 10.3.** Weight diagram for an SU(3) multiplet. The multiplicities of the points are shown, as are the symmetry axes $a$, $b$, and $c$, and the characteristic integers $p$ and $q$.

3. The *multiplicity* [number of independent substates with given $(t_3, y)$ quantum numbers] increases by one every step in from the boundary, which itself has multiplicity 1, until the hexagon becomes a triangle. Inside and on the triangle the multiplicity is constant (Fig. 10.3).

4. The *dimensionality* of an irreducible representation (number of substates of a multiplet) is given by

$$n = \tfrac{1}{2}(p+1)(q+1)(p+q+2). \tag{10.29}$$

We begin our discussion of the multiplets with the simplest nontrivial one, the *triplet* corresponding to the *defining*, or *fundamental*, *representation* that we started with [Eqs. (10.17)]. Using

$$T_3 = \begin{pmatrix} \frac{1}{2} & 0 & 0 \\ 0 & -\frac{1}{2} & 0 \\ 0 & 0 & 0 \end{pmatrix} \quad \text{and} \quad Y = \begin{pmatrix} \frac{1}{3} & 0 & 0 \\ 0 & \frac{1}{3} & 0 \\ 0 & 0 & -\frac{2}{3} \end{pmatrix}, \tag{10.30}$$

and Eqs. (10.17), we can read off the eigenvalues and write[2]

$$\begin{aligned} \hat{\mathrm{p}} &\equiv q(1) = |\tfrac{1}{2}, \tfrac{1}{3}\rangle, \\ \hat{\mathrm{n}} &\equiv q(2) = |-\tfrac{1}{2}, \tfrac{1}{3}\rangle, \\ \hat{\lambda} &\equiv q(3) = |0, -\tfrac{2}{3}\rangle. \end{aligned} \tag{10.31}$$

The corresponding weight diagram is shown in Fig. 10.4(b). We see that the triplet consists of an isospin doublet (hypercharge $\frac{1}{3}$) plus an isospin singlet (hypercharge $-\frac{2}{3}$), and that this **3** representation has $p = 1$, $q = 0$.

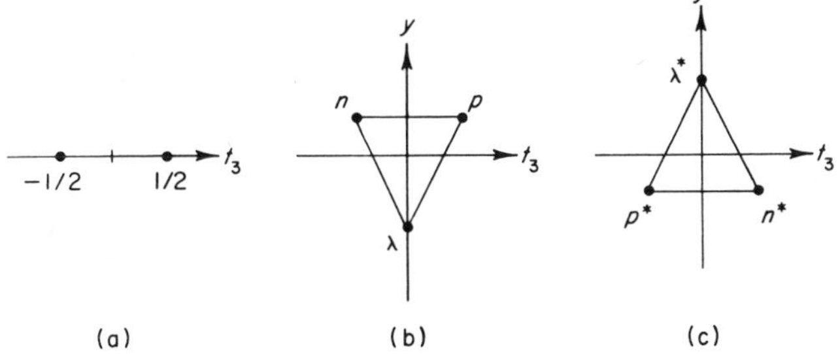


**Fig. 10.4.** Weight diagrams for the fundamental representation for (a) SU(2) and (b) SU(3); (c) shows the **3*** multiplet, which is the dual of the **3** shown in (b).

---

[2] To correspond to the notation we have used for isospin state, $|t, t_3\rangle$, we should write $|\mathbf{3}, t, t_3, y\rangle$, where the **3** indicates the three-dimensional representation. The dimension is usually clear from the context, so we omit it.

### d. Dual Representation

Now for every representation there exists a closely related one, the *dual representation*, which may or may not be distinct from the original. Consider the transformation equation

$$\psi' = D^r(\alpha)\psi, \tag{10.32}$$

where

$$D^r(\alpha) = \exp[i\alpha \cdot F^r] \tag{10.33}$$

(superscript $r$ denotes the representation). Taking the complex conjugate of these equations yields

$$\psi'^* = D^{r^*}(\alpha)\psi^*, \tag{10.34}$$

where

$$D^{r^*}(\alpha) \equiv [D^r(\alpha)]^* = \exp[-i\alpha \cdot (F^r)^*] \equiv \exp[i\alpha \cdot F^{r^*}]. \tag{10.35}$$

Clearly complex conjugation does not effect the multiplication properties of matrices; thus $D^{r^*}$ (the dual representation) is indeed a representation. Its generators are, from Eq. (10.35),

$$F_i^{r^*} \equiv -(F_i^r)^*, \tag{10.36}$$

so the diagonal matrices $T_3$ and $Y$ are reversed in sign in this new representation, and therefore so are the eigenvalues. This means that the weight diagram is reflected in the origin. As a result, any multiplet (irreducible representation) whose weight diagram is *not* a regular hexagon ($p \neq q$) has a distinct dual: $r \neq r^*$; and conversely. The diagram for **3*** is shown in Fig. 10.4(c). Analogous to Eq. (10.31), we have as basic spinors for **3***:

$$q^*(1) \equiv \begin{pmatrix} 1 \\ 0 \\ 0 \end{pmatrix} = |-\tfrac{1}{2}, -\tfrac{1}{3}\rangle, \quad \text{etc.}$$

### e. Multiplet Building

In the same way that higher spin objects can be built up out of spin one-half in SU(2), **3** and **3*** can be used to build up bigger multiplets. This is done by forming the direct product of the representations (in general $r$ and $r'$) and reducing the result to a sum of irreducible representations. Now

$$\psi^{r \times r'} \equiv \psi^r \psi^{r'}; \tag{10.37}$$

so

$$D^{\text{prod}} = D^r D^{r'}$$

or

$$D^{r \times r'}(\alpha) = \exp[i\boldsymbol{\alpha} \cdot \mathbf{F}^r] \exp[i\boldsymbol{\alpha} \cdot \mathbf{F}^{r'}]. \tag{10.38}$$

But also

$$D^{r\times r'}(\alpha) = \exp[i\boldsymbol{\alpha} \cdot \mathbf{F}^{\text{prod}}]. \tag{10.39}$$

Then, if we stick to $\psi$'s for which $T_3$ and $Y$ are diagonal, considering an infinitesimal $\alpha$ makes it obvious that

$$t_3^{r\times r'} = t_3{}^{r} + t_3{}^{r'}, \tag{10.40a}$$

$$y^{r\times r'} = y^{r} + y^{r'}. \tag{10.40b}$$

This simply states the additivity of the quantum numbers, which is all we need to find the weight diagram of the direct product.

We can use this additivity graphically as follows. Superimpose the weight diagrams with the origin of the second on a point of the first, make a mark at each of the points of the second, then move the origin of the second to another point, and continue until all the points of the first have been used. This is illustrated in Fig. 10.5 for $\mathbf{3} \otimes \mathbf{3}$. Notice that the midpoints of the sides of the

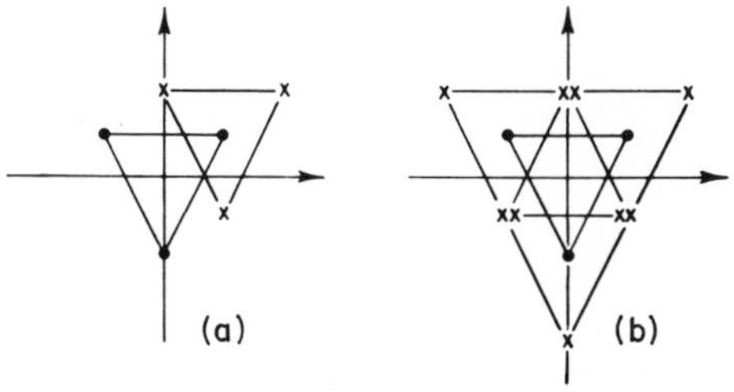


**Fig. 10.5.** Procedure for forming $\mathbf{3}\otimes\mathbf{3}$: (a) the diagram of the second $\mathbf{3}$ centered on point $(\frac{1}{2}, \frac{1}{3})$ of the first; (b) all the points marked.

large triangle are doubly occupied while the vertices are only singly occupied. In view of the preceding discussion of multiplets, this cannot be a single multiplet but is instead the sum shown in Fig. 10.6. The first diagram is just

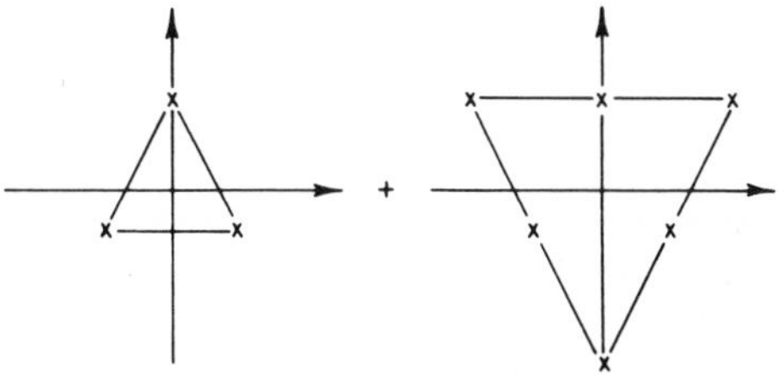

**Fig. 10.6.** Figure 10.5(b) shown as a sum of multiplets.

the $\mathbf{3}^*$, while the second is a new, six-dimensional multiplet. Thus we have found that

$$\mathbf{3} \otimes \mathbf{3} = \mathbf{3}^* \oplus \mathbf{6}. \tag{10.41}$$

Note that this equation holds for the dimensions too:

$$3 \times 3 = 9 = 3 + 6.$$

This always must be so, since it is just a matter of counting up all the (independent) combinations of eigenvectors.

Another example is $\mathbf{3} \otimes \mathbf{3}^*$. Its weight diagram and the reduction into multiplets is shown in Fig. 10.7. Here we see that

$$\mathbf{3} \otimes \mathbf{3}^* = \mathbf{1} \oplus \mathbf{8}. \tag{10.42}$$

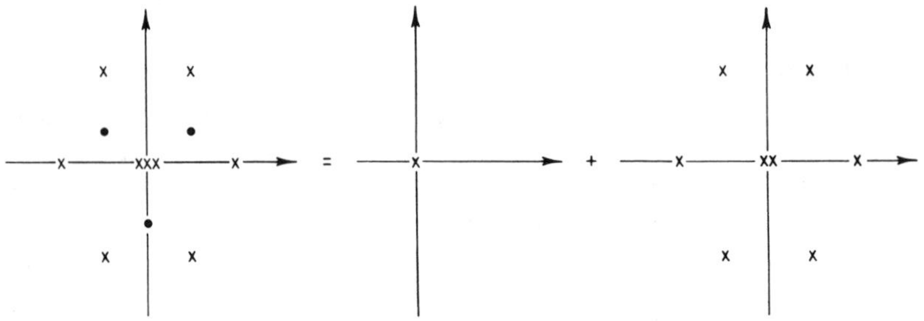

**Fig. 10.7.** The weight diagram for $\mathbf{3} \otimes \mathbf{3}^*$ and its reduction.

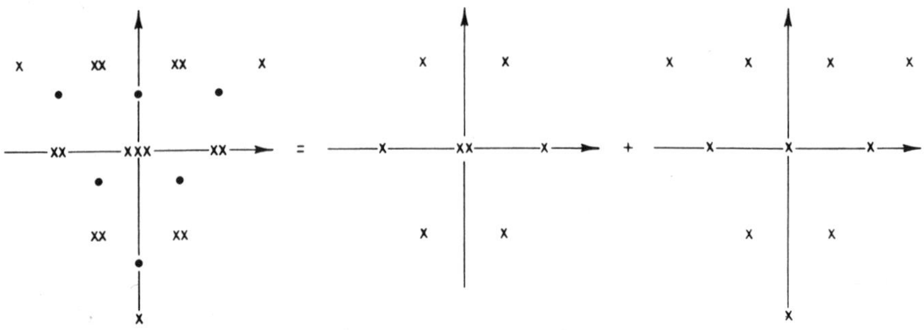

**Fig. 10.8.** The weight diagram for $\mathbf{3} \otimes \mathbf{6}$ and its reduction.

The diagrams for $\mathbf{3} \otimes \mathbf{6}$ are shown in Fig. 10.8; thus

$$\mathbf{3} \otimes \mathbf{6} = \mathbf{8} \oplus \mathbf{10}. \tag{10.43}$$

The isospin content of these multiplets is easily read off the diagrams. It is the singlet (**1**), octet (**8**), and decuplet (**10**) that are identified with the physical particles in the eightfold way; their isospin content is plotted in Fig. 10.9.

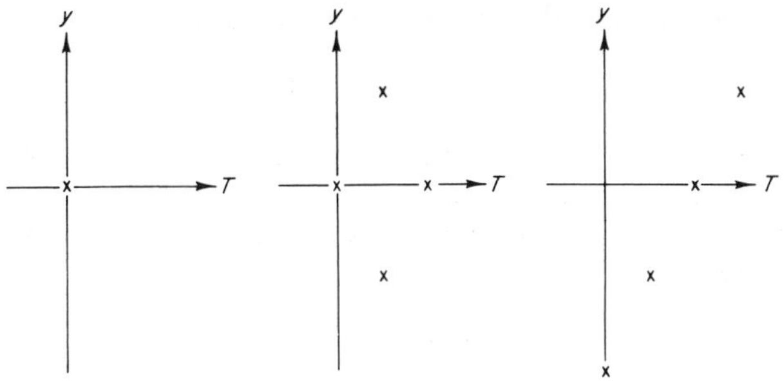


**Fig. 10.9.** $(T, y)$ diagrams for singlet, octet, and decuplet.

## *f. Tensor Method*

Instead of using this diagrammatic technique, an alternative method for reducing direct products is to construct *irreducible tensors* from elementary spinors $a, b, \ldots,$ which transform like $q$ and complex conjugate spinors $c^*, d^*, \ldots,$ which transform like $q^*$. From products of these we form tensors [cf. Eq. (1.10)]:

$$\mathcal{T} \equiv ab \cdots c^* d^* \cdots . \tag{10.44}$$

These transform under SU(3) in a very special way, since [cf. Eqs. (10.18) and (10.34)]

$$a' = Qa, \qquad c^{*\prime} = Q^* c^*, \tag{10.45}$$

$$\mathcal{T}' = QQ \cdots Q^* Q^* \cdots \mathcal{T}. \tag{10.46}$$

In terms of components, we write

$$a^i \equiv (a)_i \qquad \text{and} \qquad c_i \equiv (c^*)_i . \tag{10.47}$$

Here we use upper versus lower indices to form the necessary distinction between $q$ and $q^*$ spinors. Then

$$\mathcal{T}^{ij\cdots}_{kl\cdots} = a^i b^j \cdots c_k d_l \cdots ,$$

and [Eq. (10.46)]

$$\mathcal{T}'^{mn\cdots}_{st\cdots} = Q_{mi} Q_{nj} \cdots Q^*_{sk} Q^*_{tl} \cdots \mathcal{T}^{ij\cdots}_{kl\cdots} . \tag{10.46$'$}$$

Now a pair of upper and lower-indices (but not two upper or two lower) can be *contracted* to give a new tensor of lower rank. Since

$$\begin{aligned} a'^i c_i' &= (Q_{ij} a^j)(Q^*_{ik} c_k) \\ &= a^j ((Q^\dagger)_{ki} Q_{ij} c_k = a^j (Q^\dagger Q)_{kj} c_k \\ &= a^j c_j \end{aligned}$$

(using $Q^\dagger = Q^{-1}$), the combination $a^j c_j$ is invariant; thus

$$\mathcal{T}^{j\cdots}_{l\cdots} \equiv \mathcal{T}^{ij\cdots}_{il\cdots}$$

is indeed a tensor with one less lower and one less upper index. Then for example we can write ($\delta_j{}^i \equiv \delta_{ij}$)

$$a^i c_j = \tfrac{1}{3}\delta_j{}^i a^l c_l + (a^i c_j - \tfrac{1}{3}\delta_j{}^i a^l c_l)$$
$$\equiv A_j{}^i + B_j{}^i. \tag{10.48}$$

Each of the terms on the right-hand side is an irreducible tensor.

In addition, like indices can be symmetrized or antisymmetrized. For example,

$$a^i b^j = \tfrac{1}{2}(a^i b^j + a^j b^i) + \tfrac{1}{2}(a^i b^j - a^j b^i)$$
$$\equiv S^{ij} + A^{ij}. \tag{10.49}$$

Since there is nothing further that can be done to $S^{ij}$ and $A^{ij}$, these are irreducible tensors.

Furthermore, it can be shown that each irreducible representation corresponds to such an irreducible tensor. The fundamental representation **3**, of course, corresponds to the irreducible tensor $a^i$ and **3*** to $c_i$. The invariant combination [Eq. (10.48)] $A_j{}^i$ is the **1**, while $B_j{}^i$ being traceless has eight independent components and is thus the **8**; hence Eq. (10.48) says that

$$\mathbf{3} \otimes \mathbf{3}^* = \mathbf{1} \oplus \mathbf{8}.$$

In Eq. (10.49), $S^{ij}$ has six independent components, and so forms the **6** representation, while $A^{ij}$ has three; in fact, since $A^{ij}$ can be contracted with the constant tensor[3] $\varepsilon_{ijk}$,

$$A_i \equiv \varepsilon_{ijk} A^{jk},$$

it corresponds to the **3*** representation. Thus Eq. (10.49) performs the reduction

$$\mathbf{3} \otimes \mathbf{3} = \mathbf{3}^* \oplus \mathbf{6}.$$

## 10.3. The Eightfold Way

### *a. Baryon and Pseudoscalar Meson Octets*

The $\frac{1}{2}^+$ baryons and the pseudoscalar mesons can be displayed on a $(t_3, y)$ plot as in Fig. 10.10. (This is the same information as in Fig. 9.1.) Comparison with the octet weight diagram (as in Fig. 10.7) shows that each set of eight particles can be fitted into a single SU(3) octet. And while the members of the multiplets so formed are *not* degenerate in mass, the *mass splittings* within each multiplet follow a definite SU(3) pattern. Note that the mass splitting is $\sim m_\pi$, as compared to the isospin splitting of a mere few MeV.

---

[3] One can show that $\varepsilon_{ijk}$, $\varepsilon^{ijk}$, and $\delta_j{}^i$ transform like tensors and therefore can be used for contraction.

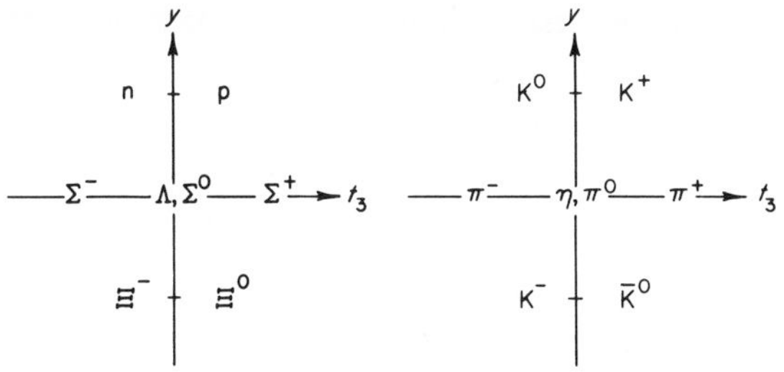


**Fig. 10.10.** The baryon and pseudoscalar meson octets.

### *b. SU(3) Symmetry Breaking. Gell-Mann–Okubo Formula*

Because of this large mass splitting, we know that the strong interactions cannot be completely SU(3) invariant but only partially so. This leads us to suppose that the strong interaction (*hadronic*) part of the Hamiltonian is composed of a *very strong* part, $H_{\text{vs}}$, which is SU(3) symmetric (i.e., invariant), and a *medium strong* part, $H_{\text{ms}}$, which is not:

$$H = H_{\text{vs}} + H_{\text{ms}}. \tag{10.50}$$

Now the mass of a particle is just the energy in the rest system; thus

$$\begin{aligned} m_a &= \langle a, \mathbf{p}_a = 0 | H | a, \mathbf{p}_a = 0 \rangle \equiv \langle a | H | a \rangle \\ &= \langle a | H_{\text{vs}} | a \rangle + \langle a | H_{\text{ms}} | a \rangle. \end{aligned} \tag{10.51}$$

Of course $\langle a | H_{\text{vs}} | a \rangle$ must be the same for all the members of a given SU(3) multiplet, since

$$m_0 \equiv \langle a | H_{\text{vs}} | a \rangle = \langle a | U^{-1} H_{\text{vs}} U | a \rangle = \langle b | H_{\text{vs}} | b \rangle,$$

using the invariance of $H_{\text{vs}}$. ($U$ is the group element that transforms substate $|a\rangle$ into substate $|b\rangle$.) What about

$$\delta m_a \equiv \langle a | H_{\text{ms}} | a \rangle ?$$

As far as its SU(3) transformation properties are concerned, *any* operator can be written as a sum of irreducible tensor operators. The eightfold way assumes that $H_{\text{ms}}$ transforms like a member of an octet, i.e., like the traceless tensor

$$q^i q_j - \tfrac{1}{3} \delta_j{}^i q^i q_k .$$

This, it turns out, implies that

$$\delta m_a = m_1 y + m_2 [T(T+1) - \tfrac{1}{4} y^2],$$

which results in the *Gell-Mann–Okubo mass formula*:

$$m_a = m_0 + m_1 y + m_2[T(T+1) - \tfrac{1}{4}y^2], \qquad (10.52)$$

where $m_0$, $m_1$, and $m_2$ are unknown constants for each multiplet.

For the baryon octet, the three unknown constants can be eliminated from the equations for the four masses to leave one mass relation:

$$\frac{m_{\mathrm{N}} + m_{\Xi}}{2} = \frac{3m_{\Lambda} + m_{\Sigma}}{4}. \qquad (10.53)$$

Inserting the observed masses in each side of this equation gives excellent agreement.

The corresponding equation for the pseudoscalar octet becomes (since $m_{\bar{\mathrm{K}}} = m_{\mathrm{K}}$)

$$m_{\mathrm{K}} = \frac{3m_\eta + m_\pi}{4}, \qquad (10.54)$$

which does *not* hold very well for the observed masses. But since in the case of mesons it is always $m^2$ rather than $m$ that appears in the Hamiltonian, there is some slight justification for writing the Gell-Mann–Okubo formula for $m^2$ rather than $m$:

$$m_a^{\ 2} = m_0^{\ 2} + m_1^{\ 2} y + m_2^{\ 2}[T(T+1) - \tfrac{1}{4}y^2]. \qquad (10.52')$$

Then instead of Eq. (10.54), we obviously have

$$m_{\mathrm{K}}^{\ 2} = \frac{3m_\eta^{\ 2} + m_\pi^{\ 2}}{4}. \qquad (10.54')$$

This formula works somewhat better than Eq. (10.54); we return to this point below.

### c. *Baryon Decuplet. The* $\Omega^-$

The greatest triumph of SU(3) has to do with the $\frac{3}{2}^+$ baryon resonances; these can be fit into a decuplet, as shown in Fig. 10.11. The existence of the

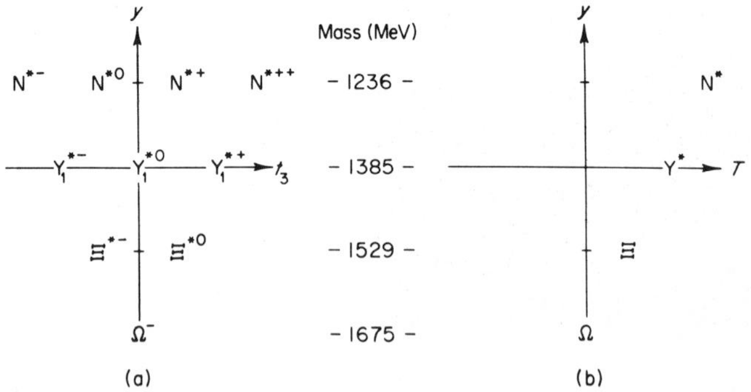


**Fig. 10.11.** (a) Weight diagram and (b) $(T, y)$ diagram for the $\frac{3}{2}^+$ baryon resonances. The mass is shown between the diagrams.

$\Omega^-$, which is needed to complete the decuplet, was undreamed of before the eightfold way, and its experimental discovery was awaited with considerable interest. The mass, which was predicted as discussed below, is less than the threshold for the strong decay

$$\Omega^- \to \Xi \overline{K}$$

and so this strangeness $-3$, spin $\frac{3}{2}$ baryon decays only weakly and lives (roughly) as long as the $\Lambda$, for example. This was predicted in 1962 and found in 1964 in the impressive (and frequently reproduced) bubble chamber photograph shown in Fig. 10.12. The chain of events is interpreted as

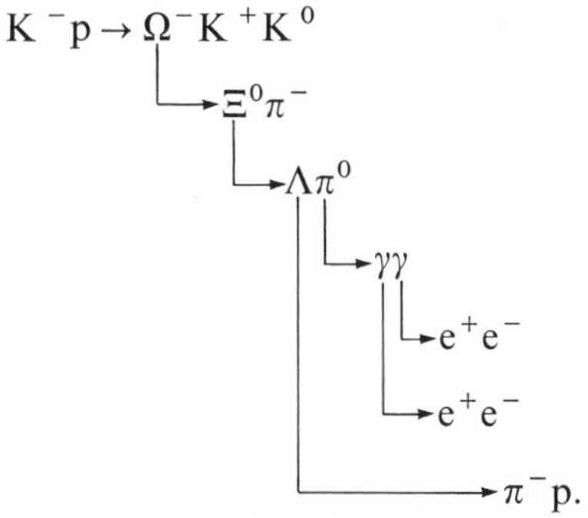


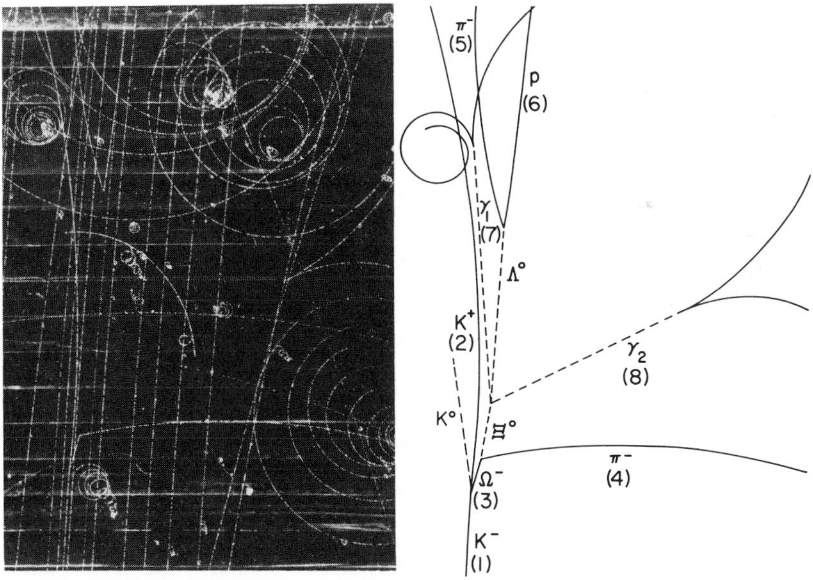


**Fig. 10.12.** First observed $\Omega^-$ decay. [From V. E. Barnes *et al.*, Observations of a hyperon with strangeness minus three. *Phys. Rev. Lett.* **12**, 204–206 (1964). Courtesy Brookhaven National Laboratory.]

The mass of the $\Omega^-$ comes out of the Gell-Mann–Okubo formula as applied to the decuplet. For any "triangular" multiplet, there is a linear relation between $T$ and $y$; for **10** (cf. Fig. 10.11),

$$T = \tfrac{1}{2}y + \text{const};$$

so Eq. (10.52) becomes (with new $m_0$ and $m_1$)

$$m_a = m_0 + m_1 y. \tag{10.55}$$

This implies the *equal spacing rule*:

$$m_{N^*} - m_{Y_1^*} = m_{Y_1^*} - m_{\Xi^*} = m_{\Xi^*} - m_{\Omega^-}. \tag{10.56}$$

Experimentally, these equations are (in MeV)

$$146.1 \pm 0.9 = 147.6 \pm 1.3 = 145 \pm 3,$$

which is astonishingly good when we consider that the SU(3) symmetry breaking is so large.

### *d. Other Multiplets*

More problematic is the octet of $1^-$ mesons: $\rho$, K*, and $\overline{\text{K}}$* fall into place in an octet (see Fig. 10.13) but the question arises of whether to insert $\phi$ or $\omega$

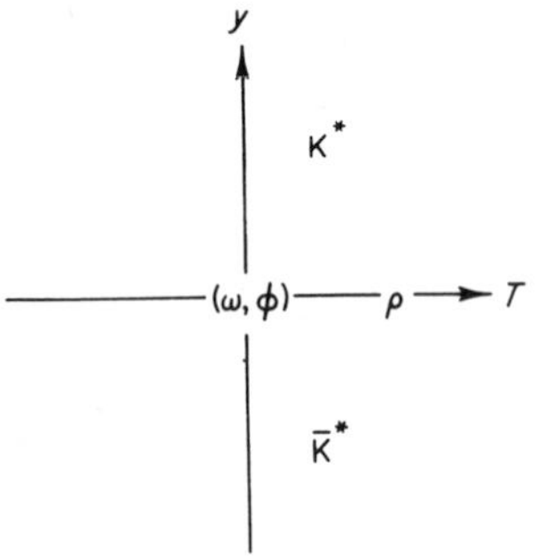


**Fig. 10.13.** $(T, y)$ diagram for the vector meson octet. The point (0, 0) is occupied by a linear combination of $\omega$ and $\phi$.

as the isospin singlet member. In fact for either choice, the mass formula fails (for either $m^2$ or $m$). To get around this, it has been proposed that, because of the SU(3) symmetry breaking, the physical particles $\phi$ and $\omega$ are actually linear combinations of unitary singlet and octet. The octet linear combination of $\phi$ and $\omega$ is then chosen to fit the mass (squared) formula; the orthogonal linear combination is the singlet.

*Nonets* of this sort, unitary singlet plus octet, are evidently the standard situation for mesons. In the pseudoscalar meson case, the ninth meson seems to be the $X^0$ [or $\eta(959)$]. Since the $\eta$ already fits the mass (squared) formula fairly well, there can only be a small amount of $X^0$ mixed in the octet, and the singlet must be nearly pure $X^0$.

Baryon and meson resonances with higher mass and spin have also been assigned to SU(3) multiplets—mesons to singlets and octets, baryons to singlets, octets, and decuplets. So we see that the eightfold way brings some much needed order to the exploding population of particles and resonances.

## 10.4. Quarks and Quark Models

### *a. Quarks*

After the eightfold way had been established, Gell-Mann and Zweig proposed that corresponding to the fundamental **3** of SU(3) there should exist a fundamental triplet of particles, $\hat{\mathrm{p}}$, $\hat{\mathrm{n}}$, and $\hat{\lambda}$. These particles, named *quarks* by Gell-Mann, are to be the building blocks which compose all of the mesons and baryons. They must have the usual triplet isospin and hypercharge, and, presumably, ordinary spin one-half; antiquarks, since they have all additive quantum numbers reversed, must correspond to **3***. Then we can suppose that the mesons (integral spin) are bound states of a quark–antiquark pair, while baryons (odd half-integral spin) are three-quark bound states. Since

$$\mathbf{3} \otimes \mathbf{3}^* = \mathbf{1} \oplus \mathbf{8},$$

and

$$\mathbf{3} \otimes \mathbf{3} \otimes \mathbf{3} = \mathbf{1} \oplus \mathbf{8} \oplus \mathbf{8} \oplus \mathbf{10},$$

this predicts singlet and octet mesons, and singlet, octet, and decuplet baryons, exactly as observed. So here we have an "explanation" of the eightfold way. But this also gives the quarks baryon number $\frac{1}{3}$ and, retaining the relation [cf. Eq. (9.1)]

$$Q = T_3 + Y/2, \tag{10.57}$$

charges $\frac{2}{3}$ and $-\frac{1}{3}$. These fractional values for baryon number and charge are certainly something new!

It is plausible to suppose that the lowest energy hadron state have quark orbital momentum zero. This means $J^P$ values of $0^-$ and $1^-$ for the mesons,[4] $\frac{1}{2}^+$ and $\frac{3}{2}^+$ for the baryons—just as we want.

### *b. SU(6) Supermultiplets*

To determine which SU(3) multiplets go with which spins, we consider the extreme model: Quarks degenerate in mass and held together by forces which are

[4] Remember that fermion and antifermion have *opposite* intrinsic parity.

both SU(3) symmetric and spin independent. We organize the spin up and spin down quarks into *six component spinors*:

$$q \equiv \begin{pmatrix} p_\uparrow \\ n_\uparrow \\ \lambda_\uparrow \\ p_\downarrow \\ n_\downarrow \\ \lambda_\downarrow \end{pmatrix}. \tag{10.58}$$

That is

$$\hat{p}_\uparrow \equiv q(1) \equiv \begin{pmatrix} 1 \\ 0 \\ 0 \\ 0 \\ 0 \\ 0 \end{pmatrix}, \ldots, \hat{\lambda}_\downarrow \equiv q(6) \equiv \begin{pmatrix} 0 \\ 0 \\ 0 \\ 0 \\ 0 \\ 1 \end{pmatrix}. \tag{10.59}$$

In this "supersymmetric" model, the strong interactions are invariant under transformations that mix up the components of Eq. (10.58). We are thus led to an SU(6) symmetry scheme; SU(6) is the group of unitary unimodular transformations on *six* objects.

In addition to identifying the quarks with the fundamental **6** multiplet of SU(6), we associate **6*** with the antiquarks. Then mesons correspond to the product **6** ⊗ **6***, which reduced is[5]

$$\mathbf{6} \otimes \mathbf{6}^* = 1 \oplus \mathbf{35}. \tag{10.60}$$

In order to see how the individual mesons fit into this scheme, we must be able to analyze SU(6) multiplets in terms of their SU(3) – SU(2) (i.e., ordinary spin) content. Introducing an [SU(3), SU(2)] notation, we write

$$\mathbf{6} \equiv (\mathbf{3}, \mathbf{2}), \tag{10.61}$$

and similarly [SU(2) multiplets being self-dual]

$$\mathbf{6}^* \equiv (\mathbf{3}^*, \mathbf{2}). \tag{10.61$'$}$$

Then, since the direct product combines the SU(3) and SU(2) parts separately,

$$\begin{aligned} \mathbf{6} \otimes \mathbf{6}^* &= (\mathbf{3}, \mathbf{2}) \otimes (\mathbf{3}^*, \mathbf{2}) = (\mathbf{3} \otimes \mathbf{3}^*, \mathbf{2} \otimes \mathbf{2}) \\ &= (\mathbf{1} \oplus \mathbf{8}, \mathbf{1} \oplus \mathbf{3}) \\ &= (\mathbf{1}, \mathbf{1}) \oplus (\mathbf{8}, \mathbf{1}) \oplus (\mathbf{1}, \mathbf{3}) \oplus (\mathbf{8}, 3), \end{aligned} \tag{10.62}$$

---

[5] Equation (10.60) can be obtained by considering the SU(6) tensor $a^i c_j$ $(i, j = 1, \ldots, 6)$ expressed as a sum of irreducible tensors [cf. Eq. (10.48)]:

$$a^i c_j = \tfrac{1}{6}\delta_j{}^i a^l c_l + (a^i c_j - \tfrac{1}{6}\delta_j{}^i a^l c_l).$$

The first term on the right-hand side has one independent component, the second 35. These irreducible tensors are in fact just the quark–antiquark spin-SU(3) wave functions.

where we have used Eqs. (10.6) and (10.42). But

$$\mathbf{1} = (\mathbf{1}, \mathbf{1}),$$

so we conclude that

$$\mathbf{35} \equiv (\mathbf{8}, \mathbf{1}) \oplus (\mathbf{1}, \mathbf{3}) \oplus (\mathbf{8}, \mathbf{3}). \tag{10.63}$$

This means that the **35** supermultiplet will accommodate a spin-zero octet, a spin-one singlet, and a spin-one octet. The **1** of course is a spin-zero singlet.

To build the baryons, we put together three quarks, and thus want the reduction of $\mathbf{6} \otimes \mathbf{6} \otimes \mathbf{6}$. It turns out that

$$\mathbf{6} \otimes \mathbf{6} \otimes \mathbf{6} = \mathbf{20} \oplus \mathbf{56} \oplus \mathbf{70} \oplus \mathbf{70}, \tag{10.64}$$

and it is the **56**, which has a completely symmetric [spin–SU(3)] wave function, that accommodates the low-lying baryon states. Now one can show that

$$\mathbf{56} \equiv (\mathbf{8}, \mathbf{2}) \oplus (\mathbf{10}, \mathbf{4}), \tag{10.65}$$

which means spin $\frac{1}{2}$ octet and spin $\frac{3}{2}$ decuplet.

The observed hadrons fit very nicely into these patterns, but of course neither the 35 mesons or the 56 baryons are degenerate in mass. In fact, as shown in Fig. 10.14,

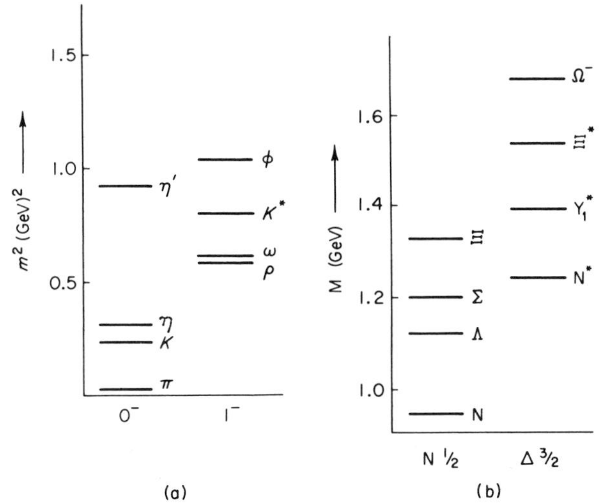


**Fig. 10.14.** (a) Masses (squared) of the $0^-$ and $1^-$ nonets; (b) masses of the $\frac{1}{2}^+$ and $\frac{3}{2}^+$ baryons.

the SU(6) mass splittings are somewhat larger than the SU(3) ones, indicating that SU(6) symmetry is even more approximate than SU(3). These mass splittings can be encompassed by an SU(6) generalization of the Gell-Mann–Okubo mass formula, but one gets more physical insight by using the quark picture. For instance, the SU(3) splittings of the pseudoscalar and vector meson octets and the $\omega$–$\phi$ mixing can successfully be ascribed to the $\hat{\lambda}$ being heavier than $\hat{\mathrm{p}}$ or $\hat{\mathrm{n}}$ (which must have the same mass for isospin invariance), while the difference between $J=0$ and 1 can be blamed on the spin-dependence of the quark–antiquark force. Things are not quite so simple for the baryons, however, because one must also introduce some SU(3) symmetry breaking into the quark interactions in order to bring about a $\Sigma$–$\Lambda$ mass splitting.

There are a host of other successful predictions that are made by various quark models (they are not all quite the same). These include relations between cross sections; strong, electromagnetic, and weak decays of hadrons; electromagnetic properties; and classification of other hadron states. The most common property of these models is the *additivity* of the interactions from each constituent quark. To illustrate some methods involved, we discuss two examples: baryon magnetic moments, where the simplest quark model works very well, and high-energy scattering, where the simplest model is not so good.

### c. Baryon Magnetic Moments

Since we are assuming zero orbital angular momenta, the hadron magnetic moments are just sums of the constituent quark magnetic moments. The simplest possible hypothesis is that the individual quark moments are proportional to the charges $Q_i$:

$$\boldsymbol{\mu}_i = \mu Q_i \boldsymbol{\sigma}_i . \tag{10.66}$$

Adding the $z$-component gives the total moment.

$$\mu_a = \mu \langle a | \sum_i Q_i \sigma_{zi} | a \rangle . \tag{10.67}$$

**Table 10.1**

*Quark Wave Functions for $\frac{1}{2}^+$ Baryons*

| Baryon | Wave function ($m=+\frac{1}{2}$) |
|---|---|
| p | $\frac{1}{18}(2p_\uparrow p_\uparrow n_\downarrow - p_\uparrow p_\downarrow n_\uparrow - p_\downarrow p_\uparrow n_\uparrow + \text{permutations})$ |
| n | $\frac{1}{18}(2n_\uparrow n_\uparrow p_\downarrow - n_\uparrow n_\downarrow p_\uparrow - n_\downarrow n_\uparrow p_\uparrow + \text{permutations})$ |
| $\Sigma^+$ | $\frac{1}{18}(2p_\uparrow p_\uparrow \lambda_\downarrow - p_\uparrow p_\downarrow \lambda_\uparrow - p_\downarrow p_\uparrow \lambda_\uparrow + \text{permutations})$ |
| $\Sigma^0$ | $\frac{1}{6}(2p_\uparrow n_\uparrow \lambda_\downarrow - p_\uparrow n_\downarrow \lambda_\uparrow - p_\downarrow n_\uparrow \lambda_\uparrow + \text{permutations})$ |
| $\Sigma^-$ | $\frac{1}{18}(2n_\uparrow n_\uparrow \lambda_\downarrow - n_\uparrow n_\downarrow \lambda_\uparrow - n_\downarrow n_\uparrow \lambda_\uparrow + \text{permutations})$ |
| $\Lambda$ | $\frac{1}{12}(p_\uparrow n_\downarrow \lambda_\uparrow - p_\downarrow n_\uparrow \lambda_\uparrow - n_\uparrow p_\downarrow \lambda_\uparrow + n_\downarrow p_\uparrow \lambda_\uparrow + \text{permutations})$ |
| $\Xi^0$ | $\frac{1}{18}(2p_\downarrow \lambda_\uparrow \lambda_\uparrow - p_\uparrow \lambda_\uparrow \lambda_\downarrow - p_\uparrow \lambda_\downarrow \lambda_\uparrow + \text{permutations})$ |
| $\Xi^-$ | $\frac{1}{18}(2n_\downarrow \lambda_\uparrow \lambda_\uparrow - n_\uparrow \lambda_\uparrow \lambda_\downarrow - n_\uparrow y_\downarrow y_\uparrow + \text{permutations})$ |

To proceed we need the quark wave function for each hadron state $a$; these are easy to find [assuming SU(6) symmetry] and are exhibited in Table 10.1. Then, for example,

$$\mu_p = \mu \langle p\uparrow | \sum_i Q_i \sigma_{zi} | p\uparrow \rangle,$$

where

$$|p\uparrow\rangle = (1/\sqrt{18})\,|2p_\uparrow p_\uparrow n_\downarrow - p_\uparrow p_\downarrow n_\uparrow - p_\downarrow p_\uparrow n_\uparrow + \text{permutations}\rangle.$$

Thus

$$\begin{aligned}
\mu_p &= \tfrac{3}{18}\mu\{\langle 2p_\uparrow p_\uparrow n_\downarrow | \Sigma\, Q_i \sigma_{zi} | 2p_\uparrow p_\uparrow n_\downarrow\rangle \\
&\quad + \langle p_\uparrow p_\downarrow n_\uparrow | \Sigma\, Q_i \sigma_{zi} | p_\uparrow p_\downarrow n_\uparrow\rangle \\
&\quad + \langle p_\downarrow p_\uparrow n_\uparrow | \Sigma\, Q_i \sigma_{zi} | p_\downarrow p_\uparrow n_\uparrow\rangle\} \\
&= \tfrac{3}{18}\mu\{4[\tfrac{2}{3} + \tfrac{2}{3} - \tfrac{1}{3}(-1)] + (\tfrac{2}{3} - \tfrac{2}{3} - \tfrac{1}{3}) + [(-\tfrac{2}{3}) + \tfrac{2}{3} - \tfrac{1}{3}]\} \\
&= \tfrac{3}{18}\mu \cdot 6 = \mu.
\end{aligned} \tag{10.68a}$$

Similarly, one finds

$$\mu_n = -\tfrac{2}{3}\mu \tag{10.68b}$$

and the other baryon octet moments shown in Table 10.2.

**Table 10.2**

*Magnetic Moments of the $\frac{1}{2}^+$ Baryons according to the Quark Model*

| Baryon | $\mu_B/\mu$ | $\mu_B/\mu_0$ | Observed value |
|---|---|---|---|
| p | 1 | 2.79 | 2.793 |
| n | $-\frac{2}{3}$ | −1.86 | −1.913 |
| $\Sigma^+$ | 1 | 2.79 | 2.6 ± 0.5 |
| $\Sigma^0$ | $\frac{1}{3}$ | 0.93 | |
| $\Sigma^-$ | $-\frac{1}{3}$ | −0.93 | |
| $\Lambda^0$ | $-\frac{1}{3}$ | −0.93 | −0.73 ± 0.16 |
| $\Xi^0$ | $-\frac{2}{3}$ | −1.86 | |
| $\Xi^-$ | $-\frac{1}{3}$ | −0.93 | |

Experimentally,

$$\mu_n/\mu_p = -1.913/2.793 = -0.685,$$

which is an amazingly good agreement.[6] It is much harder to measure the moments of the other baryons because of their short lifetimes, but the agreement so far is satisfactory. [If it were not, one could try to improve the fit by relaxing Eq. (10.66) and adjusting the three quark moments independently; this amounts to including SU(6) symmetry-breaking.]

[6] The same result was obtained earlier from a purely SU(6) model (without reference to quarks).

### d. High-Energy Scattering

The simplest application to high-energy scattering employs (1) additivity, (2) SU(6) symmetry, (3) the optical theorem, and (4) the *Pomeranchuk Theorem*. Additivity says that for hadron scattering the scattering amplitude is the sum of the constituent quark–(anti-)quark scattering amplitudes:

$$T = \sum T_{qq}, \tag{10.69}$$

where the summation is over all the q–q and q–$\bar{q}$ *pairs*. Complete SU(6) symmetry implies that the quark–(anti-)quark amplitudes are independent of which quarks are involved and their spin state; since

$$T_{qq} = T_{\bar{q}\bar{q}},$$

Eq. 10.69 becomes

$$T = n_1 T_{qq} + n_2 T_{q\bar{q}}, \tag{10.70}$$

where $n_1$ is the number of $\bar{q}$–$\bar{q}$ and q–q pairs, and $n_2$ is the number of q–$\bar{q}$ pairs. The optical theorem relates the imaginary part of the forward elastic scattering amplitudes to the total cross section:

$$\text{Im } T_{ii} = 2|\mathbf{p}|\sqrt{s}\,\sigma_T \tag{3.102}$$

which can be combined with Eq. (10.70) to give

$$\sigma = n_1 \sigma_{qq} + n_2 \sigma_{q\bar{q}}. \tag{10.71}$$

Finally the Pomeranchuk theorem states that, quite generally, total cross sections satisfy

$$\lim_{s \to \infty} (\sigma_{ab} - \sigma_{a\bar{b}}) = 0 \tag{10.72}$$

($\bar{b}$ is the antiparticle of $b$) which allows us to set

$$\sigma_{qq} = \sigma_{q\bar{q}} \tag{10.73}$$

for high enough energies. Thus Eq. (10.71) becomes

$$\sigma = n\sigma_{qq}, \tag{10.74}$$

where $n = n_1 + n_2$. Thus for a baryon–baryon scattering

$$\sigma_{BB} = \sigma_{B\bar{B}} = 9\sigma_{qq}, \tag{10.75}$$

since there are 9 pairs involved, while for meson–baryon scattering

$$\sigma_{MB} = \sigma_{\bar{M}B} = 6\sigma_{qq}. \tag{10.76}$$

For the observable cross sections, we conclude that

$$\sigma(NN) = \sigma(\bar{N}N) = \tfrac{3}{2}\sigma(\pi N) = \tfrac{3}{2}\sigma(KN) = \tfrac{3}{2}\sigma(\bar{K}N). \tag{10.77}$$

The behavior of the experimental cross sections is shown in Fig. 10.15. We see that Eqs. (10.77) are *not* well satisfied. Furthermore, we note that (1) the

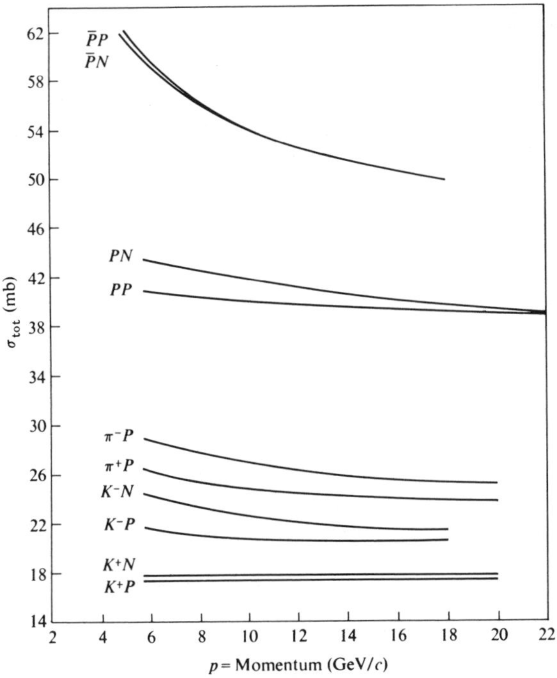


**Fig. 10.15.** Experimental total cross sections for hadron scattering at high energy; $P$ is (lab) projectile momentum. [Reprinted by permission of the publisher, from Bernard T. Feld, *Models of Elementary Particles*, © 1969 Xerox Corporation. All rights reserved.]

Pomeranchuk limit relation [Eq. (10.72)] does *not* hold at these energies, and (2) a significant amount of SU(3) symmetry breaking is apparent. One can construct less stringent models by allowing

$$\sigma_{qq} \neq \sigma_{q\bar{q}},$$

and

$$\sigma_{n\lambda} \neq \sigma_{np}, \qquad \sigma_{n\bar{\lambda}} \neq \sigma_{n\bar{p}},$$

etc., while retaining

$$\sigma_{np} = \sigma_{nn} = \sigma_{pp},$$
$$\sigma_{n\lambda} = \sigma_{p\lambda},$$

etc., to preserve isospin symmetry. This allows a much better fit to the experimental data.

### *e. Higher Resonances*

So far, we have discussed the lowest energy meson and baryon states. The next question is how to construct the higher energy hadron resonances out of quarks. There are two possibilities: (1) higher quark multiplicities, e.g., $qq\bar{q}\bar{q}$

for mesons and $qqq\bar{q}\bar{q}$ for baryons—the multiplets predicted by such models are called "*exotic*" and so far there is no experimental evidence for such resonances; (2) excited states of the familiar qqq and $q\bar{q}$ combinations, including those with nonzero orbital angular momentum. This idea seems to be successful so far, but has yet to pass a test as decisive as, say, the prediction of the $\Omega^-$ by the eightfold way.

### *f. Difficulties*

We must now discuss the two major difficulties connected with the quarks: quark statistics, and the existence of quarks as particles. Regarding statistics, the crucial point is that the **56**, which contains the $\frac{1}{2}^+$ and $\frac{3}{2}^+$ baryons, has SU(6) wave functions which are all *symmetric* under interchange of any pair of quarks. But we have been assuming zero orbital angular momentum for the quarks, which means a *symmetric* spatial wave function. This in turn implies *symmetric* overall wave functions, rather than the antisymmetric ones we expect for spin one-half! One can consider higher orbital angular momenta for the **56**, but this is unappealing for several reasons: (1) It would be peculiar to have a *ground state* spatial wave function with $l > 0$; (2) this would mess up some of the successful quark model predictions; (3) the wave function and hence the probability density would be zero for all three quarks at one point. The presence of this "hole," it turns out, implies that the *charge form factor* changes sign for some momentum transfer; experimentally this does not happen in the region of momentum transfer so far explored. Thus the rule connecting spin and statistics appears to be suspended for quarks inside a baryon.[7]

As to the existence problem, one wonders why it is so hard to find free quarks. Certainly there have been many experiments looking for particles with charge $\pm\frac{1}{3}e$ or $\pm\frac{2}{3}e$, and a natural explanation for this lack of success[8] is that the mass of a free quark is very large (>5 GeV). But this means the binding energies also must be very large, which makes the success of the simple additivity assumption puzzling. Many physicists tentatively regard the quarks as theoretical constructs rather than "physical" particles and constituents in the familiar (atomic and nuclear) sense—a sort of algebraic algorithm to an underlying and as yet unrevealed dynamics. The status of the quarks is perhaps the biggest mystery currently facing particle physics.

---

[7] But it is not suspended for quarks in different nucleons. Why?

[8] At the time of this writing there have been two experiments which claim to have found quarks, but they have been received with considerable skepticism.

## Problems

**10.1.** Compute $T_3$, $Y$, $T_\pm$, $V_\pm$, $U_\pm$ from the explicit form of the $\lambda_i$'s. Establish the commutation relations for $T_3$, $Y$, .... Using the commutation relations, show explicitly that $T_+$, $U_+$, and $V_+$ have the stepping properties claimed.

**10.2.** Show that $[Q, U_\pm] = 0$, and therefore that all members of a $U$-spin multiplet have the same charge.

**10.3.** Using the SU(3) weight diagrams, evaluate $\mathbf{3} \otimes \mathbf{8}$.

**10.4.** (a) Test the Gell-Mann–Okubo mass formula for the baryon octet by inserting the experimental values of the masses in Eq. (9.42).

(b) Do the same for the pseudoscalar octet, using first the masses [Eq. (9.43)] and then the squares of the masses [Eq. (9.44)]. Neglect any mixing of the $\eta$ with a ninth pseudoscalar meson.

**10.5.*** Derive the baryon octet quark wave-function given in Table 10.1. (Hint: Make use of the requirement of overall [spin-SU(3)] symmetry of all members of the **56**.)

**10.6.** Derive the baryon octet magnetic moments shown in Table 10.2.

### Bibliography

B. T. Feld. *Models of Elementary Particles*. Blaisdell, Waltham, Massachusetts, 1969.

W. R. Frazer. *Elementary Particles*. Chap. 6. Prentice-Hall, Englewood Cliffs, New Jersey, 1966.

M. Gell-Mann and Ne'eman (eds.). *The Eight-Fold Way*. Benjamin, New York, 1964.

M. Gourdin. *Unitary Symmetries*. North-Holland Publ. Co., Amsterdam, 1967.

J. J. J. Kokkedee. *The Quark Model*. Benjamin, New York, 1969.

F. E. Low. *Symmetries and Elementary Particles*. Gordon & Breach, New York, 1967.

CHAPTER ELEVEN

# Weak Interactions

The weak interactions are responsible for the "slow" ($\tau \geq 10^{-10}$ sec) decays of unstable particles. Weak interactions were first studied in the *beta decay* of unstable nuclei, where the charge of the nucleus changes by one unit and a *lepton pair*, i.e., an electron or positron and some kind of *neutrino*, is emitted (or an atomic electron is absorbed and a neutrino emitted). In fact, Pauli (in 1933) invented the neutrino just to preserve conservation of energy for these beta decays. Neutrinos are electrically neutral, have spin one-half, and, as far as we know, have mass zero (experiment can only give an upper limit for this mass; the present upper limit is about 60 eV). Although well-established theoretically, neutrinos remained experimentally invisible until 1960, being very hard to detect because they interact only weakly.

We can get an idea of the orbital angular momentum $l$ of an outgoing electron or neutrino in beta decay by estimating the product *momentum* $\times$ *nuclear radius*. Thus if $Q$ is the kinetic energy released and $R$ the nuclear radius, we have that

$$p \sim Q,$$

while

$$R \sim A^{1/3}/m_\pi,$$

so that

$$pR \sim (Q/m_\pi)A^{1/3}$$

($A$ being the nucleon number). Since $Q \lesssim 1$ MeV, while $m_\pi \simeq 140$ MeV, we see that

$$pR \ll 1.$$

Hence $l = 0$ is the most likely possibility for the emitted particles (this is called an *allowed transition*), while $l = 1, 2, \ldots$ are progressively less probable (*first forbidden, second forbidden, . . .*).

Now because each lepton has spin one-half, leptons can, for allowed transitions, carry off angular momentum ($J$) 0 or 1. Those with $J = 0$ are called *Fermi transitions*, while those with $J = 1$ are *Gamow–Teller transitions.* We will now discuss how these transitions can come from the sort of interaction proposed by Fermi in 1934.

## 11.1. Universal Fermi Interaction

### *a. Original Form. Neutron Decay*

Fermi proposed that beta-decay involves interaction with individual nucleons, with an interaction Hamiltonian made up of *vector currents* which are similar to the electromagnetic current:

$$\mathcal{H}_I(x) = G[\bar{\psi}_p(x)\gamma_\mu \psi_n(x)][\bar{\psi}_e(x)\gamma_\mu \psi_\nu(x)] + \text{Hermitian conjugate.} \quad (11.1)$$

One of the first things to be noticed about this type of interaction, and all of its variants which will be discussed subsequently, is that while they create and destroy both baryons and leptons (remember that $\psi_n$ both destroys neutrons and creates antineutrons, etc.), they do so in pairs in such a way that the total baryon number and the total *lepton number* are *each* conserved. Recall that the baryon number is the number of baryons minus the number of antibaryons; similarly, the lepton number is the number of negative muons, electrons, and neutrinos, minus the number of positive muons, positrons, and antineutrinos. These conservation laws are well established experimentally.

The transition matrix element is found from Eq. (11.1) by using the lowest order of perturbation theory (Born approximation). For instance, the neutron decay amplitude would then be, using the expansions (4.18) for the field operators and going over to invariant normalization,

$$T_{fi} = -G(16 m_n m_p m_e m_\nu)^{1/2}(\bar{u}_p \gamma_\mu u_n)(\bar{u}_e \gamma_\mu v_\nu), \quad (11.2)$$

for the process $n \rightarrow pe^-\bar{\nu}$ (Fig. 11.1). In the rest system of the decaying neutron, the outgoing proton is certainly nonrelativistic; so [cf. Eqs. (2.32) and (2.19)]

$$(\bar{u}_p \gamma_\mu u_n) \simeq \delta_{\mu 4}(\chi_p{}^\dagger \chi_n).$$

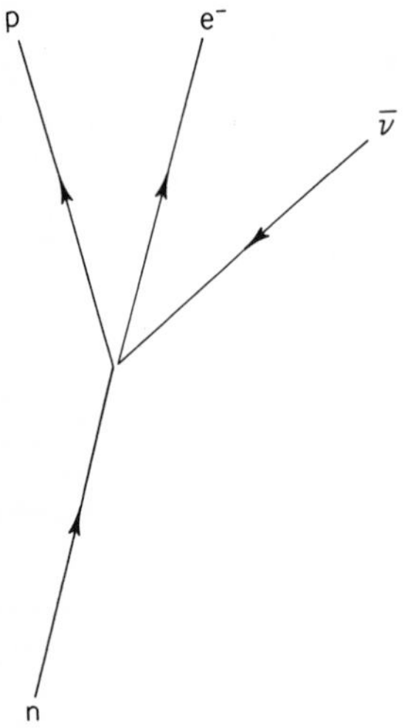


**Fig. 11.1.** The four-fermion interaction in neutron decay.

This shows that the neutron and proton spins are parallel; hence the expression *Fermi transition* for $J = 0$. It is also an allowed transition since there is no angular dependence in Eq. (11.2) to bring in an $l > 0$ contribution.

For the (e$\nu$) factor, we derive from Eq. (2.32)

$$(\bar{u}_e \gamma_4 v_\nu) = (u_e{}^\dagger v_\nu)$$
$$= \left[\frac{(E_e + m_e)(E_\nu + m_\nu)}{4m_e m_\nu}\right]^{1/2} \cdot \left(\chi_e{}^\dagger \boldsymbol{\sigma} \cdot \left[\frac{\mathbf{p}_e}{E_e + m_e} + \frac{\mathbf{p}_\nu}{E_\nu + m_\nu}\right] \chi_\nu\right).$$

Thus Eq. (11.2) becomes

$$T_{fi} = +G2m_N(\chi_p{}^\dagger\chi_n)[(E_e + m_e)E_\nu]^{1/2}(\chi_e{}^\dagger\boldsymbol{\sigma} \cdot \mathbf{K}\chi_\nu),$$

where

$$\mathbf{K} \equiv \frac{\mathbf{p}_\nu}{E_\nu} + \frac{\mathbf{p}_e}{E_e + m_e},$$

and we have taken $m_\nu \to 0$. Then

$$\sum_{\substack{p,e,\nu \\ sp}} |T_{fi}|^2 = G^2 \cdot 4m_N{}^2 \cdot 1 \cdot (E_e + m_e)E_\nu \cdot \mathrm{tr}[(\boldsymbol{\sigma} \cdot \mathbf{K})^2]$$
$$= 4G^2 m_N{}^2(E_e + m_e)E_\nu \cdot 2\mathbf{K}^2,$$

with

$$\mathbf{K}^2 = 1 + 2\,\frac{\mathbf{p}_e \cdot \mathbf{p}_\nu}{(E_e + m_e)E_\nu} + \frac{E_e - m_e}{E_e + m_e}$$
$$= \frac{2E_e E_\nu + 2\mathbf{p}_e \cdot \mathbf{p}_\nu}{(E_e + m_e)E_\nu}.$$

Therefore,

$$\sum |T_{fi}|^2 = 16G^2 m_N^2 (E_e E_\nu + \mathbf{p}_e \cdot \mathbf{p}_\nu).$$

Now the decay rate is given by Eq. (3.91):

$$\Gamma_{fi} = (2E_n)^{-1} \int (2\pi)^4 \delta^4(P_i - P_f)\left(\sum |T_{fi}|^2\right) \frac{d^3\mathbf{p}_p}{(2\pi)^3 2E_p} \frac{d^3\mathbf{p}_e}{(2\pi)^3 2E_e} \frac{d^3\mathbf{p}_\nu}{(2\pi)^3 2E_\nu},$$

and one can manipulate the three-body phase-space integral to give

$$T_3 = \frac{1}{8} \int \frac{|\mathbf{p}_e| E_\nu}{E_p} \, d\Omega_e \, d\Omega_\nu \, dE_\nu. \tag{11.3}$$

Furthermore,

$$E_p \simeq m_N \qquad \text{and} \qquad E_e \simeq Q - E_\nu,$$

where

$$Q \equiv m_n - m_p$$

are very good approximations, and

$$\int (E_e E_\nu + \mathbf{p}_e \cdot \mathbf{p}_\nu) \, d\Omega_e \, d\Omega_\nu \simeq 16\pi^2 E_e E_\nu.$$

Putting all this together,

$$\Gamma_{fi} \simeq (2m_N)^{-1} (2\pi)^{-5} (8m_N)^{-1} \int |\mathbf{p}_e| E_\nu \cdot 16G^2 m_N^2 \cdot 16\pi^2 E_e E_\nu \, dE_\nu$$

$$\simeq 4(2\pi)^{-3} G^2 \int_0^{Q-m_e} (Q - E_\nu)[(Q - E_\nu)^2 - m_e^2]^{1/2} E_\nu^2 \, dE_\nu.$$

The integral is not difficult to evaluate, and one finds $\int \simeq 1.6$. Therefore,

$$\Gamma_{fi} \simeq 4(2\pi)^{-3} G^2 m_e^5 \cdot 1.6.$$

We can use this result and the observed neutron lifetime

$$\tau \simeq 0.93 \times 10^3 \quad \text{sec}$$

to evaluate the beta-decay coupling constant:

$$(Gm_e^2)^2 \simeq \frac{(2\pi)^3}{6.4} \frac{\hbar}{m_e c^2 \tau},$$

$$Gm_e^2 \sim 10^{-11},$$

or

$$Gm_p^2 \sim 10^{-5}.$$

$G$ is indeed small.

In going from Eq. (11.1) to Eq. (11.2), we used the Born approximation. This is reasonable as far as the weak interactions are concerned because of the smallness of $G$—in fact, it is essential, since the corrections higher order in $G$ actually diverge and the theory *cannot* be renormalized! However, the neutron and proton are subject to strong interactions; since the strong interaction coupling constants are not small, there is no reason to expect that pion correction of Fig. 11.2(a), for example, is any less important than Fig. 11.1

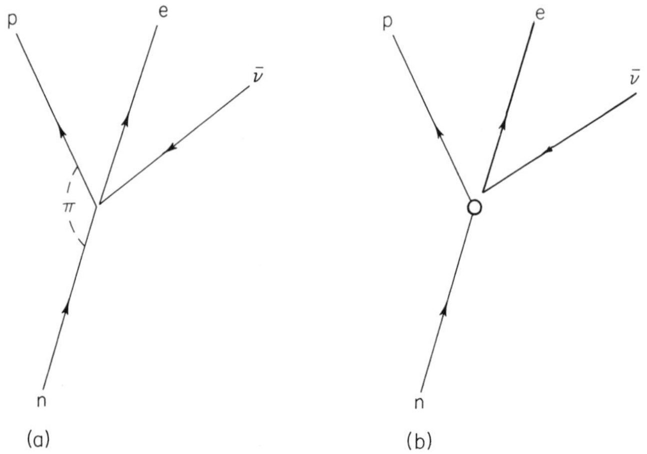


**Fig. 11.2.** The four fermion interaction as modified by the strong interactions: (a) a pion vertex correction; (b) all possible corrections to the pn vertex bring in a form factor.

itself. If we allow all possible strong interaction corrections, which of course is too much to calculate, then instead of $(\bar{u}_p \gamma_\mu u_n)$ in the matrix element we have $F \cdot (\bar{u}_p \gamma_\mu u_n)$; the lepton part $(\bar{u}_e \gamma_\mu v_\nu)$ is unaffected [Fig. 11.2(b)]. The unknown function $F$, called a *form factor*, must be an invariant function of the neutron and proton momenta, and therefore can depend only on the invariants formed from those momenta. There are only three independent invariants:

$$p_p^2, \quad p_n^2, \quad \text{and} \quad q^2 \equiv (p_p - p_n)^2.$$

Since

$$p_p^2 = -m_p^2 \qquad \text{and} \qquad p_n^2 = -m_n^2,$$

we are left with $q^2$ as the only variable. Thus $F \equiv F(q^2)$, and

$$T_{fi} = -G[16 m_n m_p m_e m_\nu]^{1/2} F(q^2)(\bar{u}_p \gamma_\mu u_n)(\bar{u}_e \gamma_\mu v_\nu). \tag{11.4}$$

Now, since the $q$ involved are all very small (compared with the hadron masses), we do not expect much variation in $F(q^2)$ as far as beta-decay is concerned, i.e.,

$$F(q^2) \simeq F(0),$$

and the experimental neutron decay rate fixes the product $F(0)G$ rather than $G$ alone.

### b. Generalized Form

Soon after Fermi's proposal of Eq. (11.1), it was pointed out that we can just as well have

$$\mathcal{H}_{\mathrm{I}} = G \sum_i C_i(\bar{\psi}_{\mathrm{p}} O_i \psi_{\mathrm{n}})(\bar{\psi}_{\mathrm{e}} O_i \psi_{\nu}) + \text{Hermitian conjugate} \qquad (11.5)$$

where the *weak currents* $(\bar{\psi} O_i \psi)$ are the bilinear covariants discussed in Chapter 2, Section 2.2d (Table 2.1). These are the combinations for which $\mathcal{H}_{\mathrm{I}}$ is invariant (including space inversion invariance). Ignoring form factors for the moment, the corresponding matrix element is

$$T_{\mathrm{fi}} = -G[16 m_{\mathrm{n}} m_{\mathrm{p}} m_{\mathrm{e}} m_{\nu}]^{1/2} \sum C_i'(\bar{u}_{\mathrm{p}} O_i u_{\mathrm{n}})(\bar{u}_{\mathrm{e}} O_i v_{\nu}). \qquad (11.6)$$

**Table 11.1**

| Type | Symbol | $O_i$ | Nonrelativistic form of $u_{\mathrm{p}} O_i u_{\mathrm{n}}$ |
|---|---|---|---|
| Scalar | S | 1 | $\chi_{\mathrm{p}}^{\dagger}\chi_{\mathrm{n}}$ |
| Vector | V | $\gamma_\mu$ | $\delta_{\mu 4} \cdot \chi_{\mathrm{p}}^{\dagger}\chi_{\mathrm{n}}$ |
| Tensor | T | $\Sigma_{\mu\nu} \begin{cases} \Sigma_{ij} \\ \Sigma_{4j} \end{cases}$ | $\varepsilon_{ijk}\chi_{\mathrm{p}}^{\dagger}\sigma_k\chi_{\mathrm{n}}$ ; 0 |
| Axial vector | A | $\gamma_\mu\gamma_5 \begin{cases} \gamma_i\gamma_5 \\ \gamma_4\gamma_5 \end{cases}$ | $\chi_{\mathrm{p}}^{\dagger}\sigma_i\chi_{\mathrm{n}}$ ; 0 |
| Pseudoscalar | P | $\gamma_5$ | 0 |

The nonrelativistic limits of the nucleon current are shown in Table 11.1. The only nonzero possibilities are

$$\bar{u}_{\mathrm{p}} O_i u_{\mathrm{n}} \simeq \begin{cases} \chi_{\mathrm{p}}^{\dagger}\chi_{\mathrm{n}}, \\ \chi_{\mathrm{p}}^{\dagger}\boldsymbol{\sigma}\chi_{\mathrm{n}}. \end{cases}$$

The first implies Fermi transitions, as we have discussed; the second, Gamow–Teller transitions $[(\chi_{\mathrm{p}}^{\dagger}\boldsymbol{\sigma}\chi_{\mathrm{n}})$ and, therefore, $(\bar{u}_{\mathrm{e}} O u_{\nu})$ are both vectors; hence the $(e\nu)$ pair must carry off $J = 1]$. This information is summarized in Table 11.2.

**Table 11.2**

| | Nonrelativistic limit of $O_i$ | Selection rule for nuclear spin |
|---|---|---|
| Fermi: S, V | 1 | $\Delta J = 0$ |
| Gamow–Teller: T, A | $\boldsymbol{\sigma}$ | $\Delta J = 0, \pm 1$ (no $0 \to 0$ transition) |

That the weak interactions involve both the Fermi and the Gamow–Teller possibilities is demonstrated by the existence of both kinds of decays: in the decay

$$O^{14} \to N^{14} e^+ \nu,$$

both nuclei have spin zero ($0 \to 0$ transition), so that the lepton pair has $J = 0$, and this is a pure Fermi transition; while for

$$He^6 \to Li^6 e^- \bar{\nu},$$

$He^6$ has spin zero, while $Li^6$ has spin one (hence a $J = 1$ transition) making this pure Gamow–Teller. Thus the complete beta decay interaction must involve (see Table 11.2) S and/or V, *and* T and/or A. Our calculation of the neutron decay rate, which assumed pure V, cannot be correct.

### c. *Time Reversal*

A constraint on these weak interaction coupling constants follows from the assumption of time-reversal invariance (Section 6.1). Thus invariance of the weak interaction Hamiltonian [Eq. (11.5)] can be shown to imply that the $GC_i$ are real, and the further assumption that the strong and electromagnetic interactions are also invariant tells us that the $GC_i'$ are nearly real. The only departure from real constants comes from any (strong or electromagnetic) scattering in the initial and/or final state and takes the form of a phase factor which is just the sum of the initial and final state phase shifts; at the low energies involved in beta-decay, these phase shifts are known to be small. A small violation of $T$-invariance has been demonstrated in the weak decay of the $K^0$, but nowhere else—so far.

### d. *Nonconservation of Parity*

The general form (11.5) has lasted remarkably well, but it took about 24 years to determine the correct combination of $O_i$ (i.e., the coefficients $C_i'$). The main impetus toward clearing up the weak interaction tangle came from the observation in 1957 of Lee and Yang that space-inversion invariance (and hence parity conservation), hitherto assumed to hold for all interactions, had never been tested for the weak interactions.

Lee and Yang were interested in the possibility of parity nonconservation, because this would resolve the *τ–θ paradox*. This refers to the two decay modes

$$\tau \rightarrow 3\pi, \qquad \theta \rightarrow 2\pi,$$

of what is now known as the $K^+$. $\tau$ and $\theta$ were established to have the same mass and lifetime, in addition to spin zero and strangeness +1. They seemed to be different decay modes of the same particle, except that, for spin zero, the $3\pi$ system must have negative parity while the $2\pi$ system has positive parity. If parity is not conserved for these weak decays, however, there is no problem.

Parity nonconservation for beta-decay was demonstrated by Wu *et al.* in their famous $Co^{60}$ experiment (1957). By placing a low temperature $Co^{60}$ sample in a very strong magnetic field, they succeeded in aligning the nuclear spins. The crucial point was the observation of a correlation between the spin polarization direction and the electron emission direction. Formally, the correlation implies a term proportional to

$$\langle \mathbf{S}_n \cdot \mathbf{p}_e \rangle$$

in the probability and thus in the transition amplitude. Since this term changes sign under space inversion, its presence means that the $S$-matrix, and hence the Hamiltonian, is *not* invariant under space inversion. Parity therefore need not be conserved, and we say that the weak interactions "violate parity."

The lack of invariance is easily allowed for in the expression for $\mathcal{H}_I$:

$$\mathcal{H}_I = \frac{G}{\sqrt{2}} \sum C_i(\bar{\psi}_p O_i \psi_n)[\bar{\psi}_e O_i(1 + \alpha_i \gamma_5)\psi_\nu] + \text{Hermitian conjugate.} \quad (11.7)$$

Correspondingly, Eq. (11.6) becomes[1]

$$T_{fi} = -\frac{G}{\sqrt{2}}(16 m_p m_n m_e m_\nu)^{1/2} \sum C_i'(\bar{u}_p O_i u_n)[\bar{u}_e O_i(1 + \alpha_i \gamma_5)u_\nu]. \quad (11.8)$$

The constants $\alpha_i$ measure the degree of "parity violation." We will not discuss in detail the experiments that have established the beta decay interaction as *equal parts* V *and* A, but only mention some of the highlights:

1. By using Eq. (11.8) to find the decay rate, one can establish the dependence of the electron energy spectrum and the electron–neutrino angular

---

[1] We write $C_i'$ rather than $C_i$ here to allow for form factors arising from strong interaction corrections to the (pn) current; these can bring in new terms as well, which we ignore here. The $1/\sqrt{2}$ is conventional.

correlation on the constants $C_i'$. The results are consistent with only $C_V$ and $C_A$ nonvanishing (and also place an upper limit on $m_\nu$, which we are simply assuming to be zero).

2. The electron asymmetry in the Wu *et al.* $Co^{60}$ experiment was found to be the maximum possible (*maximum parity violation*). This implies $\alpha_i \simeq \pm 1$.

3. The choice $\alpha_i \simeq +1$ was determined by Frauenfelder *et al.*, who found the longitudinal polarization of the $Co^{60}$ electrons to be negative.

These facts lead us to replace Eq. (11.7) simply by

$$\mathcal{H}_I = \frac{G}{\sqrt{2}}[\bar{\psi}_p \gamma_\mu (C_V + C_A \gamma_5)\psi_n][\bar{\psi}_e \gamma_\mu (1 + \gamma_5)\psi_\nu] + \text{Hermitian conjugate} \tag{11.9}$$

and

$$T_{fi} = -\frac{G}{\sqrt{2}}(16 m_p m_n m_e m_\nu)^{1/2}[\bar{u}_p \gamma_\mu (C_V' + C_A' \gamma_5) u_n][\bar{u}_e \gamma_\mu (1 + \gamma_5) v_\nu]. \tag{11.10}$$

It is advantageous at this point to adjust $G$ so that $C_V' \equiv 1$; then the nucleon part of the $T$-matrix element [Eq. (11.10)] becomes

$$\bar{u}_p \gamma_\mu (1 + r\gamma_5) u_n,$$

where $C_A'$ is now disguised as $r$. Using this matrix element in the calculation of nuclear lifetimes and comparing with experiment gives

$$G = (1.00 \pm .03) \times 10^{-5}/m_p^2 \qquad \text{and} \qquad r = 1.18 \pm 0.03. \tag{11.11}$$

## 11.2. V–A Theory

### a. Two-Component Theory of the Neutrino

This combination of vector and axial vector currents, $\gamma_\mu + \gamma_\mu \gamma_5$, gives rise to the V–A *theory* (plus becomes minus in other conventions). The V–A combination was proposed independently by Marshak and Sudarshan and Feynman and Gell-Mann in 1957. The presence of $a \equiv 1 + \gamma_5$ in the lepton current results in the *two-component* theory of the neutrino. Recall that in the representation introduced above [Eqs. (2.19)]

$$\gamma_5 = -\begin{pmatrix} 0 & 1 \\ 1 & 0 \end{pmatrix} \qquad \text{and hence} \qquad a = \begin{pmatrix} 1 & -1 \\ -1 & 1 \end{pmatrix}. \tag{11.12}$$

Then (suppressing the normalization factors)

$$au = \begin{bmatrix} \left(1 - \dfrac{\boldsymbol{\sigma}\cdot\mathbf{p}}{E_{\mathrm{p}} + m}\right)\chi \\[2ex] \left(1 - \dfrac{\boldsymbol{\sigma}\cdot\mathbf{p}}{E_{\mathrm{p}} + m}\right)\chi \end{bmatrix}.$$

For zero mass (or highly relativistic) fermions

$$\left(1 - \frac{\boldsymbol{\sigma}\cdot\mathbf{p}}{E_{\mathrm{p}} + m}\right)\chi = \begin{Bmatrix} 0 \\ 2 \end{Bmatrix}\chi \qquad \text{for} \begin{Bmatrix} \text{right-} \\ \text{left-} \end{Bmatrix} \text{handed polarization,} \tag{11.13}$$

so that $a\psi_\nu$ annihilates only left-handed neutrinos, and (since $\sigma_z\chi = +\chi$ means spin *down* for the antifermion[2]) creates only right-handed antineutrinos. Similarly $\psi_\nu{}^\dagger a$, which appears in the Hermitian conjugate part of the interaction Hamiltonian [Eq. (11.9)] creates left-handed neutrinos and destroys right-handed antineutrinos. Thus Eq. (11.10) implies that all neutrinos are left-handed and all antineutrinos are right-handed, a situation that clearly violates both space-inversion and charge-conjugation invariance.

Since the electrons in beta-decay are *not* highly relativistic, the suppression of the "wrong" handedness for the electron is only partial. The general rule is that left-handed leptons and right-handed antileptons are preferred. Note that the $a$ can be applied to either the $\psi$ or $\bar{\psi}$ (or to either $u$ or $\bar{u}$), since

$$\bar{\psi}\gamma_\mu a\psi = \psi^\dagger\gamma_4(1 - \gamma_5)\gamma_\mu\psi = \psi^\dagger a\gamma_4\gamma_\mu\psi = (a\psi)^\dagger\gamma_4\gamma_\mu\psi = \tfrac{1}{2}(a\psi)^\dagger\gamma_4\gamma_\mu a\psi.$$

It is enlightening at this point to look at the plane wave solutions of the Dirac equation using a new representation for the $\gamma$-matrices [compare Eqs. (2.19)]:

$$\boldsymbol{\gamma} = \begin{pmatrix} 0 & -i\boldsymbol{\sigma} \\ i\boldsymbol{\sigma} & 0 \end{pmatrix} \quad \text{as before} \qquad \text{and} \qquad \gamma_4 = \begin{pmatrix} 0 & 1 \\ 1 & 0 \end{pmatrix};$$

hence

$$\gamma_5 = \begin{pmatrix} 1 & 0 \\ 0 & -1 \end{pmatrix} \qquad \text{and} \qquad a = 2\begin{pmatrix} 1 & 0 \\ 0 & 0 \end{pmatrix}.$$

In this representation, the Dirac equation

$$(i\not{p} + m)u = 0,$$

[2] The spin of an antifermion is *opposite* to that of the corresponding negative energy fermion state, because the antifermion corresponds to a hole in the negative energy sea.

becomes

$$\begin{pmatrix} 0 & -E + \mathbf{p} \cdot \boldsymbol{\sigma} \\ -E - \mathbf{p} \cdot \boldsymbol{\sigma} & 0 \end{pmatrix} \begin{pmatrix} u^0 \\ v^0 \end{pmatrix} = -m \begin{pmatrix} u^0 \\ v^0 \end{pmatrix},$$

or

$$\begin{aligned} (E - \mathbf{p} \cdot \boldsymbol{\sigma}) v^0 &= m u^0, \\ (E + \mathbf{p} \cdot \boldsymbol{\sigma}) u^0 &= m v^0, \end{aligned} \tag{11.14}$$

where we have set

$$u = \begin{pmatrix} u^0 \\ v^0 \end{pmatrix}.$$

For $m = 0$, these equations (11.14) uncouple:

$$\begin{aligned} (E - \mathbf{p} \cdot \boldsymbol{\sigma}) v^0 &= 0, \\ (E + \mathbf{p} \cdot \boldsymbol{\sigma}) u^0 &= 0, \end{aligned}$$

or

$$(1 - \boldsymbol{\sigma} \cdot \mathbf{p}/E) v^0 = 0, \tag{11.15a}$$

$$(1 + \boldsymbol{\sigma} \cdot \mathbf{p}/E) u^0 = 0. \tag{11.15b}$$

We are free to set either $u^0 = 0$ or $v^0 = 0$; note that the solutions each have only *one* helicity. (These *Weyl equations* were discarded by Pauli in 1933 because of the lack of $P$-invariance.) The choice is made by $a$, since

$$a u = 2 \begin{pmatrix} u^0 \\ 0 \end{pmatrix}. \tag{11.16}$$

Thus Eq. (11.15b) is the *two-component neutrino equation.* Since

$$|\mathbf{p}| = |E|,$$

for positive energy this clearly says that the neutrino is left handed; for negative energies,

$$0 = (E + \mathbf{p} \cdot \boldsymbol{\sigma}) u^0(-) = (-|E| + \mathbf{p} \cdot \boldsymbol{\sigma}) u^0(-),$$

so the negative energy solutions, and hence the antineutrinos, are right-handed.[3]

This prediction of left-handed neutrinos has actually been checked empirically in the very beautiful experiment of Goldhaber *et al.* This involved the

[3] Because the antifermion corresponds to a hole in the negative energy sea, it has *both* momentum and spin reversed, and hence the same helicity as the corresponding negative energy solution.

K-capture reaction

$$\mathrm{Eu}^{152}\mathrm{e}^- \to \mathrm{Sm}^{152*}\ \nu$$
$$\hookrightarrow \mathrm{Sm}^{152}\ \gamma.$$

When $\nu$ and $\gamma$ come off in opposite directions, the final $\mathrm{Sm}^{152}$ has minimum recoil; the $\gamma$ then has the right energy to undergo resonant scattering from a $\mathrm{Sm}^{152}$ target. Thus resonant scattering picks out those decays with $\nu$ and $\gamma$ opposite. Now both the original $\mathrm{Eu}^{152}$ and the final $\mathrm{Sm}^{152}$ nuclei have spin zero; thus to balance the electron spin, the total $\nu$-plus $\gamma$-angular momentum must be $\frac{1}{2}$ and not $\frac{3}{2}$, and hence $\nu$ and $\gamma$ have the *same* handedness. By scattering these photons from magnetized iron, the $\gamma$- and hence $\nu$-helicity was found to be negative. The experimental arrangement is shown in Fig. 11.3.

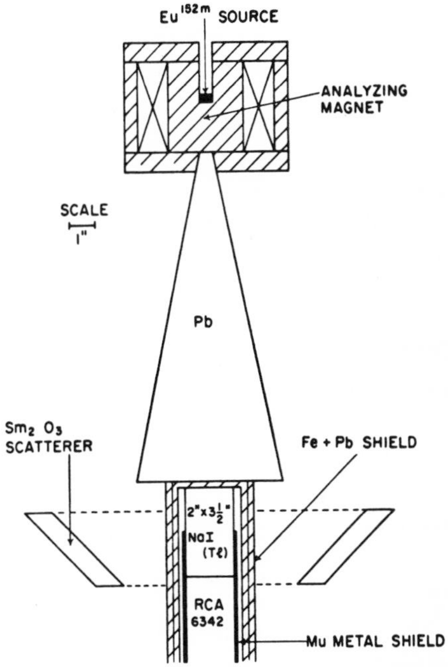


**Fig. 11.3.** Experimental arrangement for determining the helicity of the neutrino. [From M. Goldhaber, L. Grodzins, and A. W. Sunyar, Helicity of neutrinos. *Phys. Rev.* **109,** 1015–1017 (1958).]

### *b.* $\pi^{\pm}$ *Decay*

Another and very interesting test of the V–A theory is provided by charged pion decay. In addition to the muon mode (note how lepton conservation applies)

$$\pi^+ \to \mu^+ \nu, \qquad \pi^- \to \mu^- \bar{\nu},$$

V–A theory predicts a rarer electronic mode,

$$\pi^+ \to e^+ \nu, \qquad \pi^- \to e^- \bar{\nu}.$$

Since $\pi^-$, for example, is strongly coupled to the $\bar{p}n$ system, we can think of the decay as going in two steps as shown in Fig. 11.4. Although we cannot

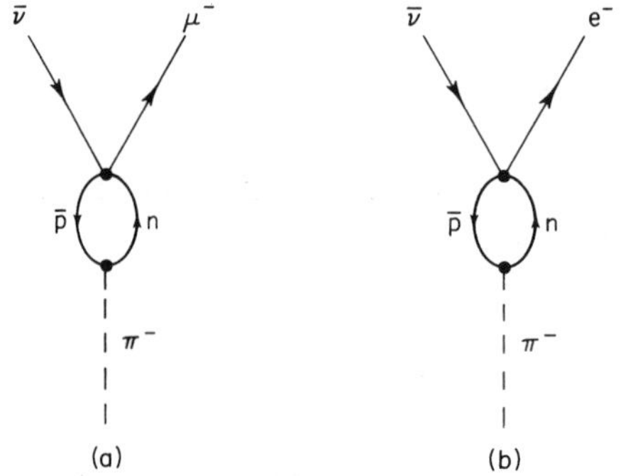


**Fig. 11.4.** Pion decay proceeds by the four fermion interaction via a virtual $N\bar{N}$ pair; (a) $\mu\nu$ mode; (b) $e\nu$ mode.

calculate the processes inside the closed loop (since they involve strong interactions), they are presumably the same for the two diagrams. The only difference in the decay amplitude is having the electron spinor replace the muon spinor—assuming, that is, that the muon and the electron are coupled to the nucleon in the same way with the *same* weak coupling constant.

Calculating the *ratio* of the decay rates is simple: The hadron part of the matrix element can only be proportional to the pion momentum $P_\mu$, there being no other four-vector available. Thus

$$\langle 0 | \bar{\psi}_p \gamma_\lambda (C_V + C_A \gamma_5) \psi_n | \pi^- \rangle \propto P_\lambda f_\pi \to i m_\pi \, \delta_{\lambda 4} f_\pi$$

in the pion rest system; here the form factor

$$f_\pi \equiv f_\pi(P^2) = f_\pi(m_\pi{}^2) = \text{const.}$$

Then

$$\begin{aligned} T_{fi} &= -(G/\sqrt{2})(4m_l m_\nu)^{1/2} P_\lambda f_\pi [\bar{u}_l \gamma_\lambda (1 + \gamma_5) v_\nu] \\ &= -i(G/\sqrt{2})(4m_l m_\nu)^{1/2} m_\pi f_\pi (u_l{}^\dagger a v_\nu). \end{aligned} \tag{11.17}$$

Using Eqs. (2.32) and (11.12) (the old representation) and choosing $\hat{z} \| \mathbf{p}_\nu$,

$$a v_\nu = \left(\frac{E_\nu}{2m_\nu}\right)^{1/2} \begin{bmatrix} -(1 - \sigma_z)\chi_\nu \\ (1 - \sigma_z)\chi_\nu \end{bmatrix};$$

hence only

$$\chi_\nu = \begin{pmatrix} 0 \\ 1 \end{pmatrix}$$

gives a nonzero contribution. As mentioned above, this means spin up (right-handed helicity) for the antineutrino. Then using Eq. (2.32) for $u_l^\dagger$, and $\mathbf{p}_l = -\mathbf{p}_\nu$,

$$u_l^\dagger a v_\nu = \left(\frac{E_l + m_l}{2m_l} \cdot \frac{E_\nu}{2m_\nu}\right)^{1/2} (-2)\chi_l^\dagger \left(1 + \frac{|\mathbf{p}_l|}{E_l + m_l}\sigma_z\right) \begin{pmatrix}0\\1\end{pmatrix},$$

so that only

$$\chi_l = \begin{pmatrix}0\\1\end{pmatrix}$$

will contribute to the decay. The lepton is thus also right handed (spin down, $\mathbf{p}_l \| -\hat{z}$); this checks with angular momentum conservation.

The resulting factor

$$\left(1 - \frac{|\mathbf{p}_l|}{E_l + m_l}\right) \to 0 \qquad \text{as} \quad \frac{m_l}{E_l} \to 0;$$

this suppression of the "wrong" helicity lepton is more nearly complete for the electron, and therefore muon emission is strongly favored over electron emission.

Putting the factors together, using Eqs. (1.32) and (1.30) and letting $m_\nu \to 0$, we find (for the contributing spin state)

$$T_{\mathrm{fi}} = +(G/\sqrt{2}) i f_\pi \cdot 2 \cdot m_l (m_\pi^2 - m_l^2)^{1/2}. \tag{11.18}$$

From Eqs. (3.91) and (1.30),

$$\Gamma_l = \frac{1}{2m_\pi}\frac{|\mathbf{p}|}{4m_\pi}\int |T_{\mathrm{fi}}|^2 \, d\Omega = \frac{m_\pi^2 - m_l^2}{16 m_\pi^3}\int |T_{\mathrm{fi}}|^2 \, d\Omega,$$

so that finally

$$\frac{\Gamma_e}{\Gamma_\mu} = \left[\frac{m_e}{m_\mu} \cdot \frac{m_\pi^2 - m_e^2}{m_\pi^2 - m_\mu^2}\right]^2 \simeq 1.3 \times 10^{-4}. \tag{11.19}$$

The rare electron decay mode has in fact been found and the experimental branching ratio agrees with Eq. (11.19). This is support for both the V–A interaction and equal coupling for muon and electron.

## 11.3. Muon Decay. Universal Weak Current

Muon decay appears to involve exactly the same *current–current* V–A type of interaction that we have become familiar with. Thus for the matrix element, we have the following.

$$T_{\mathrm{fi}} = -\frac{G'}{\sqrt{2}}(16m_\mu m_e m_\nu{}^2)^{1/2}[\bar{u}_\nu \gamma_\lambda(1+\gamma_5)u_\mu][u_e \gamma_\lambda(1+\gamma_5)v_\nu], \quad (11.20)$$

corresponding to interaction Hamiltonian (no strong interaction corrections here)

$$\mathcal{H}_\mathrm{I} = \frac{G'}{\sqrt{2}}[\bar{\psi}_\nu \gamma_\lambda(1+\gamma_5)\psi_\mu][\bar{\psi}_e \gamma_\lambda(1+\gamma_5)\psi_\nu] + \text{Hermitian conjugate.} \quad (11.21)$$

A sensitive test of Eq. (11.20) is provided by the distribution in energy of the outgoing electron. It turns out that for the more general type of interaction [as in Eq. (11.7)]

$$d\Gamma/dE_e \propto x^2[1 - x + \tfrac{2}{3}\rho(\tfrac{4}{3}x - 1)], \quad (11.22)$$

where

$$x \equiv |\mathbf{p}_e|/|\mathbf{p}_e|_{\max},$$

and $\rho$ is known as the *Michel parameter*. The V–A interaction predicts $\rho = \frac{3}{4}$, while experimentally $\rho = 0.747 \pm 0.005$.

Observation of the muon lifetime enables one to calculate the muon decay constant $G'$. Remarkably,

$$G' \simeq G. \quad (11.23)$$

(More precisely, $G'$ seems to be about 2% larger than $G$; we return to this point later.) Is it possible that *all* particles have the *same* weak interaction coupling constant? This possibility is contained in a very elegant way in the idea of a *universal weak current* which interacts with itself (*current–current interaction*) to produce the weak interaction:

$$\mathfrak{j}_\lambda{}^\mathrm{W} \equiv (\bar{\nu}e) + (\bar{\nu}\mu) + (\bar{p}n), \quad (11.24)$$

where

$$(\bar{\nu}e) \equiv \bar{\psi}_\nu \gamma_\lambda(1+\gamma_5)\psi_e, \quad \text{etc.,}$$

and

$$\mathcal{H}_\mathrm{I} = (G/\sqrt{2})\mathfrak{j}_\lambda{}^\mathrm{W}\mathfrak{j}_\lambda^{\mathrm{W}\dagger}. \quad (11.25)$$

(We are ignoring strange particles for the moment.)

[Here $\mathfrak{j}_\lambda^{\mathrm{W}\dagger}$ means the adjoint of $\mathfrak{j}_\lambda{}^\mathrm{W}$ except that we must refrain from complex conjugating the $i$ in the fourth component. Otherwise we would have, for example,

$$p_\mu p_\mu{}^\dagger = \mathbf{p}^2 + E^2,$$

which is not invariant. Formally we can introduce the symbol ‡ such that

$$\mathbf{j}^{\ddagger} \equiv \mathbf{j}^{\dagger},$$
$$j_4^{\ddagger} = -j_4^{\dagger}.$$

Then whenever we write $j_\lambda j_\lambda^{\dagger}$, we really mean $j_\mu j_\mu^{\ddagger}$. This is the price we pay for using $i$ to take care of the minus sign in the metric.]

Since

$$j_\lambda^{W\dagger} = (\bar{e}\nu) + (\bar{\mu}\nu) + (\bar{n}p),$$

$\mathcal{H}_I$ contains all the beta- and muon-decay interactions, as well as terms that can only contribute to scattering.

## 11.4. Conserved Vector Current Hypothesis

By writing Eq. (11.24) without an extra coefficient in front of the nucleon term, we have introduced an additional assumption. In Eq. (11.24) we have

$$(\bar{p}n) \equiv \bar{\psi}_p \gamma_\mu (1 + \gamma_5) \psi_n,$$

instead of the more general expression [see Eq. (11.9)]

$$\bar{\psi}_p \gamma_\mu (C_V + C_A \gamma_5) \psi_n .$$

Either form leads to

$$\bar{u}_p \gamma_\mu (C_V' + C_A' \gamma_5) u_n,$$

in the $T$-matrix element [see Eq. (11.10)]. Let us now forget about adjusting $G$ so that $C_V' = 1$, and instead fix $G$ from the muon-decay experiment. The universal weak current idea assumes that $C_V = C_A = 1$. Then, in general, we expect both that

$$C_V' \neq 1 \qquad \text{and} \qquad C_A' \neq 1$$

because of the strong interactions. The additional assumption is that

$$C_V' = 1.$$

We have already seen that this agrees with experiment ($G' \simeq G$ when $C_V' = 1$). Thus in going from interaction Hamiltonian to matrix element we have

$$\bar{\psi}_p \gamma_\mu \psi_n \to \bar{u}_p \gamma_\mu u_n$$

(just like for the lepton currents) in contrast to

$$\bar{\psi}_p \gamma_\mu \gamma_5 \psi_n \to r \bar{u}_p \gamma_\mu \gamma_5 u_n$$

(where $r \equiv C_A'$). We say that the strong interactions "renormalize" the axial-vector current, but *not* the vector current.

This assumption follows naturally from the *conserved vector current* (CVC) hypothesis, proposed by Gershtein and Zeldovitch (1956) and independently by Feynmann and Gell-Mann (1958). This hypothesis says that the hadronic part of the weak vector current is conserved in the same way as the electromagnetic current:

$$\partial_\mu j_\mu^{\text{el}} = 0.$$

But conservation of electromagnetic current implies that total charge is unaffected by the presence of strong interactions; this in turn implies that the *electric form factor* at zero momentum transfer satisfies

$$F_{\text{el}}(q^2 = 0) = 1.$$

Analogously, the CVC idea implies that the total "weak vector charge" is unaffected by the strong interactions, so that

$$F_{\text{V}}(q^2 = 0) = C_{\text{V}}'/C_{\text{V}} = 1. \tag{11.26}$$

The similarity in structure between the electromagnetic current and the weak vector current can be carried further. As we see from Eq. (11.24), the weak current increases the electric charge by one unit. Thus we can write that

$$j_\mu{}^{\text{V}} \propto T_+ \equiv T_1 + iT_2. \tag{11.27}$$

Since part (the *isotopic vector* part) of the electromagnetic current is $\propto T_3$, we expect that the two currents are related in isospin space. This idea can be used to evaluate the vector current matrix elements. For example, we can write for the matrix element between pion states

$$\begin{aligned}\langle\pi^+(\mathbf{p}_2)|\,j_\mu{}^{\text{V}}\,|\pi^0(\mathbf{p}_1)\rangle &= \langle t = 1, t_3 = 1\,|\,T_+\,|\,t = 1, t_3 = 0\rangle \\ &\quad\times\langle\pi^+(\mathbf{p}_2)|\,j_\mu^{\text{el}}\,|\pi^+(\mathbf{p}_1)\rangle \\ &= \sqrt{2}\cdot(p_2 + p_1)_\mu F_{\text{el}}^\pi(q^2)(4\omega_1\omega_2)^{-1/2}(2\pi)^{-3}e^{iqx}.\end{aligned} \tag{11.28}$$

This allows one to evaluate completely the matrix element for the very rare $\pi^+$ decay mode

$$\pi^+ \to \pi^0 e^+ \nu,$$

since there is no axial vector contribution.

[Because the space part of the axial vector current is unaffected by space inversion,

$$\begin{aligned}\langle\pi^+(\mathbf{p}_2)|\mathbf{j}^{\text{A}}|\pi^0(\mathbf{p}_1)\rangle &= \langle\pi^+(\mathbf{p}_2)|\,P^{-1}(P\mathbf{j}^{\text{A}}P^{-1})P\,|\pi^0(\mathbf{p}_1)\rangle \\ &= +\langle\pi^+(-\mathbf{p}_2)|\mathbf{j}^{\text{A}}|\pi^0(-\mathbf{p}_1)\rangle \\ &= (-\mathbf{p}_2 - \mathbf{p}_1)F_{\text{A}}{}^\pi(q^2)(4\omega_1\omega_2)^{-1/2}(2\pi)^{-}e^{iqx} \\ &= -\langle\pi^+(\mathbf{p}_2)|\mathbf{j}^{\text{A}}|\pi^0(\mathbf{p}_1)\rangle,\end{aligned}$$

and therefore,

$$\langle \pi^+(\mathbf{p}_2)|j^A|\pi^0(\mathbf{p}_1)\rangle = 0.$$

The argument for the time part is similar.]

In spite of its rarity, this decay mode has been observed and the measured branching ratio agrees with the predicted value of

$$\Gamma_{\pi^0 e^+ \nu}/\Gamma_{\mu^+\nu} = 1.02 \times 10^{-8}.$$

## 11.5. Strange Particle Decays

The weak interaction allows the strange particles to undergo *leptonic decay*, e.g.,

$$K^+ \to \mu^+ \nu,$$

or *nonleptonic* decay, e.g.,

$$K^+ \to \pi^+\pi^0, \qquad \Lambda \to \pi^- p.$$

Because there are fewer hadrons involved, leptonic decays are simpler to discuss and we confine most of our remarks to them.

### *a. Leptonic Decays*

We expect that, as above, the leptons enter through lepton currents that couple with hadron currents. The simplest way to include such interactions is to generalize the universal weak current [Eq. (11.24)] to include a *strangeness-changing part*. One is tempted simply to write, e.g.,

$$j_\lambda{}^W = (\bar{\nu}e) + (\bar{\nu}\mu) + (\bar{p}n) + (\bar{p}\Lambda). \tag{11.29}$$

Because of the strong interactions, a term like $(\bar{p}\Lambda)$ can raise the strangeness of *any* hadron and thus bring about the decay (leptonic or nonleptonic) of any strange particle.

The different kinds of weak interactions implied by this weak current in

$$\mathcal{H}_I \propto j_\lambda{}^W j_\lambda{}^{W\dagger}$$

can be indicated by the tetrahedron shown in Fig. 11.5, and possible diagrams contributing to strange particle decays are shown in Fig. 11.6.

For leptonic decays, the $(\bar{p}\Lambda)$ insertion clearly implies the rules

$$\Delta S = +\Delta Q, \tag{11.30a}$$

$$|\Delta S| = 1, \tag{11.30b}$$

$$\Delta T = \tfrac{1}{2}, \tag{11.30c}$$

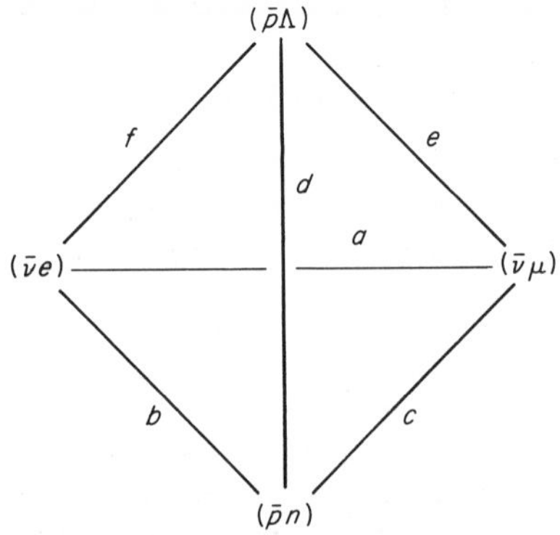


**Fig. 11.5.** Contributions to the weak current and the resulting weak interactions. Of the edges of the tetrahadron, *a* brings about muon decay, *b* beta and pion decay, *c* pion decay, *d* nonleptonic strange particle decay, *e* and *f* leptonic strange particle decay.

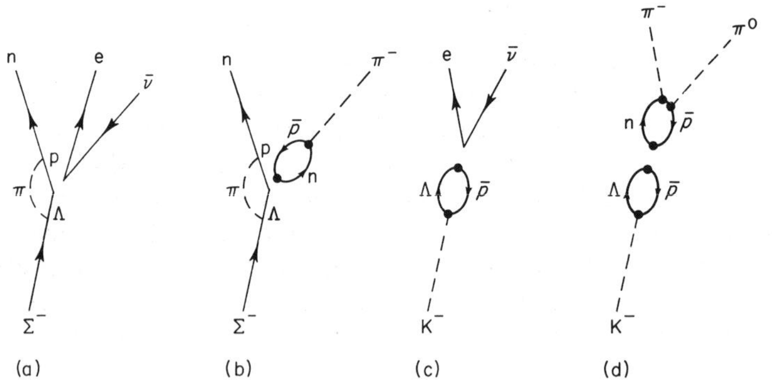


**Fig. 11.6.** Possible diagrams contributing to some strange particle decays: (a) $\Sigma^-$ leptonic, (b) $\Sigma^-$ nonleptonic, (c) $K^-$ leptonic, (d) $K^-$ nonleptonic.

which had in fact been previously established from experiment. For example, $\Delta S = \Delta Q$ allows the reaction

$$\Sigma^- \to \mathrm{n} e^- \bar{\nu},$$

while forbidding

$$\Sigma^+ \to \mathrm{n} e^+ \nu,$$

consistent with observation. The lack of a $|\Delta S| > 1$ current rules out many more decays that would otherwise go, e.g.,

$$\Xi^- \to \mathrm{n} \mu^- \bar{\nu},$$

while $\Delta T = \frac{1}{2}$ provides relations between certain decay rates, e.g., it predicts the branching ratio

$$\Gamma(\Lambda \to \pi^- p)/\Gamma(\Lambda \to \pi^0 n).$$

However, this "naive" form of universality has a significant drawback: It gives strangeness-changing leptonic decay rates that are larger by about an order of magnitude than those observed experimentally.

### *b. Cabibbo Theory*

A form of universality which is in much better accord with experiment was invented by Cabibbo in 1963. Instead of equal parts *strangeness-conserving* and *strangeness-changing* currents, Cabibbo assumes that the *hadronic part* of the weak current is composed of a more general linear combination of the two:

$$\mathfrak{J}_\mu{}^{\mathrm{H}} = \cos\theta \mathfrak{j}_\mu^{(0)} + \sin\theta \mathfrak{j}_\mu^{(1)}, \tag{11.31}$$

where we now write

$$\mathfrak{j}_\mu^{(0)} \equiv (\bar{p}n), \qquad \mathfrak{j}_\mu^{(1)} \equiv (\bar{p}\Lambda).$$

Cabibbo's model also assumes that (1) exact SU(3) symmetry holds; (2) $\mathfrak{j}_\mu^{(0)}$ and $\mathfrak{j}_\mu^{(1)}$ both transform under SU(3) like members of an octet; (3) the vector part of the hadron current, $\mathfrak{j}_\mu^{(V)}$, is equal to the electric current (isovector part) transformed by an SU(3) rotation (*generalized* CVC).

From these assumptions one can calculate all the leptonic decays of the baryons and mesons in terms of only three constants (neglecting any momentum dependence of form factors): the *Cabibbo angle* $\theta$ and two parameters which specify the "renormalization" of the axial current. The experimental data are fitted remarkably well, as Table 11.3 shows. In addition, the existence

**Table 11.3**

*Phenomenological Test of Cabibbo Theory*[a]

| | Branching ratio ($\times 10^3$) | |
|---|---|---|
| Decay | Theory | Experiment |
| $\Sigma^- \to \Lambda e^- \bar{\nu}$ | 0.062 | 0.059 ± 0.006 |
| $\Sigma^+ \to \Lambda e^+ \nu$ | 0.019 | 0.019 ± 0.004 |
| $\Lambda \to p e^- \bar{\nu}$ | 0.86 | 0.83 ± 0.08 |
| $\Lambda \to p \mu^- \bar{\nu}$ | 0.14 | 0.14 ± 0.06 |
| $\Sigma^- \to n e^- \bar{\nu}$ | 1.01 | 1.10 ± 0.05 |
| $\Sigma^- \to n \mu^- \bar{\nu}$ | 0.48 | 0.45 ± 0.05 |

[a] Data from H. Filthuth. In *Proceedings of the Topical Conference on Weak Interactions*, CERN, 1969.

of a Cabibbo angle different from zero means that the effective constant in beta decay is $G \cos \theta$ instead of $G$, which explains the small (2%) discrepancy between the nuclear and the muon constants. That the Cabibbo theory works so well is really a puzzle, since it uses exact SU(3) symmetry, while for the strong interactions SU(3) symmetry is very approximate.

## 11.6. Neutrino Scattering

Neutrino experiments are in their infancy because of the very small cross sections involved. The earliest such experiment was that of Reines *et al.* in 1960 which detected the positron and the neutron produced in the reaction

$$\bar{\nu}\mathrm{p} \to \mathrm{ne}^+.$$

A reactor was the source of antineutrinos, while an enormous volume of liquid scintillator served as both target and detector. The observed cross section agrees with the prediction of V–A theory.

### *a. Two Kinds of Neutrinos*

A more surprising result came from experiments with high-energy neutrinos. These are produced by decay in flight of a charged pion beam; the resulting neutrino beam is cleansed of everything else by passing it through a thick absorber (20 m of iron). According to Eq. (11.24), at high energy about equal numbers of electrons and muons should be produced when the neutrinos react with nuclei:

$$\begin{aligned} \nu\mathrm{n} &\to \mu^-\mathrm{p} \\ &\to \mathrm{e}^-\mathrm{p}. \end{aligned}$$

The observed branching ratio, however, is very different:

$$\sigma(\nu e)/\sigma(\nu\mu) = 0.017 \pm 0.005.$$

The most natural explanation is that there are *two distinct kinds* of neutrinos: $\nu_e$ which is coupled to the electron, and $\nu_\mu$ which is coupled to the muon; thus the weak current should be written

$$\mathfrak{j}_\lambda{}^{\mathrm{W}} = (\bar{\nu}_e e) + (\bar{\nu}_\mu \mu) + \mathfrak{j}_\lambda{}^{\mathrm{H}}. \tag{11.24'}$$

This form implies that *muon lepton number* and *electron lepton number* are conserved *separately*, and whenever we deal with neutrinos we should specify which kind. Since the predominant charged pion decay mode is

$$\pi^+ \to \mu^+ \nu_\mu,$$

the neutrino beam is overwhelmingly $\nu_\mu$, which then explains why so few electrons are produced when the neutrinos are absorbed.

### b. High-Energy Neutrino Scattering

Neutrino cross sections grow rapidly with energy according to the theory we have set out and according to the few experiments that have been done. The simplest way to see this is to realize that the scattering amplitude $T$ contains a factor of $G$. Since according to Eq. (3.96a)

$$\sigma = \frac{1}{64\pi^2}\left|\frac{\mathbf{q}}{\mathbf{p}}\right|\frac{1}{s}\int |T|^2\, d\Omega,$$

$T$ is dimensionless. But

$$G \simeq 10^{-5}/m_{\mathrm{p}}{}^2,$$

so $T$ must have an additional factor of dimension $M^2$. At high energy, we do not expect any of the particle masses to enter, so the only available factor is $E^2 \equiv s$. Thus

$$|T| \sim GE^2 = Gs, \qquad (11.32)$$

and, therefore,

$$\sigma \sim G^2E^2 = G^2 s. \qquad (11.33)$$

If this growth with increasing energy held for arbitrarily high energy, weak scattering would eventually dominate all other kinds. It is clear, however, that this energy behavior *cannot* persist because it would violate unitarity. For high energies the statement of unitarity [Eq. (3.103′)] reads

$$\operatorname{Im} T_{\mathrm{ii}} \simeq s\sigma_{\mathrm{T}}. \qquad (11.34)$$

Then since

$$\sigma_{\mathrm{T}} \geq \sigma \qquad \text{and} \qquad |T| \geq \operatorname{Im} T_{\mathrm{ii}},$$

we have

$$|T| \geq s\sigma \sim G^2 s^2 \sim |T|^2,$$

which is clearly impossible behavior for $|T| > 1$.

Hence $T$ must depart from this behavior while it still satisfies

$$|T| \lesssim 1, \qquad (11.35)$$

i.e., for $E \lesssim E_c$, where from Eq. (11.32),

$$E_c \equiv G^{-1/2} \simeq 10^{5/2} m_{\mathrm{p}} \simeq 300 \quad \mathrm{GeV}.$$

Indeed the calculation of $T$ from the weak Hamiltonian is limited to the lowest order of perturbation theory (Born approximation), for otherwise we are presented with infinities that cannot be handled by renormalization. But Born approximation can only be trusted when the scattering is small, i.e., $|T| \ll 1$.

So far, no clear way has been found around these difficulties, and our understanding of the weak interactions, while satisfying in some respects, is certainly incomplete. One idea, which has received a lot of attention, both theoretical and experimental, is that the weak interactions are actually mediated by an *intermediate vector boson* (W). In this picture the weak currents couple to the W field just the way the electric current couples to the photon field:

$$\mathcal{H}_I = g_W W_\lambda(x) \mathfrak{j}_\lambda^W(x).$$

Then instead of four fermions interacting at one point, we have the Feynman diagram of Fig. 11.7, and

$$G/\sqrt{2} \simeq g_W^2/M_W^2.$$

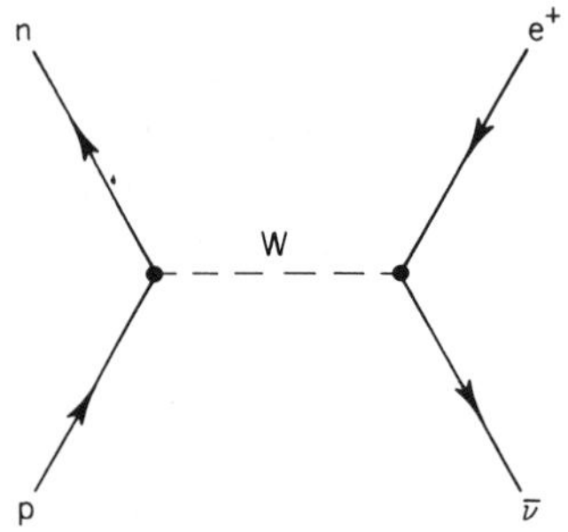


**Fig. 11.7.** Weak interactions *via* intermediate vector boson exchange.

The boson propagator moderates the unphysical high-energy behavior already discussed. So far there has been no experimental confirmation of the existence of such a particle, but that might be because it is very massive; the current experimental lower limit to $M_W$ is only about 2.5 GeV. The search will continue. Furthermore, high-energy neutrino scattering experiments will eventually be pursued into the 300 GeV range, which should help considerably in clarifying the situation.

## 11.7. Neutral K Decay

### *a. Effects of* $K^0 \rightleftarrows \overline{K}^0$

While looking for further tests of the strangeness idea, Gell-Mann and Pais (in 1955) predicted some very unexpected and beautiful phenomena in $K^0$ decay. These arise basically because (1) $K^0$ and $\overline{K}^0$ differ only in strangeness and (2) strangeness is not conserved in their (weak) decay. Thus two successive weak interactions, each with $|\Delta S| = 1$, can convert a $K^0$ into a $\overline{K}^0$ (or vice versa); since the principle decay mode is to two pions, the conversion can be pictured as in Fig. 11.8. The amplitude for the process cannot be calculated reliably, but we can do much analysis simply using invariance arguments.

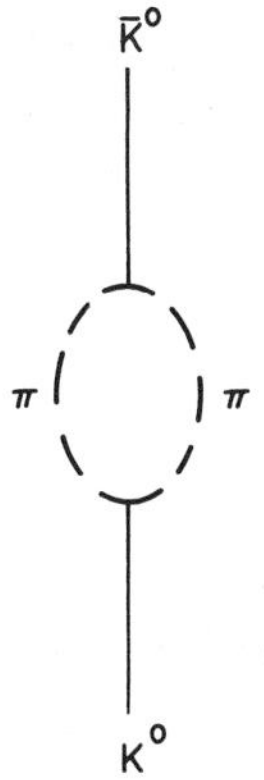


**Fig. 11.8.** The $K^0 \rightleftarrows \bar{K}^0$ transition, mediated by the two pion intermediate state.

The useful invariance here is that under the combined operation *CP* [which, assuming the *CPT* theorem (Chapter 6, Section 6.4) is equivalent to *T* invariance]. It is now known that there is actually a small violation of *CP* invariance in $K^0$ decay (it seems to hold elsewhere); to simplify the analysis, we ignore this violation for the moment. Now, neither $K^0$ nor $\bar{K}^0$ can be a eigenstate of *CP*, but $\pi^+\pi^-$ and $\pi^0\pi^0$ can be. Thus in the CM system

$$\begin{aligned} CP|\pi^+(\mathbf{p}_1)\pi^-(\mathbf{p}_2)\rangle &= C|\pi^+(\mathbf{p}_2)\pi^-(\mathbf{p}_1)\rangle \\ &= |\pi^+(\mathbf{p}_1)\pi^-(\mathbf{p}_2)\rangle, \end{aligned} \tag{11.36}$$

since $\mathbf{p}_2 = -\mathbf{p}_1$; this is illustrated in Fig. 11.9. Thus $\pi^+\pi^-$ is an eigenstate of *CP* with eigenvalue $+1$, and similarly for $\pi^0\pi^0$.

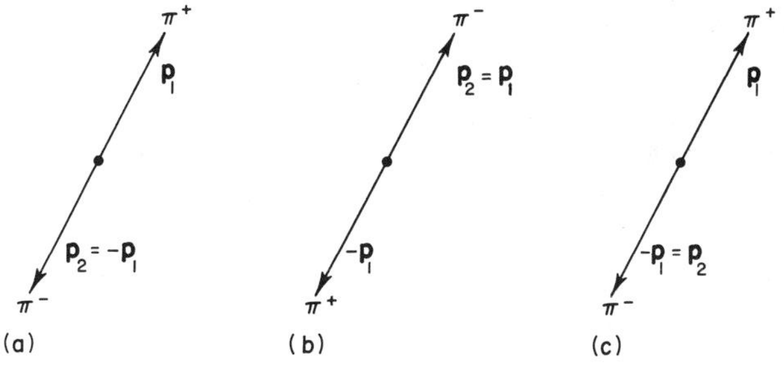


**Fig. 11.9.** *CP* acting on a $\pi^+\pi^-$ state (CM system): (a) the original state; (b) after applying *P*; (c) after applying *C*.

Now using the convention that

$$C|K^0\rangle = +|\bar{K}^0\rangle, \tag{11.37}$$

we have

$$\begin{aligned} CP|K^0\rangle &= -|\bar{K}^0\rangle, \\ CP|\bar{K}^0\rangle &= -|K^0\rangle \end{aligned} \tag{11.38}$$

($K^0$ rest system). Form *CP eigenstates*:

$$\begin{aligned}|K_1\rangle &\equiv (1/\sqrt{2})(|K^0\rangle - |\overline{K}^0\rangle),\\ |K_2\rangle &\equiv (1/\sqrt{2})(|K^0\rangle + |\overline{K}^0\rangle).\end{aligned} \tag{11.39}$$

Hence

$$\begin{aligned}CP|K_1\rangle &= +|K_1\rangle,\\ CP|K_2| &= -|K_1\rangle.\end{aligned} \tag{11.40}$$

Then, *CP* conservation (invariance) implies that while

$$\begin{aligned}&K_1 \to 2\pi,\\ &K_2 \nrightarrow 2\pi;\end{aligned}$$

because it is forbidden the $2\pi$ mode, $K_2$ can only decay into $3\pi$ or $\pi l v$; this gives it a *much longer lifetime* than $K_1$ has. (In fact, $\tau_1 \simeq 0.9 \times 10^{-10}$ sec, while $\tau_2 \simeq 5 \times 10^{-8}$ sec.) In addition, the masses of $K_1$ and $K_2$ must differ slightly, since there is a $2\pi$ self-energy contribution to $K_1$ but not to $K_2$.

Suppose we start with a $K^0$ state at time $t = 0$,

$$|K^0\rangle = (1/\sqrt{2})(|K_1\rangle + |K_2\rangle). \tag{11.41}$$

At a later time $t \geq$ several $\tau_1$, most of the $K_1$ will be gone to leave a nearly pure $K_2$ state, which is a mixture of $K^0$ and $\overline{K}^0$! Thus a pure $K^0$ beam becomes a 50–50 mixture of $K^0$ and $\overline{K}^0$. The $\overline{K}^0$'s are easy to distinguish from the $K^0$'s, since, as mentioned above, their interactions with matter are so different; e.g., the $\overline{K}^0$'s can produce hyperons:

$$\overline{K}^0 N \to \Lambda\pi,$$

etc.

We can discuss the evolution of a $K^0$ state more quantitatively by considering the time dependence explicitly:

$$|\psi(t)\rangle = (1/\sqrt{2})[|K_1\rangle \exp(-im_1 t - t/2\tau_1) + |K_2\rangle \exp(-im_2 t - t/2\tau_2)].$$

Let us compute, for example, the $\overline{K}^0$ component of this state as a function of time:

$$\begin{aligned}\langle \overline{K}^0|\psi(t)\rangle &= (1/\sqrt{2})(\langle K_2| - \langle K_1|)|\psi(t)\rangle\\ &= \tfrac{1}{2}[\exp(-im_2 t - t/2\tau_2) - \exp(-im_1 t - t/2\tau_1)],\end{aligned}$$

so that

$$\begin{aligned}&|\langle \overline{K}^0|\psi(t)\rangle|^2\\ &\quad = \tfrac{1}{4}\{\exp(-t/\tau_1) + \exp(-t/\tau_2) - 2\exp(-t/2\tau_1 - t/2\tau_2)\cos(\Delta m \cdot t)\}.\end{aligned} \tag{11.42}$$

This time variation of the $\bar{K}^0$ probability is shown in Fig. 11.10. Observing this time variation, or rather the corresponding variation with distance along the beam, is one of the ways to measure the mass difference $\Delta m$; the currently accepted value is

$$\Delta m \equiv m_2 - m_1 \simeq 0.47 \times 1/\tau_1,$$

which means that

$$\Delta m/m \sim 10^{-14}!$$

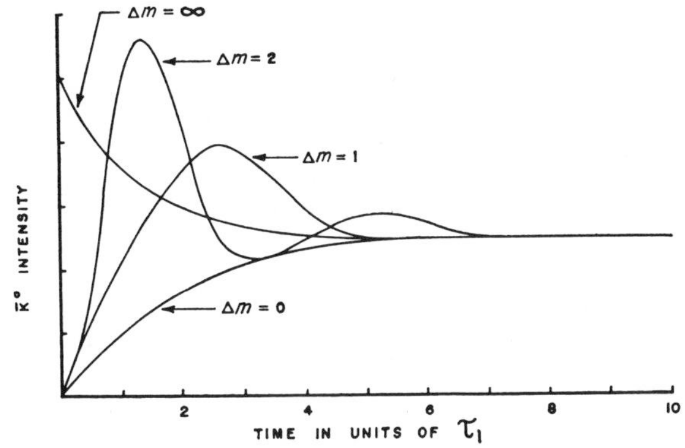


**Fig. 11.10.** Time dependence of the $\bar{K}^0$ intensity for several values of the $K_1$–$K_2$ mass difference. [From U. Camerini *et al.*, *$K_1{}^0$–$K_2{}^0$ mass difference*. *Phys. Rev.* **128**, 362–367 (1962).]

### *b.* $K_1$ *Regeneration*

Another interesting phenomenon is the *regeneration* of $K_1$ mesons in a beam which has become pure $K_2$. This is accomplished by passing the $K_2$ through matter; since the $K^0$ and $\bar{K}^0$ components scatter differently, one gets a new linear combination which is a mixture of $K_1$ and $K_2$. Now because it is coherent over the entire target, forward scattering is the most important. We start with

$$|K_2\rangle = (1/\sqrt{2})(|K^0\rangle + |\bar{K}^0\rangle);$$

so after scattering we have

$$(1/\sqrt{2})(f|K^0\rangle + \bar{f}|\bar{K}^0\rangle),$$

$f$ and $\bar{f}$ being the forward scattering amplitudes for $K^0$ and $\bar{K}^0$. The $K_1$ component of this scattered state is then

$$f_{21} = \langle K_1|(1/\sqrt{2})(f|K^0\rangle + \bar{f}|\bar{K}^0\rangle) = \tfrac{1}{2}(f - \bar{f}). \tag{11.43}$$

Now suppose the scatterer is a plate (thickness $L$, number density of scatterers $N$) placed in the beam. Then the scattering amplitude from a layer of thickness $dx$ is $\propto f_{21}N\,dx$. However, we must take into account the fact that the *phase* of the $|K_1\rangle$ is going to depend upon where in the sample it was produced. Thus we have a phase factor of

$$e^{ip_2 x}$$

for the propagation from 0 to $x$, and a factor of

$$e^{ip_1(L-x)}$$

for the propagation from $x$ to $L$ ($p_1 \equiv |\mathbf{p}_1|$, etc.). Since for coherent scattering the plate as a whole takes up the recoil, and energy and momentum conservation require

$$\Delta E \equiv E_1 - E_2 = (\Delta p)^2/2M = \Delta p(\Delta p/2M) \ll \Delta p \equiv p_1 - p_2,$$

we can take $\Delta E = 0$, and therefore can ignore $e^{-iEt}$ factors. ($\Delta E$, $\Delta p$, and $M$ are the kinetic energy, momentum, and mass of the *plate*.) Furthermore, $E_1 = E_2$ implies

$$p_1{}^2 - p_2{}^2 = m_2{}^2 - m_1{}^2,$$

or

$$\Delta p \equiv p_1 - p_2 = m\,\Delta m/p,$$

where

$$m \equiv (m_1 + m_2)/2, \qquad p \equiv (p_1 + p_2)/2.$$

Finally, we insert the factor that expresses the exponential decay of the $K_1$ component

$$\exp[-(L-x)/2v\gamma\tau_1]$$

(this factor for the $K_2$ component can be neglected). Putting all this together, integrating over $x$, and taking the absolute square, one finds for intensity of $K_1$'s emerging from the plate,

$$I \propto \frac{4\,|N\Lambda f_{21}|^2}{1 + 4(\tau_1\,\Delta m)^2}\,|e^{-L/2\Lambda} - e^{i\Delta p L}|^2, \tag{11.44}$$

where $\Lambda \equiv v\gamma\tau_1$ is the *decay length*. (Notice the factor of $\gamma$ in $\Lambda$, which expresses *time dilation*.) This result has been used to measure $\Delta m$ by observing the dependence of $I$ on plate thickness $L$; no knowledge of $f_{21}$ is required.

### c. CP Violation

In 1964, Christensen *et al.* looked for and found a small amount of the forbidden $2\pi$ decay of $K_2$; they found the branching ratio

$$\frac{\Gamma(K_2 \to \pi^+\pi^-)}{\Gamma(K_2 \to \text{all charged modes})} = (2.0 \pm 0.4) \times 10^{-3}.$$

By looking far enough "downstream," they insured that the $2\pi$ events represented $K_2$ rather than $K_1$ decay. This, in fact, is an extremely sensitive test of the *CP* invariance idea, since, as we have discussed, *CP* conservation forbids this decay. As a result of this *CP* violation, the long- and short-lived components are not precisely $K_2$ and $K_1$, but linear combinations of them, i.e.,

$$\begin{aligned} |K_L\rangle &\simeq |K_2\rangle + \varepsilon|K_1\rangle, \\ |K_S\rangle &\simeq |K_1\rangle - \varepsilon|K_2\rangle, \end{aligned} \tag{11.45}$$

with

$$|\varepsilon| \simeq 2 \times 10^{-3}.$$

Further experiments to determine the parameters of *CP* violation in $K^0$ decay have been and are being performed, but any underlying "cause" of the violation remains hidden. *CP* violation plus the *CPT* theorem, of course, implies a violation of *T*-invariance and has triggered many investigations of *T*-invariance in the weak interactions of other particles and in the strong and electromagnetic interactions as well. So far, the experimental evidence indicates that $T(CP)$ violation is peculiar to $K^0$ decay.

## Problems

**11.1.** Derive Eq. (11.3).

**11.2.*** Carry out the calculation of the neutron decay rate using the V–A interaction [Eq. (11.10)].

**11.3.** Check that the Dirac matrices introduced in Section 11.2a satisfy the anticommutation relations, Eq. (2.13).

**11.4.** Verify Eqs. (11.18) and (11.19).

**11.5.** Verify Eq. (11.28) using Eqs. (2.7) and (4.13).

**11.6.** Calculate the

$$\Gamma(\Lambda \to \pi^- p)/\Gamma(\Lambda \to \pi^0 n)$$

ratio implied by the $\Delta T = \frac{1}{2}$ rule.

**11.7.** Evaluate the ratio

$$\Gamma(K^+ \to \mu^+ \nu)/\Gamma(\pi^+ \to \mu^+ \nu),$$

using the fact that the *only* difference in the matrix elements, apart from kinematic factors, lies in the fact that $K^+$ decay involves the $\mathfrak{j}^{(1)}$ current, while $\pi^+$ decay involves the $\mathfrak{j}^{(0)}$ current.

**11.8.** Verify Eq. (11.44).

**11.9.*** The *phase* $\phi_{+-}$ of the relative amplitude

$$\eta_{+-} \equiv \frac{\langle \pi^+\pi^- | K_2 \rangle}{\langle \pi^+\pi^- | K_1 \rangle} \equiv |\eta_{+-}|\, e^{i\phi_{+-}}$$

can be measured by observing the interference of $K_2 \to \pi^+\pi^-$ with $K_1 \to \pi^+\pi^-$, where the $K_1$ is produced by passing the $K_2$ beam through a thin slab of material. Find an expression for $I(\pi^+\pi^-)$ in terms of $f_{21}$, plate thickness, etc. How should the parameters be chosen for determining $\phi_{+-}$?

## Bibliography

B. T. FELD. *Models of Elementary Particles.* Blaisdell, Waltham, Massachusetts, 1969.

W. R. FRAZER. *Elementary Particles.* Chap. 8. Prentice-Hall, Englewood Cliffs, New Jersey, 1966.

G. KÄLLÉN. *Elementary Particle Physics.* Addison-Wesley, Reading, Massachusetts, 1964.

H. MUIRHEAD. *Physics of Elementary Particles.* Chap. 12. Pergamon, Oxford, 1965.

L. B. OKUN. *Weak Interactions of Elementary Particles.* Pergamon, Oxford, 1965.

CHAPTER TWELVE

# Hadron Dynamics

Because of the lack of a substitute for perturbation theory for making explicit calculations of strong interaction amplitudes, physicists cannot be sure that the world of hadrons is governed by a relativistic quantum field theory at all. Thus because field theory is (1) hard and (2) possibly irrelevant, many have adopted the *S-matrix theory* point of view, which concentrates on using various general properties (like *analyticity*, *crossing symmetry*, *unitarity*) to derive equations (*dispersion relations*) for hadron amplitudes. Here we will try to indicate the ideas involved, beginning with a discussion of *nucleon form factors*.

## 12.1. Nucleon Form Factors

Electron–nucleon elastic scattering has been pursued as a means of determining the electromagnetic "shape" of the proton and neutron. First we discuss how scattering from a charge distribution gives rise to a form factor.

### *a. Nonrelativistic Scattering*

Let us consider a point charge $e$ that scatters from a fixed charge distribution $\rho(\mathbf{r})$. The interaction energy is as follows.

$$V(\mathbf{r}) = e\Phi(\mathbf{r}) = e\int \frac{\rho(\mathbf{r}')}{4\pi|\mathbf{r}-\mathbf{r}'|}\, d^3\mathbf{r}', \tag{12.1}$$

where $\Phi(\mathbf{r})$ is the scalar potential and satisfies Poisson's equation:

$$\nabla^2 \Phi(\mathbf{r}) = -\rho(\mathbf{r}). \tag{12.2}$$

Then

$$H' = \int V(\mathbf{r})\psi^\dagger(\mathbf{r})\psi(\mathbf{r})\, d^3\mathbf{r}$$

is the (nonrelativistic) interaction Hamiltonian. The scattering amplitude in Born approximaıion is ($\delta$-function normalization)

$$\begin{aligned} S_{1\mathrm{fi}} &= -i2\pi\delta(E_\mathrm{f} - E_\mathrm{i})\int d^3\mathbf{r}\, \phi_\mathrm{f}^*(\mathbf{r})V(\mathbf{r})\phi_\mathrm{i}(\mathbf{r}) \\ &= -i(2\pi)^{-3}2\pi\delta(E_\mathrm{f} - E_\mathrm{i})\int d^3\mathbf{r}\, \exp[i(\mathbf{k}_\mathrm{i} - \mathbf{k}_\mathrm{f})\cdot \mathbf{r}]\, e\Phi(\mathbf{r}), \end{aligned}$$

or

$$\overline{M}_{\mathrm{fi}} = -(2\pi)^{-3}e\int d^3\mathbf{r}\, e^{i\mathbf{q}\cdot\mathbf{r}}\, \Phi(\mathbf{r}), \tag{12.3}$$

with $\mathbf{q} \equiv \mathbf{k}_\mathrm{i} - \mathbf{k}_\mathrm{f}$. We are able to express the integral in terms of the charge density:

$$\begin{aligned} \int d^3\mathbf{r}\, e^{i\mathbf{q}\cdot\mathbf{r}}\Phi(\mathbf{r}) &= -1/\mathbf{q}^2 \int d^3\mathbf{r}\, \nabla^2\, (e^{i\mathbf{q}\cdot\mathbf{r}})\Phi(\mathbf{r}) \\ &= -1/\mathbf{q}^2 \int d^3\mathbf{r}\, e^{i\mathbf{q}\cdot\mathbf{r}}\nabla^2\Phi(\mathbf{r}) \\ &= 1/\mathbf{q}^2 \int d^3\mathbf{r}\, e^{i\mathbf{q}\cdot\mathbf{r}}\rho(\mathbf{r}), \end{aligned} \tag{12.4}$$

using some integrations by parts and Poisson's equation. Thus

$$\overline{M}_{\mathrm{fi}} = -(2\pi)^{-3}e[F(\mathbf{q})/\mathbf{q}^2], \tag{12.5}$$

where the form factor $F(\mathbf{q})$ is simply the Fourier transform of the charge density,

$$F(\mathbf{q}) \equiv \int d^3\mathbf{r}\, e^{i\mathbf{q}\cdot\mathbf{r}}\rho(\mathbf{r}), \tag{12.6}$$

and is a function only of the magnitude $|\mathbf{q}|$ if $\rho$ is spherically symmetric: $\rho \equiv \rho(r)$. For a point charge, $\rho(\bar{r}) = Q\delta^3(\mathbf{r})$, which means that the form factor is constant: $F(\mathbf{q}) = Q$.

### *b. Relativistic Electron–Nucleon Scattering*

Now we look at relativistic electron–proton scattering (lowest order in $e$). We start with the minimal electromagnetic coupling expression (cf. Chapter 4, Section 4.2c) for the interaction of the proton with the electromagnetic field:

$$\mathcal{H}' = -\,eA_\mu(x)\bar{\psi}(x)\gamma_\mu\psi(x) = -\,eA_\mu(x)j_\mu(x). \qquad (12.7)$$

Then if we neglect all strong interaction corrections, i.e., assume the proton, like the electron, is a point Dirac particle [Fig. 12.1(a)] the relativistic Feynman

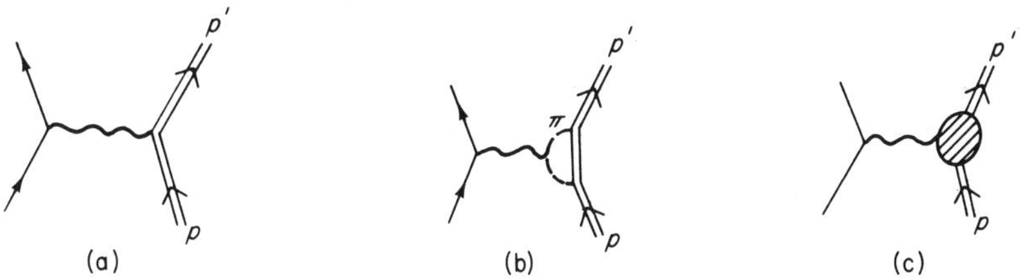


**Fig. 12.1.** Electron–proton scattering via one photon exchange: (a) with no strong interaction corrections; (b) pion correction; (c) all possible strong interaction corrections.

rules will have a factor [using Eqs. (4.18), (4.19), and (12.7) and invariant normalization]

$$e\langle \mathbf{p}'|j_\mu(0)|\mathbf{p}\rangle = e\cdot 2Mi\bar{u}'\gamma_\mu u, \qquad (12.8)$$

for the photon–proton vertex ($M$ = proton mass). With the strong interactions "turned on," however, we expect corrections like that of Fig. 12.1(b) to be important. The sum of all the strong interaction corrections produces the incalculable blob of Fig. 12.1(c). Instead of Eq. (12.8), we now have

$$\langle \mathbf{p}'|j_\mu(0)|\mathbf{p}\rangle = 2Mi\bar{u}'\Gamma_\mu u, \qquad (12.9)$$

where the operator $\Gamma_\mu$ can be expressed in terms of two form factors:

$$\Gamma_\mu = F_1(q^2)\gamma_\mu + F_2(q^2)(2M)^{-1}\,\Sigma_{\mu\nu}q_\nu, \qquad (12.10)$$

with $q \equiv p - p'$ and $\Sigma_{\mu\nu} \equiv -i/2(\gamma_\mu\gamma_\nu - \gamma_\nu\gamma_\mu)$. These form factors express the fact that the charge and current densities in a proton are spread out over a finite volume. For a point Dirac particle, on the other hand, we must have $F_1(q^2) = 1$, $F_2(q^2) = 0$, to agree with Eq. (12.8).

To see where Eq. (12.10) comes from, we start with the most general vector form we can think of for $\Gamma_\mu$:

$$\Gamma_\mu = A(q^2)q_\mu + B(q^2)P_\mu + C(q^2)\gamma_\mu + D(q^2)\Sigma_{\mu\nu}q_\nu + E(q^2)\Sigma_{\mu\nu}P_\nu. \qquad (12.11)$$

Here $P \equiv p + p'$. The coefficients $A(q^2)$, etc., must be invariant and therefore functions only of invariants; since

$$p^2 = p'^2 = -M^2,$$

$q^2$ is the only invariant variable. Because $u$ and $u'$ satisfy the Dirac equation

$$(i\not{p} + M)u = 0, \qquad \bar{u}'(i\not{p}' + M) = 0,$$

one can show that

$$\bar{u}'\Sigma_{\mu\nu}P_\nu u = +i\bar{u}'q_\mu u \tag{12.12a}$$

and

$$\bar{u}'\Sigma_{\mu\nu}q_\nu u = +2M\bar{u}'\gamma_\mu u + i\bar{u}'P_\mu u. \tag{12.12b}$$

This allows Eq. (12.11) to be rewritten as

$$\Gamma_\mu = A'q_\mu + B'\gamma_\mu + C'\Sigma_{\mu\nu}q_\nu.$$

Current conservation, $\partial_\mu \mathfrak{j}_\mu(x) = 0$, is then used to eliminate the $q_\mu$ term: Since [from Eqs. (3.33) and (3.37)]

$$\begin{aligned}\mathfrak{j}_\mu(x) &= e^{-iPx}\mathfrak{j}_\mu(0)e^{iPx},\\ 0 &= \langle \mathbf{p}' | \partial_\mu \mathfrak{j}_\mu(0) | \mathbf{p}\rangle\\ &= \partial_\mu \langle \mathbf{p}' | \mathfrak{j}_\mu(x) | \mathbf{p}\rangle_{x=0}\\ &= \partial_\mu \langle \mathbf{p}' | e^{-i(p'-p)x}\mathfrak{j}_\mu(0) | \mathbf{p}\rangle_{x=0}\\ &= i(p-p')_\mu \langle \mathbf{p}' | \mathrm{A}_\mu(0) | \mathbf{p}\rangle\\ &= iq_\mu \cdot 2Mi\bar{u}'\Gamma_\mu u;\end{aligned}$$

so we must have

$$\bar{u}'q_\mu\Gamma_\mu u = \bar{u}'(A'q^2 + B'\not{q} + C'\Sigma_{\mu\nu}q_\mu q_\nu)u = 0.$$

The third term is clearly zero ($\Sigma_{\mu\nu}$ is antisymmetric under $\mu \rightleftarrows \nu$) and, because the Dirac equation implies

$$\bar{u}'\not{q}u = \bar{u}'(\not{p} - \not{p}')u = 0,$$

so is the second. But $q^2 \neq 0$; so we must set

$$A'(q^2) \equiv 0.$$

Then changing notation and introducing $(2M)^{-1}$ for convenience, we have Eq. (12.10). Furthermore, since $\mathfrak{j}$ is Hermitian, $F_1$ and $F_2$ must be *real.*

Equations (12.9) and (12.10) can be used with the Feynman rules to find the electron–proton scattering matrix element [Fig. 12.1(c)]:

$$\begin{aligned}S_{2\mathrm{fi}} &= (-i)^2(-ie)(+ie)\\ &\quad\times 4m_e M(2\pi)^4\delta(p' + k' - p - k)\; \bar{u}_e'\gamma_\mu u_e \bar{u}'\Gamma_\mu u \cdot [-i/(q^2 - i\varepsilon)],\end{aligned}$$

or

$$T_{2\mathrm{fi}} = -e^2 \cdot 4m_e M\bar{u}_e'\gamma_\mu u_e \bar{u}'\Gamma_\mu u \cdot t^{-1}, \tag{12.13}$$

where $t \equiv -q^2 \equiv -(p-p')^2$. Notice that $t$ is *negative* in the physical region for electron scattering. The standard methods (cf. Chapter 5, Section 5.1a) then lead to the *Rosenbluth formula* (lab system):

$$\frac{d\sigma}{d\Omega} = \sigma_M(\theta)\left\{F_1{}^2 + \frac{(-t)}{4M^2}\left[2(F_1+F_2)^2\tan^2\frac{\theta}{2} + F_2{}^2\right]\right\}, \tag{12.14}$$

where

$$\sigma_M(\theta) = \alpha^2\frac{1}{4E^2}\frac{\cos^2\theta/2}{\sin^4\theta/2}\cdot\frac{1}{1+(2E/M)\sin^2\theta/2}$$

(*Mott scattering*) is what one gets for electron scattering (initial energy $E$) from a point proton with no magnetic moment. (In these formulas a very good approximation, $m_e \to 0$, has been used.) Thus most of the energy and angle dependence come from $\sigma_M(\theta)$, but measuring the deviations of the experimental cross section from $\sigma_M(\theta)$ can both test the Rosenbluth formula and determine the form factors.

Analysis of the experiments is somewhat simpler in terms of the *electric* and *magnetic* form factors:

$$G_E(t) \equiv F_1(t) + (t/4M^2)F_2(t), \tag{12.15a}$$

$$G_M(t) \equiv F_1(t) + F_2(t). \tag{12.15b}$$

The $G$'s can be expressed as Fourier transforms of the charge ($G_E$) and magnetization ($G_M$) distributions of the proton.

In spite of their *total* charge being zero, neutrons can also interact with the electromagnetic field and scatter electrons; thus we introduce the *neutron form factors* $F_1{}^n$, $F_2{}^n$, and $G_E{}^n$, $G_M{}^n$. (The proton form factors will henceforth be written $F_1{}^p$, etc.)

To discuss the low-energy behavior of the form factors, we first use Eqs. (12.9), (12.10), and (12.12b) to rewrite the matrix element of the current:

$$\langle \mathbf{p}'|j_\mu(0)|\mathbf{p}\rangle = \bar{u}'[F_1 P_\mu + i(F_1+F_2)\Sigma_{\mu\nu}q_\nu]u. \tag{12.16}$$

Here the first term on the right-hand side represents the charge and convective current, while the second is the magnetic moment term. Now for low energy (or for small momentum transfer at *any* energy), only the *total* charge can contribute to the first term, so that (the $e$ has already been factored out)

$$F_1{}^p(0) = 1 = G_E{}^p(0), \qquad F_1{}^n(0) = 0 = G_E{}^n(0).$$

Similarly, the second term gives the *total static magnetic moment* (cf. Section 4.2c):

$$\mu = [F_1(0) + F_2(0)](e/2M) = G_M(0)\cdot(e/2M)$$

($M$ is now the *nucleon* mass, the neutron–proton mass difference being neglected). Now if the nucleons were point Dirac particles, we would have

$$F_1{}^{\mathrm{p}} \equiv 1, \qquad F_2{}^{\mathrm{p}} \equiv F_1{}^{\mathrm{n}} \equiv F_2{}^{\mathrm{n}} \equiv 0,$$

so the $F_2(0)$ part is called the *anomalous* (or *Pauli*) magnetic moment. The known values of the static moments give us

$$G_M{}^{\mathrm{p}}(0) = 1 + F_2{}^{\mathrm{p}}(0) = 2.79,$$
$$G_M{}^{\mathrm{n}}(0) = F_2{}^{\mathrm{n}}(0) = -1.91.$$

Information on

$$\left.\frac{dF_1{}^{\mathrm{n}}(t)}{dt}\right|_{t=0}$$

is provided by the scattering of thermal neutrons from atomic electrons. The result is usually expressed in terms of the *neutron charge radius*

$$\langle \mathbf{r}^2 \rangle_1{}^{\mathrm{n}} \equiv 6[dF_1{}^{\mathrm{n}}(0)/dt] = (0 \pm 0.006) \quad \text{fermi}^2.$$

At higher energies, en scattering cannot be measured directly but must be inferred, with the aid of theoretical models, by combining ep and ed data.

The scattering data collected so far are consistent with the Rosenbluth formula, which means that the one-photon-exchange approximation holds. Experimental results for one of the form factors are shown in Fig. 12.2. Very surprisingly, in the range so far explored ($-t \lesssim 25\ \text{GeV}^2 \simeq 600\ \text{fermi}^{-2}$) the experimental points satisfy the relations

$$G_E{}^{\mathrm{p}}(t) \simeq \frac{G_M{}^{\mathrm{p}}(t)}{2.79} \simeq -\frac{G_M{}^{\mathrm{n}}(t)}{1.91} \simeq G(t), \tag{12.17a}$$

where

$$G(t) = \left[1 + \frac{(-t)}{0.71\ \text{GeV}^2}\right]^{-2}. \tag{12.7b}$$

This *dipole fit* has been so successful that the data are now usually presented in terms of the deviations from it. [$G_E{}^{\mathrm{n}}(t)$ is small and only poorly determined by the data.] Equation (12.17b) is the sort of thing one gets from an exponential distribution of charge and magnetization,

$$\rho(r) \propto e^{-ar}.$$

However, it is very mystifying from the point of view of the usual theoretical analysis of form factors, to which we now turn.

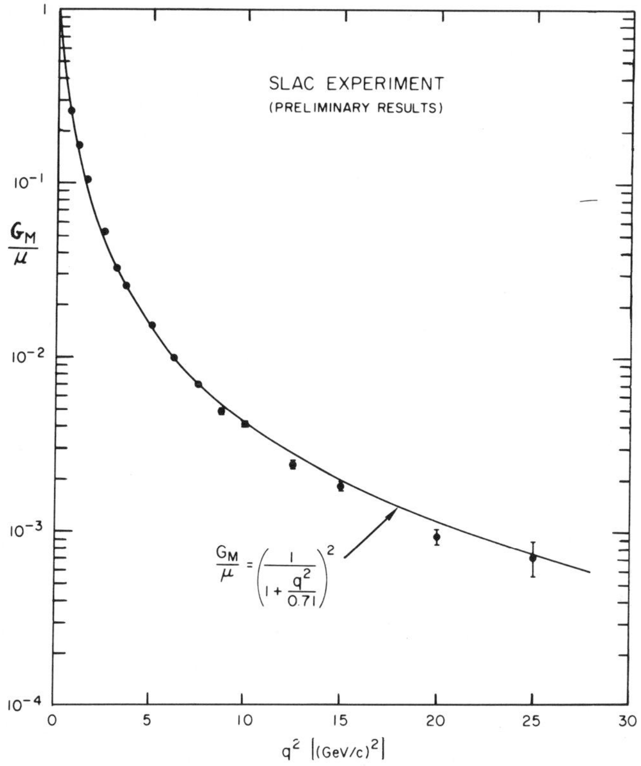


**Fig. 12.2** Experimental data on $G_M{}^p(t)$. [From W. K. H. Panofsky, Electromagnetic interactions. In *Proceedings of the Heidelberg International Conference on Elementary Particles*, H. Filthuth, ed. North Holland, Amsterdam, 1968.]

## 12.2. Dispersion Relations for Form Factors

Elastic eN scattering defines the form factors for $t < 0$, i.e., $q$ *spacelike* [$t \equiv -q^2 = -(\mathbf{p} - \mathbf{p}')^2$ in the CM system]. If we instead consider the annihilation–creation process $e\bar{e} \to N\bar{N}$ and *timelike* $q$, we have just the same Feynman graphs (Fig. 12.1) and perturbation theory contributions (except for two of the spinors for which $u' \to v$). This leads us to expect the *same* form factors (functions of $t$) at the photon–nucleon vertex, but for a different range of $t$. The physical region is now $t \geq 4M^2$. However, in the theoretical analysis of the form factors, an important role is played by the *unphysical region* $0 < t < 4M^2$. To see how this comes about, we first discuss analytic properties in general, and then their application to the nucleon form factors.

***a. Analyticity. Dispersion Relations***

The basic idea to be exploited is that there exists a (complex) function of *complex variable* $z$, $F(z)$, which for $z$ real and negative is a scattering form factor, and for $z$ real and $>4M^2$ is the corresponding pair creation form factor. (This is an expression of crossing symmetry, to be discussed shortly.) Of course, in our ignorance, we usually do not know the explicit form of this function and must therefore proceed on the basis of general properties.

One of the most important of these is the *domain of analyticity*, i.e., the region of the complex $z$-plane that is free of *singularities* (poles, branch points, etc.). While there have been rigorous derivations of analyticity domains from relativistic quantum field theory, one more often relies on perturbation theory for this. Indeed, perturbation theory seems to be quite reliable for giving the *singularity structure*, is more generous as far as the size of the analyticity domain is concerned, and is very much simpler than the rigorous theory. For the form factors, perturbation theory tells us that the domain of analyticity is the entire $z$-plane except for the *positive real axis*.

It is important to realize that once the function is specified along a line (within the analyticity domain) it is determined, by a process of *analytic continuation*,[1] throughout the analyticity domain. Thus, in principle, the form factor can be continued from the physical scattering region—the negative $z$-axis—throughout the $z$-plane (cut along the positive real axis). This region is known as the *physical sheet*. Furthermore, the physical pair creation form factor is obtained by approaching the positive real axis from *above*. (This prescription can be shown to follow from the $m^2 \to m^2 - i\varepsilon$ prescription for Feynman propagators.)

Now since the form factor $F(t)$ is real for (real) negative $t$, its analytic continuation automatically satisfies

$$[F(z)]^* = F(z^*). \tag{12.18}$$

This property is often called *Hermitian analyticity*. Suppose that

(A) $F(z) \to 0$, as $|z| \to \infty$.

Then from Cauchy's theorem,

$$F(z) = \frac{1}{2\pi i} \int_C \frac{F(u)\, du}{u - z}. \tag{12.19}$$

---

[1] For example, it can be determined by power series expansion. But such analytic continuation requires that the function be known along the line *with arbitrary accuracy*—that is the rub, as far as using experimental data is concerned.

The contour $C$, shown in Fig. 12.3, has been chosen so that $F(u)$ is analytic inside and on it. Condition (A) implies that the contribution of the circular part of the contour vanishes as its radius $\to\infty$. Hence we use contour $C'$ (Fig. 12.4) and write

$$F(z) = \frac{1}{2\pi i}\int_0^\infty \frac{[F(u+i\varepsilon) - F(u-i\varepsilon)]}{u-z}\,du, \tag{12.20}$$

with $\varepsilon \to 0^+$. Then, using Eq. (12.18), the *discontinuity* of $F(u)$ is given by

$$\text{discontinuity} \equiv [F(u+i\varepsilon) - F(u-i\varepsilon)] = 2i\,\text{Im}\,F(u+i\varepsilon) \equiv 2i\,\text{Im}\,F(u). \tag{12.21}$$

Hence Eq. (12.20) becomes

$$F(z) = \frac{1}{\pi}\int_0^\infty \frac{\text{Im}\,F(u)}{u-z}\,du. \tag{12.22}$$

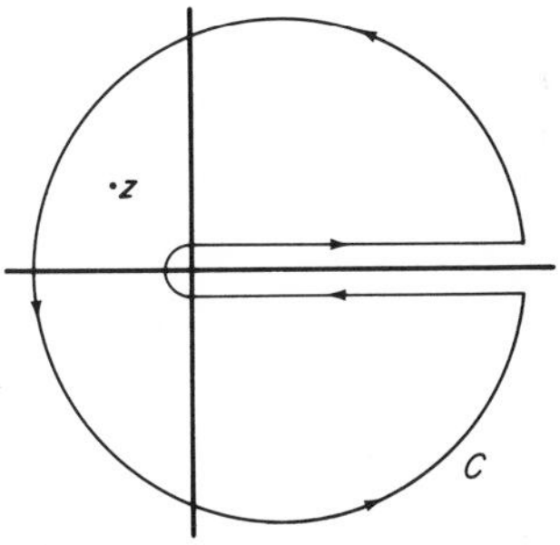


**Fig. 12.3.** The contour $C$.

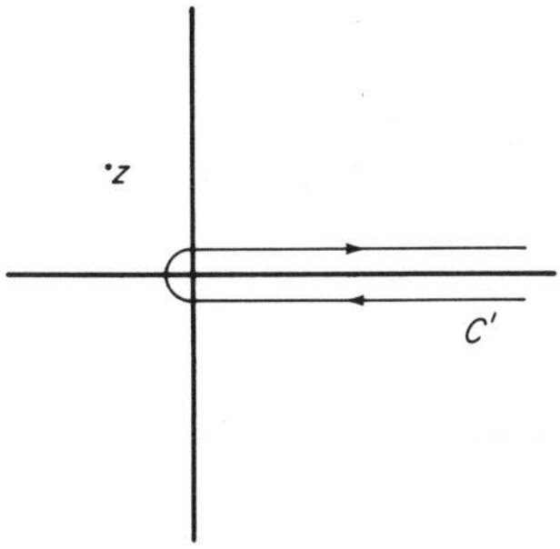


**Fig. 12.4.** The contour $C'$.

Hence we have the function $F(z)$ being given *anywhere* in the $z$-plane by an integral of Im $F$ over a line. Equation (12.22) is known in mathematics as a *Hilbert transform* and in physics as a *dispersion relation*, and the Im $F(u)$ in the integrand is the *spectral function*.[2] Analyticity and the existence of such dispersion relations can be shown in many cases to follow from causality.

If instead of condition (A) we have the weaker condition

(B) $F(z)/z \to 0$, as $|z| \to \infty$,

we can go through the same procedure for the function

$$[F(z) - F(z_0)]/(z - z_0)$$

and arrive at the (once-) *subtracted* dispersion relation

$$F(z) = F(z_0) + \frac{z - z_0}{\pi} \int_0^\infty \frac{\operatorname{Im} F(u)}{(u - z_0)(u - z)}\, du. \tag{12.23}$$

### *b. Singularities. Discontinuities*

Before analyzing the dispersion integral of Eq. (12.22) more closely, we point out that perturbation theory yields a lot more than the fact that the singularities of $F(z)$ are confined to the positive real axis; it also tells us the *position* and *nature* of the singularities. For example, consider the Feynman graphs shown in Fig. 12.5. Analysis of the corresponding perturbation terms

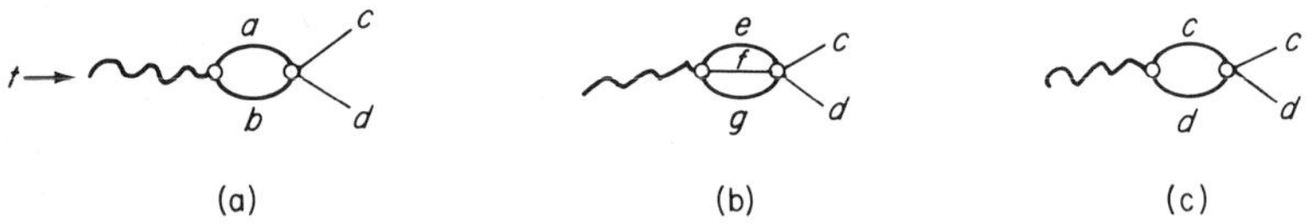


**Fig. 12.5.** Graphs giving rise to thresholds in $t$: (a) two-particle threshold; (b) three-particle threshold; (c) elastic threshold.

shows that (1) they have *branch points* at the *intermediate particle thresholds*,

$$t_1 = (m_a + m_b)^2, \qquad t_2 = (m_e + m_f + m_g)^2, \qquad t_3 = (m_e + m_d)^2,$$

etc., and (2) the discontinuities across the corresponding *branch cuts* are obtained by replacing, in the intermediate line propagators, $-i(k^2 + m^2 - i\varepsilon)^{-1}$ by $2\pi\delta_+(k^2 + m^2)$. This replacement puts the intermediate lines *on the mass shell* and converts the closed loop integration(s) to an integration over *real intermediate states*. Because it exactly reproduces the statement of unitarity for scattering amplitudes, (2) is often known as *generalized unitarity*.

To see how (2) works, we write the perturbation theory term corresponding to Fig. 12.5(a) in the following form.

---

[2] An alternative version for real $x$, $\operatorname{Re} F(x) = (1/\pi)P\int_0^\infty [\operatorname{Im} F(u)/(u - x)]\, du$, is obtained from Eq. (11.22) by letting $z = x + i\varepsilon$, $\varepsilon \to 0^+$. $P$ means *principal value*.

$$F_{ab}(t) = i \int f_{\gamma ab} \frac{-i}{k_a^{\,2} + m_a^{\,2} - i\varepsilon} \frac{-i}{k_b^{\,2} + m_b^{\,2} - i\varepsilon} f_{ab,\,cd}$$

$$\times (2\pi)^4 \delta(k_a + k_b - q) \frac{d^4k_1}{(2\pi)^4} \frac{d^4k_2}{(2\pi)^4}.$$

The functions $f_{\gamma ab}$ and $f_{ab,\,cd}$ correspond to the blobs of Fig. 12.5(a) and are the perturbation theory $\gamma \to ab$ and $ab \to cd$ amplitudes. Then for the discontinuity across the *ab* branch cut

$$[F(t)]_{ab} = i \int f_{\gamma ab} f_{ab,\,cd} (2\pi)^4 \delta(k_a + k_b - q) 2\pi\delta_+ (k_a^{\,2} + m_a^{\,2}) \frac{d^4k_a}{(2\pi)^4}$$

$$\times\, 2\pi\delta_+(k_b^{\,2} + m_b^{\,2}) \frac{d^4k_b}{(2\pi)^4}, \tag{12.24}$$

which exhibits the familiar phase space structure [c.f. Eq. (3.72′)]. But in line with the *S*-matrix theory point of view, now we say that Eq. (12.24) (and those like it) also holds for the *actual* functions, $F, f_{\gamma ab}$, and $f_{ab,\,cd}$, not merely for the perturbation theory terms. The only alteration required is to replace $f_{\gamma ab}$ by $f^*_{\gamma ab}$.[3]

The cuts attached to these branch points can be arranged as shown in Fig. 12.6. Now the dispersion integral of Eq. (12.22) calls for the discontinuity $[F(u + i\varepsilon) - F(u - i\varepsilon)]$, which includes a contribution from *each* cut whose branch point lies to the left of $u$ (cf. Fig. 12.6). Thus the discontinuity includes the effects of *all* possible real intermediate states (with the correct quantum numbers—otherwise one of the blob functions will vanish). One-particle

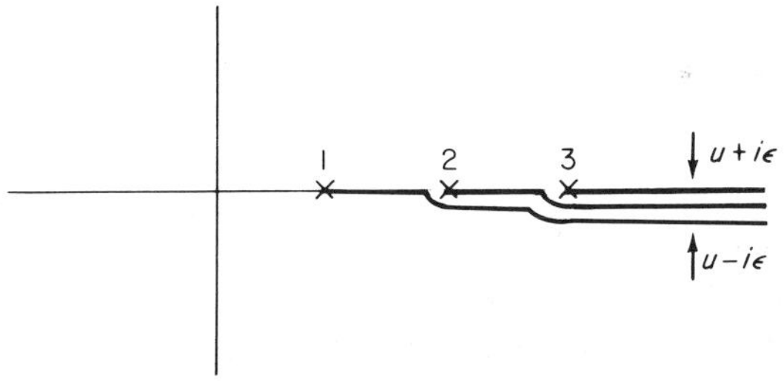


**Fig. 12.6.** Cuts attached to the threshold branch points.

[3] This is because now we are dealing with *interacting* states, and it is necessary to distinguish between states which are composed of plane waves in the remote past (*in* states) and those which have plane waves in the remote future (*out* states). Since $ab \to cd$ means *in* states for *ab*, we want the sum over states in Eq. (12.24) to be over *in* states. However, $f_{\gamma ab}$ means *out* states for *ab*, so instead we want $f^*_{\gamma ab}$ which has the *ab* going in.

states can also contribute (Fig. 12.7) but give rise to *poles* instead of branch points; for stability, the particle mass must be smaller than the lowest multiparticle threshold. These poles contibute $\delta$-functions to Im $F$. Then, except for the points corresponding to the poles, Im $F$ *vanishes* below the lowest multiparticle threshold ($\Delta$), and Eq. (12.22) becomes (for a single pole)

$$F(z) = \frac{A_i}{m_i^2 - z} + \int_\Delta^\infty \frac{\text{Im } F(u)}{u - z}\, du. \qquad (12.22')$$

In effect, the contour in Fig. 12.4 becomes that of Fig. 12.8.

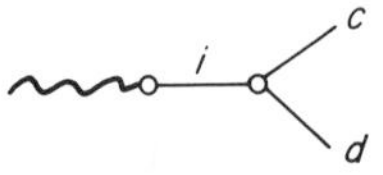

**Fig. 12.7.** Pole contribution graph.

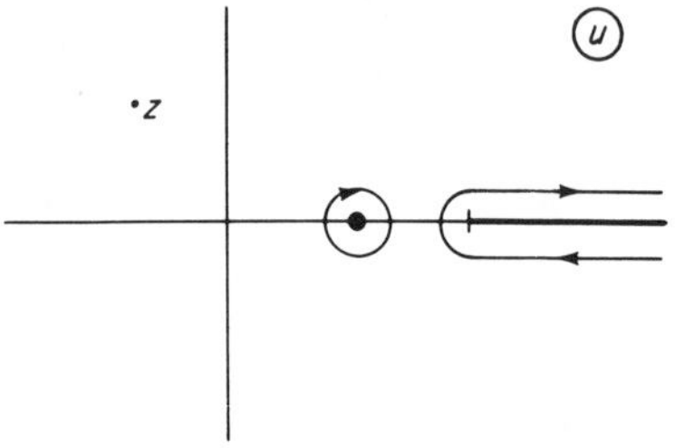

**Fig. 12.8.** Contour for a single pole below the lowest threshold.

### c. *Application to Nucleon Form Factors*

Here multipion intermediate states are expected to make the most important contributions to Im $F$. The analysis is simplified by using isospin. Because there is only *one* photon line attached to the hadrons, and

$$Q = T_3 + Y/2,$$

we write

$$F_i = F_i^{\text{s}} + \tau_3 F_i^{\text{v}},$$

where $F_i^{\text{s}}$ and $F_i^{\text{v}}$, the *isoscalar* and *isovector* form factors, are isospin invariant. By applying this expression to proton and neutron isospinors, we see that

$$F_i^{\text{p}} = F_i^{\text{s}} + F_i^{\text{v}}, \qquad F_i^{\text{n}} = F_i^{\text{s}} - F_i^{\text{v}}.$$

Since the photon is odd under charge conjugation, the isoscalar form factor has $G$-parity $(-1)$, while the isovector must have $G = +1$. Thus Im $F_i^{\text{s}}$ will have contributions from *odd* numbers of $\pi$'s only, and Im $F_i^{\text{v}}$ *even*, so that

$$F_2^{\text{s}}(z) = \frac{1}{\pi} \int_{9m_\pi^2}^\infty \frac{\text{Im } F_2^{\text{s}}(u)}{u - z}\, du, \qquad (12.25\text{a})$$

and

$$F_2^{\mathrm{v}}(z) = \frac{1}{\pi}\int_{4m_\pi^2}^{\infty} \frac{\mathrm{Im}\, F_2^{\mathrm{v}}(u)}{u-z}\, du. \qquad (12.25\mathrm{b})$$

For the $F_1$'s, on the other hand, it is usual to write the dispersion relations in subtracted form [with $z_0 = 0$; using Eq. (12.23) and $F_1^{\mathrm{p}}(0) = 1$, $F_1^{\mathrm{n}}(0) = 0$]:

$$F_1^{\mathrm{s}}(z) = \frac{1}{2} + \frac{z}{\pi}\int_{9m_\pi^2}^{\infty} \frac{\mathrm{Im}\, F_1^{\mathrm{s}}(u)}{u(u-z)}\, du, \qquad (12.26\mathrm{a})$$

$$F_1^{\mathrm{v}}(z) = \frac{1}{2} + \frac{z}{\pi}\int_{4m_\pi^2}^{\infty} \frac{\mathrm{Im}\, F_1^{\mathrm{v}}(u)}{u(u-z)}\, du. \qquad (12.26\mathrm{b})$$

This amounts to attempting to predict the static anomalous magnetic moments of the nucleons, but not the total charges.

Focusing on the isovector form factors and neglecting the $4\pi$, $6\pi$, etc., intermediate states, one can infer from Fig. 12.9 that [cf. Eq. (12.24)]

$$\mathrm{Im}\, F_i^{\mathrm{v}}(t) \propto \left(\frac{t-4m_\pi^2}{t}\right)^{1/2} f_{\gamma\pi\pi}^*(t) f_{\pi\pi,\,\mathrm{NN}}^{\mathrm{i}}(t). \qquad (12.27)$$

**Fig. 12.9.** Lowest mass contributions to the nucleon form factors: (a) isovector; (b) isoscalar.

Here $[(t-4m_\pi^2)/t]^{1/2}$ comes from the two-body phase space factor [cf. Eq. (3.95]) $f_{\gamma\pi\pi}(t)$ is the *pion* electromagnetic isovector form factor, and the $f_{\pi\pi,\,\mathrm{NN}}^{\mathrm{i}}(t)$ are $J^P = 1^-$, $T = 1$, $\pi\pi \to \mathrm{N}\bar{\mathrm{N}}$ scattering amplitudes (related by crossing symmetry to the $\pi\mathrm{N} \to \pi\mathrm{N}$ amplitudes). The simplest assumption, $f_{\gamma\pi\pi} =$ constant (corresponding to a point $\pi$), was found not to work, because the remaining factors increase smoothly with $t$ [Fig. 12.10(a)]; this results in a much more gradual $t$-dependence of $F_i^{\mathrm{v}}(t)$ for negative $t$ than is observed experimentally.

In order to get the required peak in the spectral function $\mathrm{Im}\, F_i^{\mathrm{v}}(t)$, Frazer and Fulco in 1959 proposed the existence of a $J^P = 1^-$, $T = 1$, $2\pi$ resonance (the $\rho$, which has the required properties, was discovered later), which enters by way of $f_{\gamma\pi\pi}$. To see how, consider that $f_{\gamma\pi\pi}$ also satisfies a dispersion relation:

$$f_{\gamma\pi\pi}(z) = 1 + \frac{z}{\pi}\int_{4m_\pi^2}^{\infty} \frac{\mathrm{Im}\, f_{\gamma\pi\pi}(u)}{u(u-z)}\, du, \qquad (12.28)$$

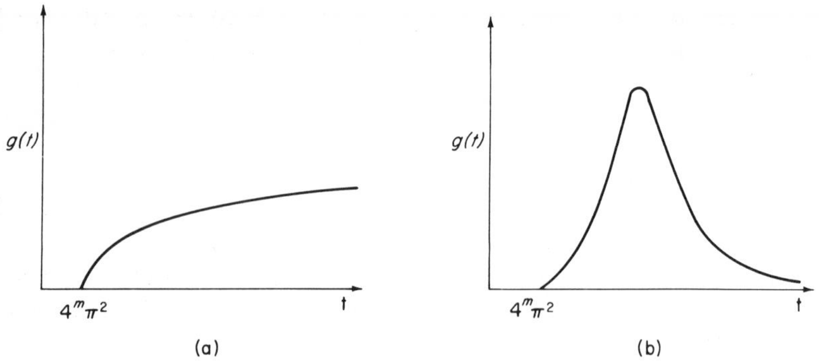


**Fig. 12.10.** Spectral function (a) for $f_{\gamma\pi\pi} = \text{const}$; (b) with a contributing $\pi\pi$ resonance. [From S. Bergia and A. Stanghellini, Electromagnetic formfactors of the nucleon and pion–pion interaction. *Phys. Rev. Lett.* **6,** 367–371 (1961).]

and the spectral function in this case is given by

$$\operatorname{Im} f_{\gamma\pi\pi} \propto \left(\frac{t - 4m_\pi^{\,2}}{t}\right)^{1/2} f^*_{\gamma\pi\pi} f_{\pi\pi,\,\pi\pi}\,, \tag{12.29}$$

where $f_{\pi\pi,\,\pi\pi}$ is the $J^P = 1^-$, $T = 1$, $\pi\pi$ elastic scattering amplitude. The resonance maximum in $f_{\pi\pi,\,\pi\pi}$ works its way back through Eqs. (12.29) and (12.28) to produce the peak in Im $F_i^{\,\mathrm{v}}$ [Fig. 12.10(b)].

## d. Vector Meson Dominance

A simple approximate method of including the effect of such a vector meson resonance is to neglect its width, i.e., treat it like a stable particle. The resonance contribution can then be pictured as in Fig. 12.11(a); this means a pole in $F_i^{\,\mathrm{v}}$ and a $\delta$-function in the spectral function, so that

$$F_1^{\,\mathrm{v}}(t) \simeq \frac{1}{2}\left[1 + \frac{a_1}{m_\rho^{\,2}}\,\frac{t}{m_\rho^{\,2} - t}\right], \tag{12.30a}$$

$$F_2^{\,\mathrm{v}}(t) \simeq \frac{a_2}{2}\,\frac{1}{m_\rho^{\,2} - t}. \tag{12.30b}$$

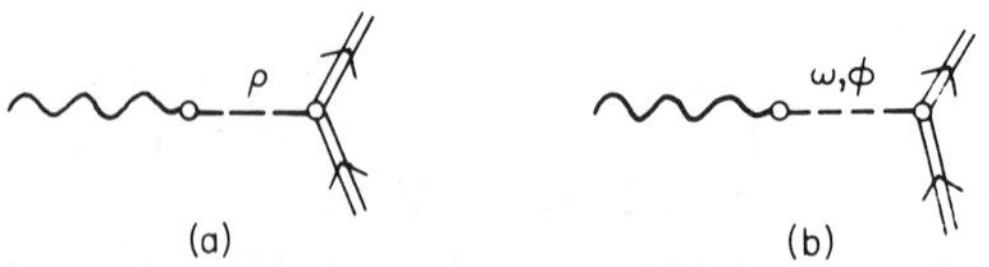


**Fig. 12.11.** Vector meson contributions to (a) isovector, (b) isoscalar form factors.

The constants $a_1$ and $a_2$ depend on the $\gamma$–$\rho$ and $\rho$–N couplings and must be adjusted empirically. In terms of the $G$'s, these equations become

$$G_E^{\text{v}}(t) \simeq \frac{1}{2}\left(1 - a + \frac{a}{1 - t/m_\rho^2}\right), \tag{12.31a}$$

$$G_M^{\text{v}}(t) \simeq \frac{6}{2}\left(1 - c + \frac{c}{1 - t/m_\rho^2}\right). \tag{12.31b}$$

Similarly, the *isoscalar* form factors might be dominated by the $\omega$ and $\phi$ $3\pi$ resonances, leading to analogous diagrams [Fig. 12.11(b)] and formulas for $G_E^{\text{s}}$, $G_M^{\text{s}}$;

$$G_E^{\text{s}}(t) \simeq \frac{1}{2}\left[1 - a' - a'' + \frac{a'}{1 - t/m_\omega^2} + \frac{a''}{1 - t/m_\phi^2}\right], \tag{12.32a}$$

$$G_M^{\text{s}}(t) \simeq \frac{b'}{2}\left[1 - c' - c'' + \frac{c'}{1 - t/m_\omega^2} + \frac{c''}{1 - t/m_\omega^2}\right]. \tag{12.32b}$$

However, these formulas provide an *accurate* fit to the electron scattering data only if certain liberties are taken with them; for example, it is usual to put a second pole into the isovector form factors (corresponding to a non-existent (?) $\rho'$ meson, and to adjust $b$ and $c$ independently in $G_M^{\text{v}}$. Such fits really do not tell us much more than that the vector mesons play a role in the form factors, but are not the whole story.

More interesting data about what is going on in the $t > 4m^{\pi 2}$ region is now beginning to emerge from clashing $e^+e^-$ beam experiments and will no doubt play an important role in future discussions of the form factors in the timelike region.

## 12.3. Dispersion Relations for Scattering Amplitudes

Scattering amplitudes, since they depend on (at least) *two* variables, energy and scattering angle or $s$ and $t$, are harder to analyze than form factors. Dispersion relations were first derived for forward elastic scattering ($t = 0$) (*forward dispersion relations*); these have been very useful for analyzing $\pi$N scattering, for example.

Many other applications employ *partial wave dispersion relations*, i.e., dispersion relations for $f_l(W)$, which are comparatively easy to handle. Crossing symmetry, on the other hand, is exhibited most clearly in the *double dispersion relations* introduced by Mandelstam.

Here we will limit ourselves to the basic ideas, and begin by discussing crossing symmetry and analyticity as applied to "two in–two out" scattering.

### a. Crossing Symmetry

For simplicity, we consider spinless particles so that reaction (I) $ab \to cd$ (called the $s$-channel, since $s$ is here the energy variable) is described by a single invariant amplitude $T(s, t)$; as usual (Fig. 12.12)

$$s = -(p_a + p_b)^2, \qquad t = -(p_a + p_c)^2, \qquad u = -(p_a + p_d)^2, \tag{1.24}$$

and

$$s + t + u = m_a^2 + m_b^2 + m_c^2 + m_d^2. \tag{1.26}$$

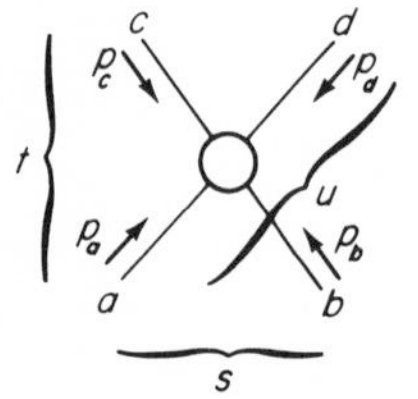


**Fig. 12.12.** Kinematic variables for two-body scattering.

Then crossing symmetry says that reactions (II) $a\bar{c} \to \bar{b}d$ ($t$-channel) and (III) $a\bar{d} \to \bar{b}c$ ($u$-channel) are described by the same function $T(s, t)$, but with different values for the arguments $s$ and $t$.[4] To simplify the kinematics, we let all the particles have the same mass $m$, Then the physical region for (I) $ab \to cd$ is

$$\text{I:} \quad s \geq 4m^2, \qquad 4m^2 - s \leq t \leq 0, \qquad 0 \geq u \geq 4m^2 - s,$$

while those for (II) and (III) are

$$\text{II:} \quad 4m^2 - t \leq s \leq 0, \qquad t \geq 4m^2, \qquad 0 \geq u \geq 4m^2 - t,$$

and

$$\text{III:} \quad 4m^2 - u \leq s \leq 0, \qquad 0 \geq t \geq 4m^2 - u, \qquad u \geq 4m^2.$$

The physical regions are shown in Fig. 12.13. In order to be able to use crossing symmetry, we must be able to make an analytic continuation from region I to region II or III.

If we have an *explicit* expression for $T$, then all we have to do is to substitute in the new values of $s$ and $t$. For example, if we have just the pole term corresponding to Fig. 12.14, then

$$T(s, t) \simeq gg'/(t - m_e^2 + i\varepsilon) \tag{12.33}$$

holds in all three regions.

---

[4] Both time-reversal invariance and the *CPT* theorem are assumed to hold in the strong interactions. Time-reversal relates (I) to $cd \to ab$, etc., while the *CPT* theorem relates (I) to $\bar{c}\bar{d} \to \bar{a}\bar{b}$, etc., both at the *same* $s$, $t$ values.

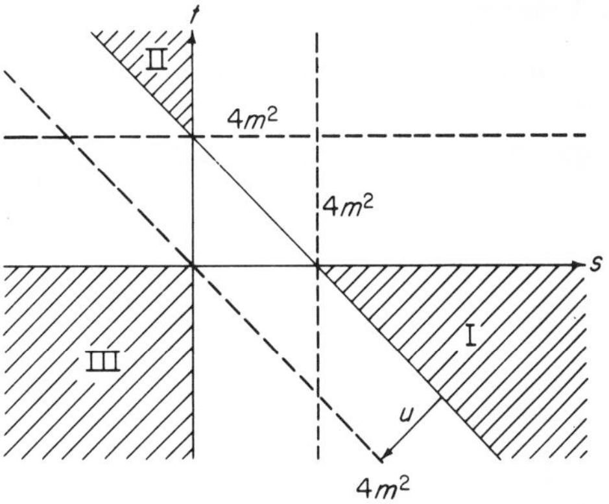


**Fig. 12.13.** The three physical regions.

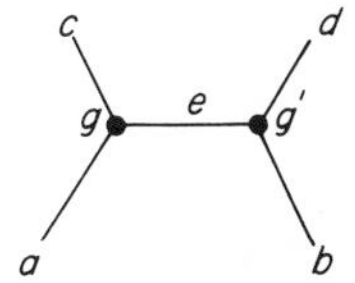

**Fig. 12.14.** Single-particle exchange contribution to $T(s, t)$.

Sometimes it happens that two of the processes I, II, and III are really the same; then crossing symmetry places a *constraint* on the amplitude. For instance, suppose $\bar{b} = c$. Then processes I and II are the same, and since the exchange $s \rightleftarrows t$ takes us from a point in region I to the equivalent point in region II, we must have the same value for the amplitude, i.e.,

$$T(s, t) = T(t, s), \tag{12.34}$$

which is symmetry with respect to the line $s = t$ in Fig. 12.13. So here, crossing symmetry is seen to be a generalization of Bose symmetry. Equation (12.33) does not possess this symmetry, but that is fair enough, since $\bar{b} = c$ means that instead of just the graph of Fig. 12.14. we should have the *two* graphs of Fig. 12.15. Hence

$$T(s, t) \simeq gg'[(t - m_e{}^2 + i\varepsilon)^{-1} + (s - m_e{}^2 + i\varepsilon)^{-1}], \tag{12.35}$$

which has the required symmetry.

If I and III are the identical channels ($\bar{b} = d$), then writing

$$T(s, t, u) \equiv T(s, t)$$

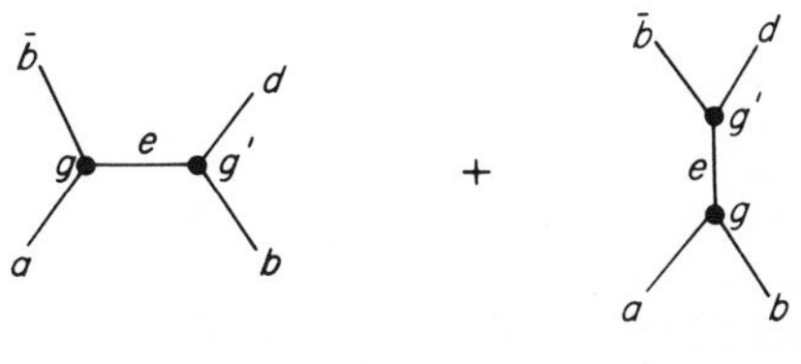


**Fig. 12.15.** Single-particle contributions when $c = \bar{b}$.

we have

$$T(s, t, u) = T(u, t, s), \tag{12.36}$$

or

$$T(s, t) = T(4m^2 - s - t, t), \tag{12.36'}$$

since I → III means $s \rightleftarrows u$. [To exhibit the resulting symmetry around the line $s = u$ graphically, we would have to change Fig. 12.13 to a triangular plot, as in Fig. 8.5(b)]

### b. Analyticity

Let us now look closely at the analytic continuation involved in crossing. The general statement (which follows from quantum field theory) is that the physical scattering amplitudes in the three physical regions are *boundary values* of a single complex function (of two independent variables) $T(s, t) \equiv T(s, t, u)$. This function has (at least) physical branch cuts starting at the ($4m^2$) threshold in each of the three variables, and the proper boundary value is chosen when a given energy variable approaches its physical cut from *above*. Thus for the three physical amplitudes ($s$, $t$, $u$ real, $\lim_{\varepsilon \to 0^+}$ understood)

$$T_{\mathrm{I}}(s, t, u) = T(s + i\varepsilon, t, u - i\varepsilon),$$

$$T_{\mathrm{II}}(s, t, u) = T(s - i\varepsilon, t + i\varepsilon, u),$$

$$T_{\mathrm{III}}(s, t, u) = T(s, t - i\varepsilon, u + i\varepsilon).$$

The *negative* imaginary parts, which have here been inserted arbitrarily to satisfy the constraint $s + t + u = 4m^2$, are not significant, because in each physical region $T(s, t, u)$ is *analtyic* (no cut) in whichever momentum transfer variable is taken to be the independent variable.

A path in the complex $s$-plane (for fixed real negative $t$) which takes us from the $s$-channel (I) physical amplitude to that of the $u$-channel (III) is shown in Fig. 12.16. The *left-hand cut* in $s$ occurs because $s$ real, $\leq -t$, implies $u$ real, $\geq 4m^2$; $s$ must approach the left-hand cut from *below* in order that $u$

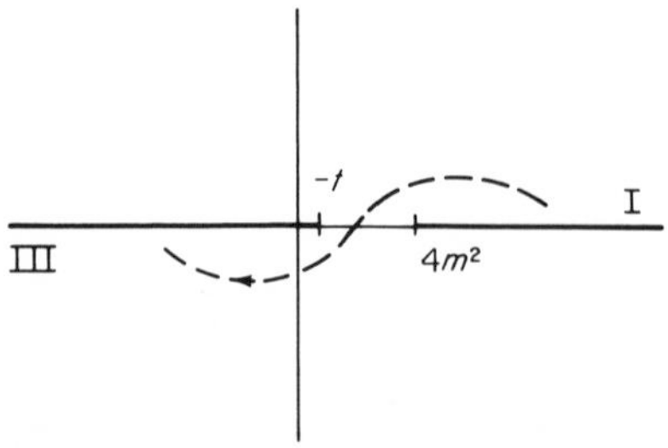


**Fig. 12.16.** Path in $s$-plane to get from region I to region III.

approach its physical cut from above. Clearly, the existence of a gap between the two cuts is necessary for making the analytic continuation and hence for using the crossing property. Furthermore, $T(s, t, u)$ can be shown to be real in the gap (real $s$ as well as $t$), so that $T$ has the Hermitian analyticity property: $T^*(s, t, u) = T(s^*, t, u^*)$ (complex $s$, real $t$).

Now if $\bar{b} = d$, so that I and III are the same process, we must have (real $s, t, u$)

$$T_{\text{I}}(s, t, u) = T_{\text{III}}(u, t, s).$$

Hence

$$T(s + i\varepsilon, t, u - i\varepsilon) = T(u - i\varepsilon, t, s + i\varepsilon),$$

so that even for complex $s, t, u$

$$T(s, t, u) = T(u, t, s). \tag{12.36}$$

### c. *Mandelstam Representation*

In 1958 Mandelstam introduced a very convenient and symmetric way of exhibiting the analyticity and crossing properties of $T$. This *Mandelstam representation* can be written

$$\begin{aligned} T(s, t, u) = {} & \frac{A}{s - m_e^2} + \frac{B}{t - m_f^2} + \frac{C}{u - m_g^2} \\ & + \int_{4m^2}^{\infty} \frac{\rho_1(s')}{s' - s}\, ds' + \int_{4m^2}^{\infty} \frac{\rho_2(t')}{t' - t}\, dt' + \int_{4m^2}^{\infty} \frac{\rho_3(u')}{u' - u}\, du' \\ & + \iint_{4m^2}^{\infty} \frac{\rho_{12}(s', t')}{(s' - s)(t' - t)}\, ds'\, dt' + \iint_{4m^2}^{\infty} \frac{\rho_{23}(t', u')}{(t' - t)(u' - u)}\, dt'\, du' \\ & + \iint_{4m^2}^{\infty} \frac{\rho_{31}(u', s')}{(u' - u)(s' - s)}\, du'\, ds'. \end{aligned} \tag{12.37}$$

(Subtractions can complicate this expression.) Here we have assumed a single bound state (or stable particle) pole in each channel, as shown in Fig. 12.17 ($m_e^2$, $m_f^2$, $m_g^2$

**Fig. 12.17.** Pole contributions to $T$, assuming a single particle in each channel.

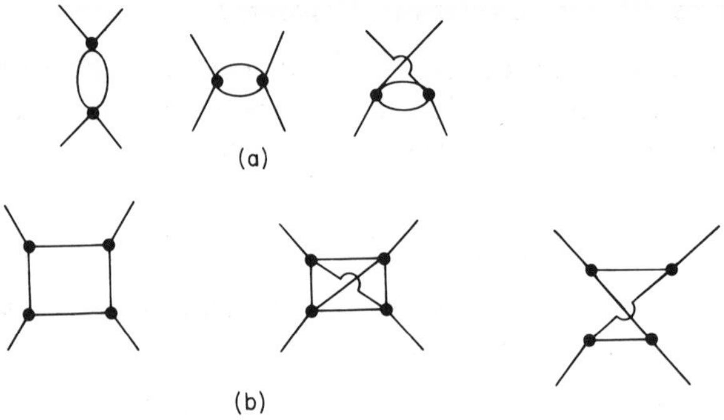


**Fig. 12.18.** Diagrams contributing to (a) the single integrals of Eq. (12.37), (b) the double integrals of Eq. (12.37).

$< 4m^2$). The three single integral terms arise in the same way as for the form factors, and correspond for example to the diagrams of Fig. 12.18(a) (assuming no lighter pairs of particles contribute). The double integral terms come from diagrams like those shown in Fig. 12.18(b), and more complicated ones. The *double spectral functions* $\rho_{12}(s', t')$ etc., are nonvanishing in the regions sketched in Fig. 12.19.

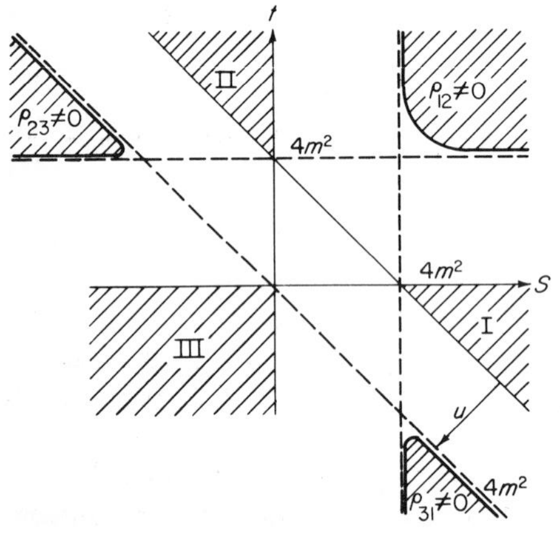


**Fig. 12.19.** Regions where the double spectral functions do not vanish.

By using generalized unitarity, one can express the various spectral functions in terms of amplitudes for getting to the particular real intermediate states involved, exactly as discussed for the form factors. Similarly, the residue of the poles are simply products of coupling constants:

$$A = g_{abc}\, g_{cde}\,, \quad \text{etc.}$$

Even with the neglect of the more complicated diagrams, the resulting nonlinear integral equations are sufficiently difficult so that one is not sure whether any or many solutions exist.

***d. Partial Wave Dispersion Relations. Bootstraps***

The Mandelstam representation is often used in deriving single variable and partial wave dispersion relations. For the latter, one forms the partial wave projection of $T(s, t)$ (in a particular channel):

$$T_l(s) = \int_{-1}^{1} d(\cos\theta) T(s, t) P_l(\cos\theta). \tag{12.38}$$

Because of the integration, the poles (and branch points) in $t$ and $u$ give rise to branch points and cuts in $s$. For example,

$$\int_{-1}^{1} \frac{P_l(\cos\theta)}{t - m_f^2}\, d(\cos\theta) = \int_{-1}^{1} \frac{P_l(x)}{-2\mathbf{k}^2[1 + (m_f^2/2\mathbf{k}^2) - x]}\, dx,$$

which produces a (logarithmic) branch point in $\nu \equiv \mathbf{k}^2 \equiv s/4 - m^2$ when the denominator vanishes at an end point of the integration, i.e., at $\nu = -m_f^2/4$. The branch point furthest to the right and hence closest to the ($s$-channel) physical region, corresponds to the lightest particle that can be exchanged. These *left-hand cuts* can be arranged as shown in Fig. 12.20, and represent the

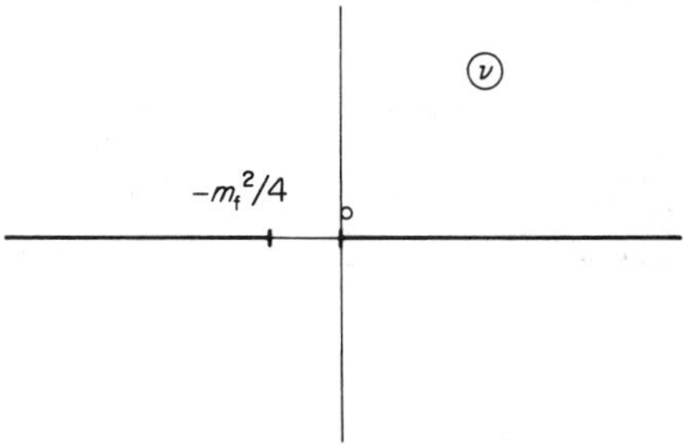


**Fig. 12.20.** Cuts for partial wave dispersion relations.

effect of the exchanged particles or sets of particles. Exactly the same structure appears in potential scattering: the left-hand cut is said to express the forces between the scattering particles.

This approach is widely useful; one of its uses is to formulate the *bootstrap* idea. The simplest example of this is the $\pi\pi$ system. Here the exchange of the $\rho$ contributes a strong attractive force to the $J = 1$, $T = 1$ $\pi\pi$ system, which gives rise to the $\rho$ as a $\pi\pi$ resonance (Fig. 12.21). Using the partial wave

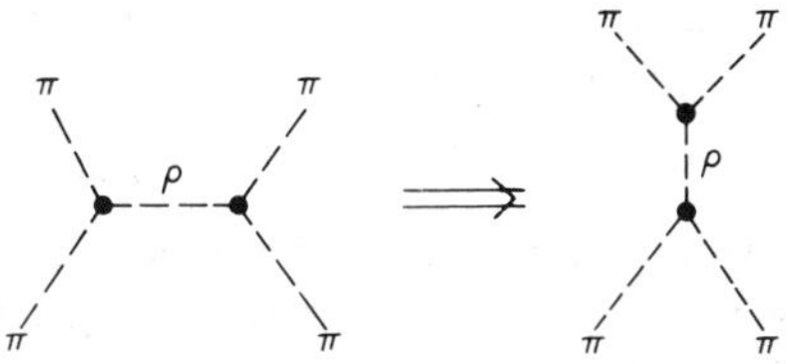


**Fig. 12.21.** Bootstrap of the $\rho$.

dispersion relation (neglecting higher order thresholds and exchanges) and "input "values of $m_\rho$ and $g^2_{\rho\pi\pi}$, ideally one can calculate the $\pi\pi$ scattering and hence "output" values of $m$ and $g^2_{\rho\pi\pi}$ (position and width of the resonance). When input and output match, *self-consistency* has been achieved, and the result can be compared with experiment. Unfortunately, the equations involved are singular (because of the spin of the $\rho$) and a completely quantitative test of the idea has not yet been possible.

Similary, the N*(1236) can be understood as a $\pi$N resonance arising mainly from nucleon exchange, while the nucleon itself is viewed as a $\pi$N bound state arising mainly from N* exchange (Fig. 12.22). Chew and

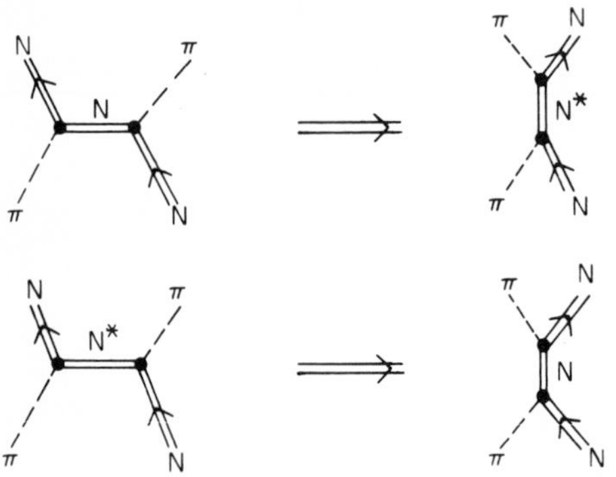


**Fig. 12.22.** Bootstrap of N and N*.

Frautschi have suggested that *all* hadrons are bound states or resonances of other hadrons, and hence that *none* are elementary. Again while the idea seems qualitatively to work, definitive quantitative tests have yet to be achieved.

## 12.4. Regge Poles

The *Regge Pole* idea, and analysis in *complex angular momentum*, is playing a very important role in hadron physics, especially in the analysis of high-energy (greater than several GeV) scattering. This activity was triggered by Regge's analysis (in 1959, 1960) of nonrelativistic potential scattering.

### *a. Regge Poles in Potential Scattering*

The starting point is the partial wave expansion for the scattering of spinless particles:

$$f(\theta) \equiv f(\mathbf{k}, \theta) = 1/2ik \sum_{l=0}^{\infty} (2l + 1)a_l(k)P_l(\cos\theta). \tag{3.56}$$

This expression is rather limited as far as analytic continuation into the complex $z \equiv \cos\theta$ plane is concerned, because the series converges only within an ellipse with foci $\pm 1$. But the $\sum_l$ above can be rewritten as $\int dl$ in the *complex l-plane* (*Watson–Sommerfeld transform*):

$$f(k, \theta) = \frac{i}{2k} \int_C dl \, \frac{(2l + 1)a(l, k)P_l(-z)}{\sin \pi l}, \tag{12.39}$$

with the contour shown in Fig. 12.23. Since $\sin \pi l$ has poles at each integer, Cauchy's theorem gives exactly the $\sum_l$. The function $a(l, k)$ is defined for complex $l$ by the partial wave Schrödinger equation. Then, using properties of

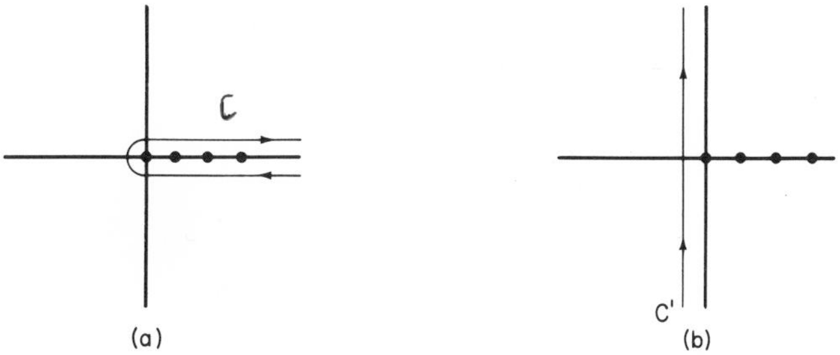


**Fig. 12.23.** (a) $l$-plane contour for Sommerfeld–Watson transform; (b) after swinging the contour.

$a(l, k)$ and $P_l(-z)$, the contour can be distorted as shown in Fig. 12.23(b). In addition to the $\int_{C'}$ (*background integral*) we now have contributions from the poles of $a(l, k)$, at $l = \alpha_n(k)$; these are the Regge poles. Thus

$$f(k, \theta) = \frac{i}{2k} \int_{-\frac{1}{2}-i\infty}^{-\frac{1}{2}+i\infty} dl \, \frac{(2l + 1)a(l, k)P_l(-z)}{\sin \pi l}$$

$$- \frac{\pi}{k} \sum_n \frac{(2\alpha_n + 1)\beta_n(k)P_{\alpha_n}(-z)}{\sin \pi\alpha_n}, \tag{12.40}$$

the $\beta_n(k)$ are the residues of $a(l, k)$ at each Regge pole.

The usefulness of this expression lies in the information it gives about the complex $z$ behavior of the scattering amplitude. For $z \to \infty$, it turns out that

$$\left| \int_{C'} \right| \sim |z|^{-1/2},$$

while $P_\alpha(z) \sim z^\alpha$. Thus $|P_{\alpha_n}(z)| \sim |z|^{\mathrm{Re}\,\alpha_n}$, so that in this limit, the dominant contribution to Eq. (12.40) is the term corresponding to the Regge pole lying furthest to the right (largest Re $\alpha_n$), and we can write

$$f(k, z) \sim g(k) z^{\alpha_1(k)}. \tag{12.41}$$

This fact is very useful for dispersion theory.

As $k$ is varied, each Regge pole $\alpha_n(k)$ moves and traces out a *Regge trajectory*. For $k^2 < 0$ (below threshold), it turns out that Im $\alpha_n(k) = 0$, and if $\alpha_n(k)$ passes through a (nonnegative) integer, we have a pole in $f(k, \theta)$ from the sin $\pi\alpha_n$ denominator; these are the *bound state* poles. For $k^2 > 0$, on the other hand, Im $\alpha_n(k) > 0$. Thus while $|f|$ has a maximum as Re $\alpha_n$ increases through an integer, the poles now are located below the real $k^2$ axis; these are the *resonance* poles. Hence there can be *several* bound states and resonances associated with a *single* Regge trajectory.

### b. Generalization to Relativistic Scattering

Very soon after Regge's work on potential scattering, these ideas were applied to relativistic scattering. For simplicity we consider spinless, equal mass particles. Referring to Eqs. (1.27a) and (1.27b′), we have

$$s = (2E)^2 = 4(k^2 + m^2)$$

and

$$t = -2k^2(1 - \cos\theta) = -\frac{(s - 4m^2)}{2}(1 - z).$$

Then $z \to \infty$ at fixed energy means $t \to \infty$ at fixed $s$ and $z \propto t$. Therefore, in this limit the Regge pole with largest Re $\alpha$ can be expected to dominate the invariant amplitude:

$$A(s, t) \sim G(s)t^{\alpha_1(s)}. \tag{12.42}$$

Now $t \to \infty$ is unphysical for $s$-channel scattering, but for $s \leq 0$ this is just the $t$-channel physical region. Thus high-energy scattering is dominated by the (crossed-channel) Regge pole with largest Re $\alpha$. Similarly, for large $s$ and $t \leq 0$, we expect

$$A(s, t) \sim G(t)s^{\alpha_1(t)}. \tag{12.43}$$

If there are several Regge poles to the right of Re $l = -\frac{1}{2}$, we can include them all and write

$$A(s, t) \sim \sum G_n(t)s^{\alpha_n(t)}. \tag{12.43′}$$

Analogous to the exchange of particles, we now have the scattering amplitude being dominated by the exchange of Regge poles (Fig. 12.24).

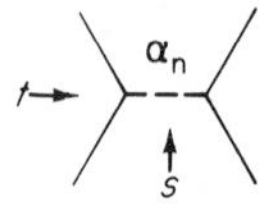


**Fig. 12.24.** Regge pole exchange.

Immediate information on the leading Regge pole for elastic scattering comes from the empirical fact that hadron total cross sections seem to become constants at high enough energy. Then from unitarity [cf. Eq. (3.102′)]

$$A(s, 0) \sim s, \tag{12.44}$$

which implies that the leading trajectory satisfies $\alpha_P(0) = 1$. Because it provides an immediate proof of the Pomeranchuck theorem, this is called the *Pomeranchuck trajectory*, and since it evidently can be exchanged by any hadron pair, it is assigned the quantum numbers (except for angular momentum) of the vacuum. No trajectory can have a greater Re $\alpha(0)$, because this would imply a total cross section increasing as a power of $s$ for large $s$, which leads to a violation of unitarity.

### c. Diffraction Peak

The "classic" prediction of Regge pole theory is the *shrinkage* of *diffraction peaks*. For very high-energy hadron scattering there are many final state channels open, and most of the scattering is into inelastic channels. Thus the incident wave is largely absorbed when the particles come together, and elastic scattering occurs because of this depletion of the incident wave. In effect, we have scattering from a black sphere; hence the term diffraction scattering. Most of the elastic scattering is then confined to the forward diffraction peak.

Let us pursue this idea. The expected angular width of the diffraction peak in the laboratory is simply

$$\Delta\theta \sim \lambda/R,$$

where $\lambda$ is the (lab) wavelength of the projectile and $R$ the sphere radius. This amounts to solid angle $\Delta\Omega_{\text{diff}} \sim \pi\lambda^2/R^2$. Now

$$\lambda = 2\pi/|\mathbf{p}_1{}^{\text{L}}|,$$

and from Eqs. (1.36) and (1.39) for large $s$,

$$(\mathbf{p}_1{}^{\text{L}})^2 \sim \frac{s^2}{4m_2{}^2},$$

$$\cos\theta_{\text{L}} \sim 1 - \frac{2m_2{}^2(t - m_1{}^2 - m_3{}^2)}{su}.$$

Hence,

$$(\Delta\Omega)_{\text{diff}} \sim \frac{4\pi^3}{s^2/4m_2{}^2 \cdot R} = \frac{16\pi^3 m_2{}^2}{s^2 R^2},$$

while

$$d\Omega_{\text{L}} = 2\pi\, d\cos\theta_{\text{L}} \sim (4m_2{}^2/s^2)\, dt,$$

so that the width in $t$ of the diffraction peak is given by

$$(4\pi m_2{}^2/s^2)(\Delta t)_{\text{diff}} \sim 16\pi^3 m_2{}^2/s^2 R^2,$$

or

$$(\Delta t)_{\text{diff}} \sim 4\pi^2/R^2, \tag{12.45}$$

*independent of s.*

To find the Regge pole prediction for diffraction scattering, we first note that at high energies Eq. (3.96b) gives

$$d\sigma/dt \sim s^{-2}\,|A|^2.$$

Then for elastic scattering which is supposed to be dominated by the Pomeranchuck pole,

$$d\sigma/dt \sim f(t)(s/s_0)^{2\alpha_{\text{P}}(t)-2} \tag{12.46}$$

($s_0$ is a scale factor introduced for convenience and usually taken to be 1 GeV). Writing

$$\alpha_{\text{P}}(t) \simeq \alpha_{\text{P}}(0) + t\alpha_{\text{P}}'(0) \qquad [\alpha'(0) \equiv (d\alpha/dt)\,|_{t=0}]$$

we have

$$d\sigma/dt \sim f(t)e^{at}(s/s_0)^{2\alpha_{\text{P}}(0)-2}, \tag{12.47}$$

with

$$a \equiv 2\alpha_{\text{P}}'(0)\ln(s/s_0).$$

Most of the $t$ dependence is in the $e^{at}$, and since experimentally $d\sigma/dt$ decreases with increasing $(-t)$, $\alpha_{\text{P}}'(0)$ must be positive. But then the $\ln(s/s_0)$ in $a$ implies that $d\sigma/dt$ falls off faster as a function of $(-t)$ as $s$ increases, i.e., the diffraction peak narrows as the energy goes up:

$$(\Delta t)_{\text{diff}} \sim a^{-1} \sim [\ln(s/s_0)]^{-1}, \tag{12.48}$$

in contrast to the black sphere predictions.

In actual fact, only PP and $\pi^+$p elastic scattering exhibit this diffraction peak shrinkage. Its failure in all the other cases, can be blamed on other contributing Regge poles, and/or the presence of branch points and cuts in the $l$-plane. Indeed, the relativistic problem turns out to be horrendously more complicated than potential scattering (especially when spins and unequal masses are taken into account), and the complex angular momentum plane seems (from theoretical analysis) to harbor such diverse monsters as branch points, essential singularities, and infinite families of Regge poles. In spite of these unwelcome complications, analysis of data in terms of complex $l$- (or $j$-) plane properties proceeds apace and is remarkably successful.

## d. Signature

One complication that occurs in even the nonrelativistic problem is that of *signature.* That is, when an exchange potential is present (so that, in effect, the potential is different for even and odd $l$) the scattering amplitude becomes a sum of *even* and *odd signature* parts; each of these has its own Regge pole contributions, which have the form [compare Eq. (12.40)]

$$\frac{(2\alpha_n + 1)\beta_n P_{\alpha_n}(-z)}{\sin \pi\alpha_n} \cdot \frac{1}{2}(1 \pm e^{-i\pi\alpha_n}),$$

for $\left\{\begin{smallmatrix}\text{even}\\\text{odd}\end{smallmatrix}\right\}$ signature. The *signature factor,* $\frac{1}{2}(1 \pm e^{-i\pi\alpha_n})$, removes alternate bound state or resonance poles, leaving only the $l$-$\left\{\begin{smallmatrix}\text{even}\\\text{odd}\end{smallmatrix}\right\}$ poles for $\left\{\begin{smallmatrix}\text{even}\\\text{odd}\end{smallmatrix}\right\}$ signature. Thus, for example, the Pomeranchuck trajectory is supposed to be of even signature, so that $\alpha_P(0) = 1$ does *not* predict a physical spin-one zero mass particle.

## e. πN *Charge-Exchange Scattering*

The clearest example of dominance by a single Regge pole is provided by the charge-exchange scattering $\pi^- p \to \pi^0 n$. Here the exchanged object (Fig. 12.25)

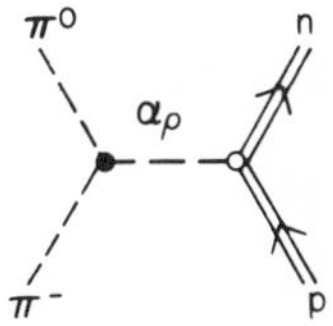


**Fig. 12.25.** $\pi^-$p charge exchange dominated by the $\rho$-trajectory.

must have the quantum numbers of the $\rho$ (except for the spin which is variable), so the trajectory is called $\alpha_\rho(t)$. Then analogous to Eq. (12.46), we have

$$(d\sigma/dt)_{ce} \sim f_{ce}(t)(s/s_0)^{2\alpha_\rho(t)-2}. \tag{12.49}$$

The predicted shrinkage of the forward peak (the argument being exactly analogous to that of Section 12.4d) shows up very clearly, and data can be used to determine $\alpha_\rho(t)$ (for negative $t$). This is shown in Fig. 12.26; the straight-line trajectory gives a good fit both to the scattering and to the $\rho$ mass (positive $t$). To be associated with the $\rho$, the trajectory must have odd signature.

A similar fit to $(d\sigma/dt)_{ce}$ is shown in Fig. 12.27. The dip at $t = -0.6$ GeV$^2$, which corresponds to $\alpha_\rho = 0$ (*wrong signature point*), is additional evidence for the Regge pole model. The dip occurs because *one* of the two charge exchange amplitudes contains a factor of $\alpha_\rho$, so it vanishes at $\alpha_\rho = 0$.

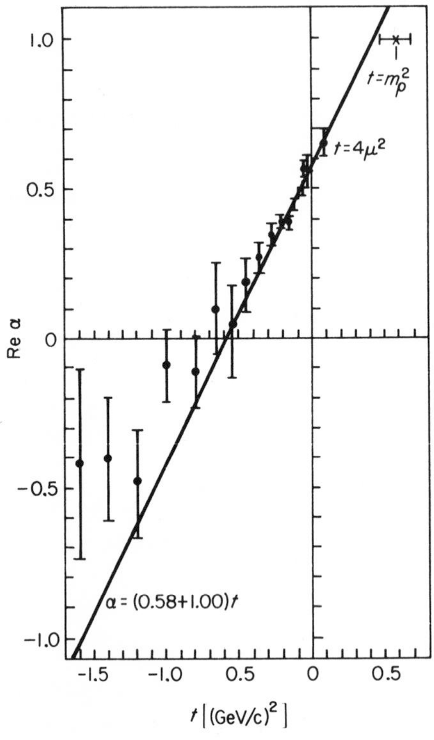


Fig. 12.26. The $\rho$-trajectory from $\pi^- p \to \pi^0 n$ data. [From G. Höhler *et al.*, *Phys. Lett* **20**, 79 (1966).]

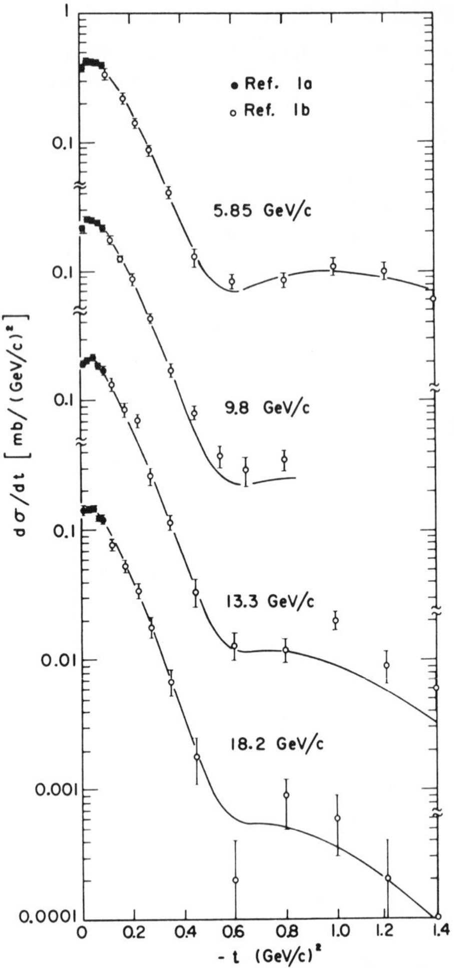


**Fig. 12.27.** Angular distributions for $\pi^- p \to \pi^0 n$ showing the single Regge pole fits and the dip at $t = -0.6$ GeV$^2$. [From F. Arbab and C. B. Chiu, Association between the dip in the $\pi^- p \to \pi^0 n$ high-energy angular distribution and the zero of the $\rho$ trajectory. *Phys. Rev.* **147**, 1045–1047 (1966).]

## *f. Baryon Trajectories. Chew–Frautschi Plot*

If we consider *backward* $\pi$N scattering, $(-t)$ is large and $u$ small. Then the poles in $u$ are the closest to the physical region, and we expect objects exchanged in the *u-channel* to dominate. This means baryon exchange (Fig. 12.28) and baryon Regge

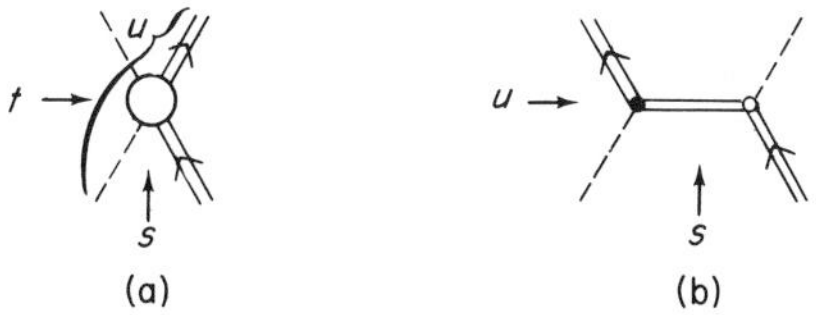


**Fig. 12.28.** (a) The Mandelstam variables for $\pi$N scattering; (b) the $u$-channel (baryon) exchange pole.

trajectories, which have $j = \alpha + \frac{1}{2}$. Analogous to the above, dominance of the backward $\pi^-$p scattering (denoted by $\pi^- p \rightarrow p\pi^-$) by a single Regge pole leads to

$$d\sigma/du \sim f(u)(s/s_0)^{2\alpha(u)-2}.$$

The exchanged object must have charge +2, and indeed a linear trajectory (denoted by $\Delta_\delta$) fitted to the resonances with $N^{*++}(1236)$ quantum numbers agrees very well with the scattering data. This is shown in Fig. 12.29.

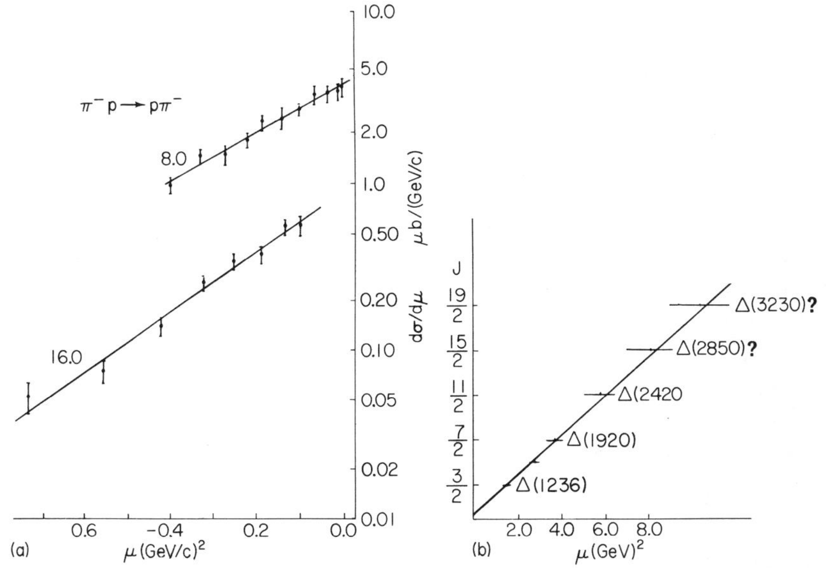


**Fig. 12.29.** (a) Fit to $d\sigma/du$ for backward $\pi^-$p scattering from the $\Delta_\delta$ trajectory; (b) Chew-Frautschi plot showing this trajectory. [From H. J. Lubatti, Regge poles in high-energy scattering. In *Theory and Phenomenology*, Pt. B, A. Zichichi, ed. Academic Press, New York, 1969.]

The plot of $j$ versus $M^2$ [like Fig. 12.29(b)], with resonances (or particles) at steps of $\Delta j = 2$, is known as a *Chew–Frautschi diagram*. Chew–Frautschi diagrams for $\pi$N resonances are shown in Fig. 12.30. The evident linearity of the Regge trajectories was a surprise, and is very different from potential theory behavior. When we take SU(3) symmetry into account, we expect SU(3) multiplets of trajectories; these are shown in Fig. 12.31.

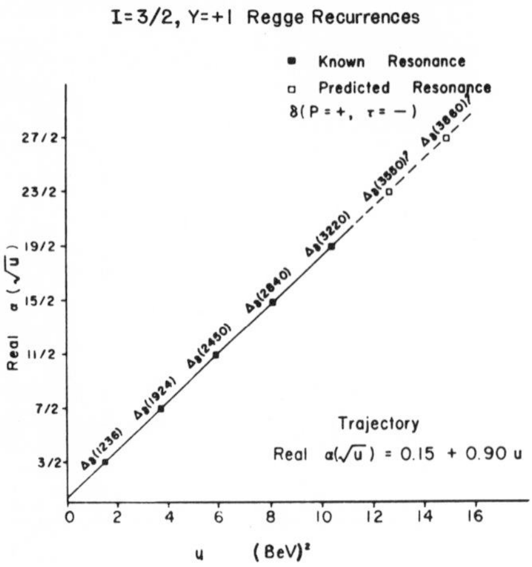


**Fig. 12.30.** Chew-Frautschi plots for $\pi$N resonances. [From G. E. Hite, Recent developments of the Regge pole model. *Rev. Mod. Phys.* **41,** 669–723 (1969).]

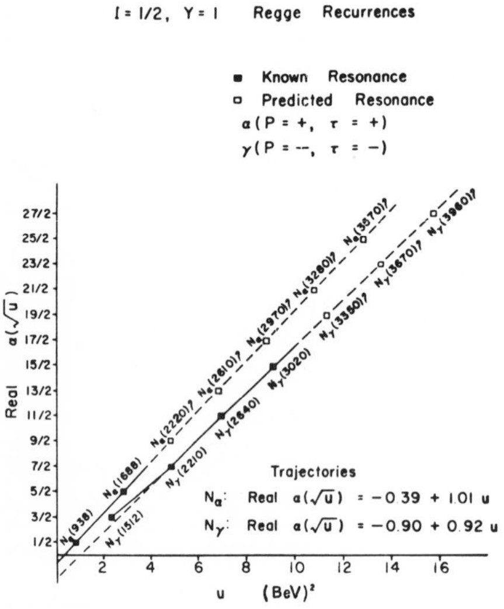


Since particles lying on Regge trajectories are evidently composite (bound states or resonances) it has been suggested that this be used to tell "composite particles" from "elementary particles," that is, an elementary particle is one that does *not* lie on a Regge trajectory. But all of the hadrons, as far as we know, lie on Regge trajectories, so none are any more elementary than any others. This idea is often known as *nuclear democracy.*

---

**Fig. 12.31.** Chew-Frautschi plots for SU(3) baryon multiplets. [From G. E. Hite, Recent developments of the Regge pole model. *Rev. Mod. Phys.* **41,** 669–723 (1969).]

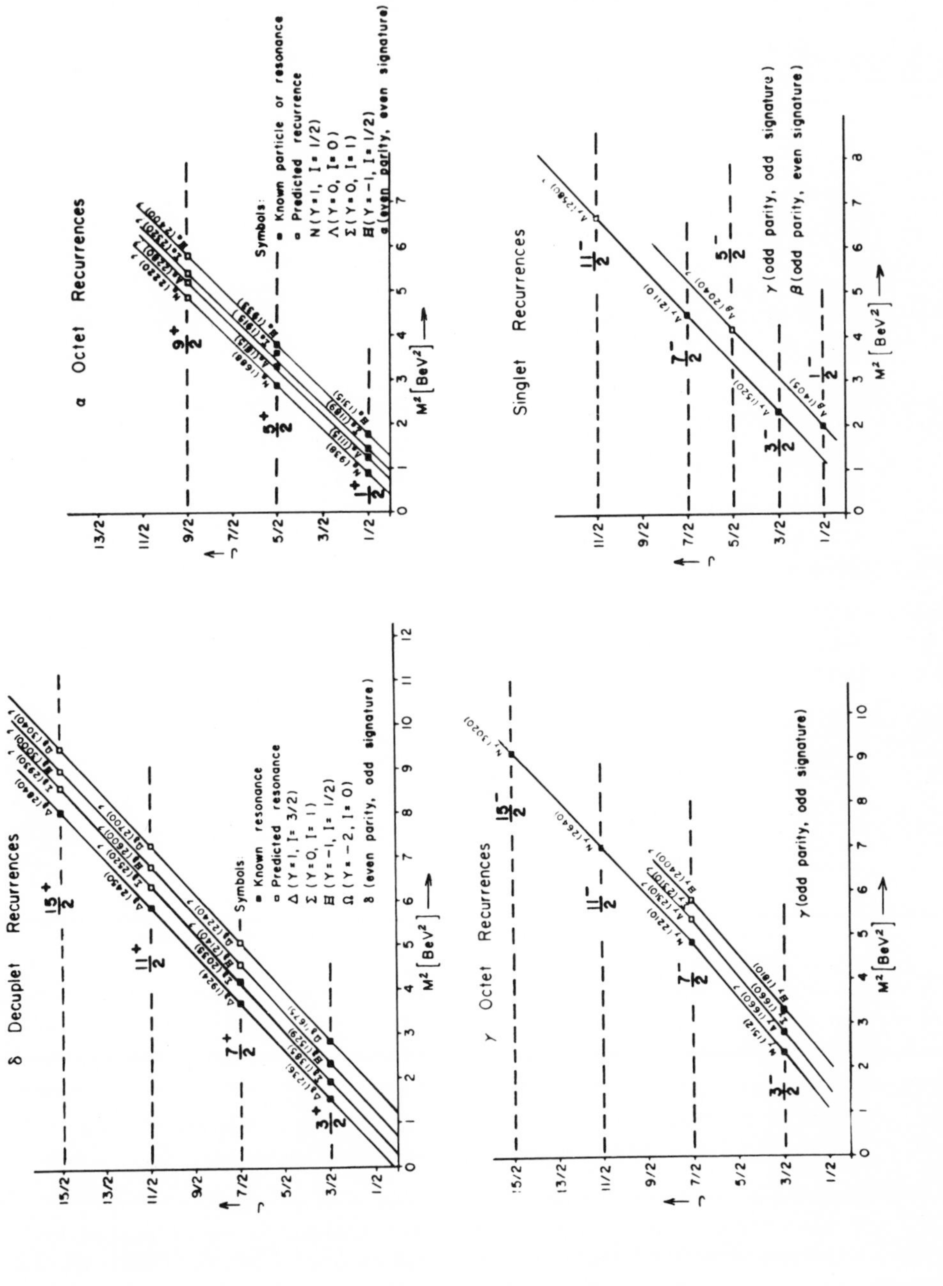
δ Decuplet Recurrences
Symbols:
■ Known resonance
□ Predicted resonance
Δ (Y = 1, I = 3/2)
Σ (Y = 0, I = 1)
Ξ (Y = −1, I = 1/2)
Ω (Y = −2, I = 0)
δ (even parity, odd signature)
α Octet Recurrences
Symbols:
■ Known particle or resonance
□ Predicted recurrence
N (Y = 1, I = 1/2)
Λ (Y = 0, I = 0)
Σ (Y = 0, I = 1)
Ξ (Y = −1, I = 1/2)
α (even parity, even signature)
γ Octet Recurrences
γ (odd parity, odd signature)
Singlet Recurrences
γ (odd parity, odd signature)
β (odd parity, even signature)
M² [BeV²]

**Problems**

**12.1.** (a) Show that for a spherically symmetric $\rho(\mathbf{r})$ and small $q^2$,

$$F(\mathbf{q}) \equiv F(q^2) \cong Q[1 - \tfrac{1}{6}\langle r^2\rangle q^2],$$

where $\langle r^2\rangle$, the mean-square radius, is given by

$$\langle r^2\rangle \equiv 4\pi Q^{-1} \int \rho(r)\, r^4\, dr,$$

and the total charge $Q$ by

$$Q \equiv 4\pi \int \rho(r) r^2\, dr.$$

(b) Find $\langle r^2\rangle$ for $\rho(r) \propto e^{-r/R}$.
(c) Find $F(q^2)$ exactly for this $\rho(r)$.

**12.2.** Verify Eq. (12.8).

**12.3.** Verify Eqs. (12.12).

**12.4.** Verify Eq. (12.16).

**12.5.** Verify Eq. (12.23).

**12.6.** Show that, unless $G_E = G_M$ at $t = 4M^2$, $F_1$ and $F_2$ will have poles at this point.

**12.7.*** From the double dispersion relation, Eq. (12.37), derive a single variable dispersion relation, e.g., in $s$ for fixed $t$ (ignore subtractions).

**12.8.** Verify Eq. (12.39).

**12.9.** Use the relation

$$\int_{-1}^{1} dz\, P_l(z) P_\alpha(z) = \frac{2}{\pi} \frac{\sin \pi\alpha}{(l - \alpha)(l + \alpha + 1)},$$

to show that each Regge pole contributes a term

$$a_l^{(n)} = \frac{\beta_n}{(l - \alpha_n)(l + \alpha_n + 1)}$$

to a partial wave amplitude $a_l$.

**Bibliography**

G. F. Chew. *S-Matrix Theory of Strong Interactions.* Benjamin, New York, 1969.
G. F. Chew. *Analytic S-Matrix.* Benjamin, New York, 1966.
B. T. Feld. *Models of Elementary Particles.* Blaisdell, Waltham, Massachusetts, 1969.
W. R. Frazer. *Elementary Particles.* Chap. 7. Prentice-Hall, Englewood Cliffs, New Jersey, 1966.
H. Muirhead. *Physics of Elementary Particles,* Chap. 10. Pergamon, Oxford, 1965.
W. S. C. Williams. *An Introduction to Elementary Particles* (2nd ed.). Chaps. 12, 13. Academic Press, New York, 1971.

# Appendix

# Tables of Particle Properties*

## Stable Particle Table

For additional parameters, see Addendum to this table.

*Quantities in italics have changed by more than one (old) standard deviation since April 1971.*

| Particle | $I^G(J^P)C_n$ | Mass (MeV) Mass$^2$ (GeV)$^2$ | Mean life (sec) $c\tau$ (cm) | Partial decay mode: Mode | Fraction[a] | p or $p_{max}$[b] (MeV/c) |
|---|---|---|---|---|---|---|
| $\gamma$ | $0,1(1^-)^-$ | $0(<2)10^{-21}$ | stable | stable | | |
| $\nu$ $\nu_e$ | $J=\frac{1}{2}$ | $0(<60\ eV)$ | stable | stable | | |
| $\nu_\mu$ | | $0(<1.2)$ | | | | |
| e | $J=\frac{1}{2}$ | 0.5110041 | stable | stable | | |
| | | ±.0000016 | $(>2\times10^{21}y)$ | | | |
| $\mu$ | $J=\frac{1}{2}$ | 105.6594 | *$2.1994\times10^{-6}$* | $e\nu\bar{\nu}$ | 100 | 53 |
| | | ±.0004 | *±.0006* S=1.1* | $e\gamma\gamma$ | $(<1.6\quad)10^{-5}$ | 53 |
| | | $m^2=$ 0.0112 | $c\tau=6.593\times10^4$ | 3e | $(<6\quad)10^{-9}$ | 53 |
| | $m_\mu - m_{\pi^\pm}$ | = −33.917 | | $e\gamma$ | $(<2.2\quad)10^{-8}$ | 53 |
| | | ±.011 | | | | |
| $\pi^\pm$ | $1^-(0^-)$ | 139.576 | $2.6024\times10^{-8}$ | $\mu\nu$ | 100 % | 30 |
| | | ±.011 | ±.0024 | $e\nu$ | $(1.24\pm0.03)10^{-4}$ | 70 |
| | | $m^2=$ 0.0195 | $c\tau=780.2$ | $\mu\nu\gamma$ | c$(1.24\pm0.25)10^{-4}$ | 30 |
| | | | $(\tau^+-\tau^-)/\bar{\tau}=$ | $\pi^0 e\nu$ | $(1.02\pm0.07)10^{-8}$ | 5 |
| | | | $(0.05\pm0.07)\%$ | $e\nu\gamma$ | c$(3.0\pm0.5)10^{-8}$ | 70 |
| | | | (test of CPT) | $e\nu e^+e^-$ | $(<3.4\quad)10^{-8}$ | 70 |
| $\pi^0$ | $1^-(0^-)^+$ | 134.972 | $0.84\times10^{-16}$ | $\gamma\gamma$ | $(98.84\pm0.04)\%$ | 67 |
| | | ±.012 | ±.10 S=2.1* | $\gamma e^+e^-$ | $(1.16\pm0.04)\%$ | 67 |
| | | $m^2=$ 0.0182 | $c\tau=2.5\times10^{-6}$ | $\gamma\gamma\gamma$ | $(<5\quad)10^{-6}$ | 67 |
| | $m_{\pi^\pm}-m_{\pi^0}$ | = 4.6043 | | $e^+e^-e^+e^-$ | d$(3.47\quad)10^{-5}$ | 67 |
| | | ±.0037 | | | | |

* From *Review of Particle Properties*, Particle Data Group, reprinted from *Phys. Lett.* **39B**, April 1972.

| Particle | $I^G(J^P)C_n$ | Mass (MeV) Mass² (GeV)² | Mean life (sec) cτ (cm) | Partial decay mode: Mode | Fraction[a] | p or $p_{max}$[b] (MeV/c) |
|---|---|---|---|---|---|---|
| $K^{\pm}$ | $\frac{1}{2}(0^-)$ | 493.84 | $1.2371\times10^{-8}$ | $\mu\nu$ | (63.77±0.28)% S=1.1* | 236 |
| | | ±0.10 | ±.0026 S=1.9* | $\pi\pi^0$ | (20.92±0.29)% S=1.2* | 205 |
| | | | cτ=370.8 | $\pi\pi^-\pi^+$ | (5.58±0.03)% S=1.1* | 126 |
| | | $m^2$=0.244 | $(\tau^+-\tau^-)/\bar{\tau}$= | $\pi\pi^0\pi^0$ | (1.68±0.04)% | 133 |
| | | | (.11±.09)% | $\mu\pi^0\nu$ | (3.20±0.11)% S=1.8* | 215 |
| | | | (test of CPT) | $e\pi^0\nu$ | (4.86±0.07)% S=1.1* | 228 |
| | | | S=1.2* | $e\pi^0\pi^0\nu$ | ($1.8^{+2.4}_{-0.6}$) $10^{-5}$ | 207 |
| | | $m_{K^\pm}-m_{K^0}$ =−3.95 | | $\pi\pi^\mp e^\pm\nu$ | (*3.7 ±0.2*) *$10^{-5}$* | 204 |
| | | ±0.13 | | $\pi\pi^\pm e^\mp\nu$ | (<5) $10^{-7}$ | 204 |
| | | S=1.1* | | $\pi\pi^\mp\mu^\pm\nu$ | (0.9 ±0.4) $10^{-5}$ | 151 |
| | | | | $\pi\pi^\pm\mu^\mp\nu$ | (<3) $10^{-6}$ | 151 |
| | | | | $e\nu$ | (1.30±0.18) $10^{-5}$ | 247 |
| | | | | $e\nu\gamma$ | c(<7) $10^{-5}$ | 247 |
| | | | | $\pi\pi^0\gamma$ | c(2.2 ±0.7) $10^{-4}$ | 205 |
| | | | | $\pi\pi^+\pi^-\gamma$ | c(10 ±4) $10^{-5}$ | 126 |
| | | | | $\pi e\nu\gamma$ | c(3.7 ±1.4) $10^{-4}$ | 227 |
| | | | | $\pi e^+e^-$ | (< 0.4) $10^{-6}$ | 227 |
| | | | | $\pi^\mp e^\pm e^\pm$ | (< 1.5) $10^{-5}$ | 227 |
| | | | | $\pi\mu^+\mu^-$ | (< 2.4) $10^{-6}$ | 172 |
| | | | | $\pi\gamma\gamma$ | c(< 3.5) $10^{-5}$ | 227 |
| | | | | $\pi\gamma\gamma\gamma$ | c(< 3) $10^{-4}$ | 227 |
| | | | | $\pi\nu\bar{\nu}$ | (< 1.2) $10^{-6}$ | 227 |
| | | | | $\pi\gamma$ | (< 4) $10^{-6}$ | 227 |
| $K^0$ | $\frac{1}{2}(0^-)$ | 497.79 | 50% $K_{Short}$, 50% $K_{Long}$ | | | |
| | | ±0.15 | | | | |
| $K^0_S$ | $\frac{1}{2}(0^-)$ | S=1.1* | $0.862\times10^{-10}$ | $\pi^+\pi^-$ | (68.85 ±0.31)% S=1.1* | 206 |
| | | $m^2$=0.248 | ±.006 S=1.2* | $\pi^0\pi^0$ | (31.15)% | 209 |
| | | | cτ=2.58 | $\mu^+\mu^-$ | (< .7) $10^{-5}$ | 225 |
| | | | | $e^+e^-$ | (<35) $10^{-5}$ | 249 |
| | | | | $\pi^+\pi^-\gamma$ | c(2.3 ±0.8) $10^{-3}$ | 206 |
| | | | | $\gamma\gamma$ | (< 2) $10^{-3}$ | 249 |
| $K^0_L$ | $\frac{1}{2}(0^-)$ | | $5.172\times10^{-8}$ | $\pi^0\pi^0\pi^0$ | (21.4 ±0.7)% S=1.1* | 139 |
| | | | ±0.042 | $\pi^+\pi^-\pi^0$ | (12.6 ±0.3)% | 133 |
| | | | cτ=1550 | $\pi\mu\nu$ | (26.8 ±0.6)% | 216 |
| | | | | $\pi e\nu$ | (39.0 ±0.6)% | 229 |
| | | | | $\pi e\nu\gamma$ | c(1.3 ±0.8)% | 229 |
| | | $m_{K_L}-m_{K_S}$ = $0.5402\times10^{10}$ ħ sec$^{-1}$ | | $\pi^+\pi^-$ | (0.157±0.005)% | 206 |
| | | ± 0.0035 | | $\pi^0\pi^0$ | (0.094±0.019)% S=1.5* | 209 |
| | | | | $\pi^+\pi^-\gamma$ | c(< 0.4) $10^{-3}$ | 206 |
| | | $\frac{\Gamma(K_S\to\pi^+\pi^-\pi^0)}{\Gamma(K_L\to\pi^+\pi^-\pi^0)}$ <0.45 | | $\pi^0\gamma\gamma$ | (< 2.4) $10^{-4}$ | 231 |
| | | | | $\gamma\gamma$ | (*4.9 ±0.4*) *$10^{-4}$* | 249 |
| | | (test of CP) | | $e\mu$ | (< 1.6) $10^{-9}$ | 238 |
| | | | | $\mu^+\mu^-$ | (< 1.9) $10^{-9}$ | 225 |
| | | | | $e^+e^-$ | (< 1.6) $10^{-9}$ | 249 |
| $\eta$ | $0^+(0^-)^+$ | 548.8 | Γ=(2.63±0.58)keV | $\gamma\gamma$ | (38.0 ±1.0)% S=1.2* | 274 |
| | | ±0.6 | Neutral decays | $\pi^0\gamma\gamma$ | e(3.1 ±1.1)% S=1.2* | 258 |
| | | S=1.4* | 71.1% | $3\pi^0$ | (30.0 ±1.1)% S=1.1* | 180 |
| | | $m^2$=0.301 | | $\pi^+\pi^-\pi^0$ | (24.0 ±0.6)% S=1.1* | 175 |
| | | | | $\pi^+\pi^-\gamma$ | (4.9 ±0.2)% S=1.1* | 236 |
| | | | Charged decays | $\pi^0e^+e^-$ | (< 0.04)% | 258 |
| | | | 28.9% | $\pi^+\pi^-e^+e^-$ | (0.1 ±0.1)% | 236 |
| | | | | $\pi^+\pi^-\pi^0\gamma$ | (< 0.2)% | 175 |
| | | | | $\pi^+\pi^-\gamma\gamma$ | (< 0.2)% | 236 |
| | | | | $\mu^+\mu^-$ | (2.2 ±0.8) $10^{-5}$ | 253 |
| | | | | $\mu^+\mu^-\pi^0$ | (< 5) $10^{-4}$ | 211 |
| p | $\frac{1}{2}(\frac{1}{2}^+)$ | 938.2592 | stable | | | |
| | | ±0.0052 | (> $2\times10^{28}$y) | | | |
| | | $m^2$=0.8803 | | | | |
| n | $\frac{1}{2}(\frac{1}{2}^+)$ | 939.5527 | e(0.935±0.014)$10^3$ | $pe^-\nu$ | 100 % | 1 |
| | | ±0.0052 | cτ = $2.80\times10^{13}$ | | | |
| | | $m^2$=0.8828 | | | | |
| | | $m_p-m_n$= −1.29344 | | | | |
| | | ±0.00007 | | | | |

| Particle | $I^G(J^P)C_n$ | Mass (MeV) Mass$^2$ (GeV)$^2$ | Mean life (sec) $c\tau$ (cm) | Partial decay mode: Mode | Fraction[a] | p or $p_{max}$[b] (MeV/c) |
|---|---|---|---|---|---|---|
| $\Lambda$ | $0(\frac{1}{2}^+)$ | 1115.59 | $2.521\times10^{-10}$ | $p\pi^-$ | $(64.2 \pm 0.5)\%$ | 100 |
| | | ±0.05 | ±.021 S=1.2* | $n\pi^0$ | $(35.8 \pm 0.5)\%$ | 104 |
| | | S=1.2* | $c\tau$ = 7.56 | $pe\nu$ | $(8.13\pm0.29)10^{-4}$ | 163 |
| | | $m^2$= 1.245 | | $p\mu\nu$ | $(1.62\pm0.35)10^{-4}$ | 131 |
| $\Sigma^+$ | $1(\frac{1}{2}^+)$ | 1189.42 | $0.800\times10^{-10}$ | $p\pi^0$ | $(51.6 \pm 0.7)\%$ | 189 |
| | | ±0.11 | ±.006 | $n\pi^+$ | $(48.4 \pm 0.7)\%$ | 185 |
| | | S= 1.7* | $c\tau$ = 2.40 | $p\gamma$ | c$(1.24\pm0.18)10^{-3}$ S=1.4* | 225 |
| | | $m^2$= 1.415 | | $n\pi^+\gamma$ | c$(1.31\pm0.24)10^{-4}$ | 185 |
| | | $m_{\Sigma^+}-m_{\Sigma^-}$= -7.92 | | $\Lambda e^+\nu$ | $(2.02\pm0.47)10^{-5}$ | 72 |
| | | ±.13 | $\frac{\Gamma(\Sigma^+\to\ell^+n\nu)}{\Gamma(\Sigma^-\to\ell^-n\nu)}<.035\leftarrow$ | $n\mu^+\nu$ | $(<2.4)10^{-5}$ | 202 |
| | | S= 1.6* | | $ne^+\nu$ | $(<1.0)10^{-5}$ | 224 |
| | | | | $pe^+e^-$ | $(<7)10^{-6}$ | 225 |
| $\Sigma^0$ | $1(\frac{1}{2}^+)$ | 1192.48 ±0.11 | $<1.0\times10^{-14}$ | $\Lambda\gamma$ | 100 % | 74 |
| | | S= 1.3* | $c\tau<3\times10^{-4}$ | $\Lambda e^+e^-$ | d$(5.45)10^{-3}$ | 74 |
| | | $m^2$=1.422 | | | | |
| $\Sigma^-$ | $1(\frac{1}{2}^+)$ | 1197.34 | $1.484\times10^{-10}$ | $n\pi^-$ | 100 % | 193 |
| | | ±0.10 | ±.019 S=1.6* | $ne^-\nu$ | $(1.10\pm0.05)10^{-3}$ | 230 |
| | | S= 1.5* | $c\tau$ = 4.45 | $n\mu^-\nu$ | $(0.45\pm0.04)10^{-3}$ | 210 |
| | | $m^2$ = 1.434 | | $\Lambda e^-\nu$ | $(0.60\pm0.06)10^{-4}$ | 79 |
| | | $m_{\Sigma^0}-m_{\Sigma^-}$=-4.86 | | $n\pi^-\gamma$ | c$(1.0 \pm0.2)10^{-4}$ | 193 |
| | | ±.06 | | | | |
| $\Xi^0$ | $\frac{1}{2}(\frac{1}{2}^+)$[f] | 1314.7 | $3.03\times10^{-10}$ | $\Lambda\pi^0$ | 100 % | 135 |
| | | ±0.7 | ±.18 | $p\pi^-$ | $(<0.9)10^{-3}$ | 299 |
| | | $m^2$= 1.729 | $c\tau$ = 9.08 | $pe^-\nu$ | $(<1.3)10^{-3}$ | 323 |
| | | $m_{\Xi^0}-m_{\Xi^-}$ =-6.6 | | $\Sigma^+e^-\nu$ | $(<1.5)10^{-3}$ | 119 |
| | | ±.7 | | $\Sigma^-e^+\nu$ | $(<1.5)10^{-3}$ | 112 |
| | | | | $\Sigma^+\mu^-\nu$ | $(<1.5)10^{-3}$ | 64 |
| | | | | $\Sigma^-\mu^+\nu$ | $(<1.5)10^{-3}$ | 49 |
| | | | | $p\mu^-\nu$ | $(<1.3)10^{-3}$ | 309 |
| $\Xi^-$ | $\frac{1}{2}(\frac{1}{2}^+)$[f] | 1321.30 | $1.660\times10^{-10}$ | $\Lambda\pi^-$ | 100 % | 139 |
| | | ±0.15 | ±.037 S=1.1* | $\Lambda e^-\nu$ | $(0.70\pm0.21)10^{-3}$ | 190 |
| | | $m^2$= 1.746 | $c\tau$ = 4.98 | $\Sigma^0e^-\nu$ | $(<0.5)10^{-3}$ | 123 |
| | | | | $\Lambda\mu^-\nu$ | $(<1.3)10^{-3}$ | 163 |
| | | | | $\Sigma^0\mu^-\nu$ | $(<0.5)\%$ | 70 |
| | | | | $n\pi^-$ | $(<1.1)10^{-3}$ | 303 |
| | | | | $ne^-\nu$ | $(<1.0)\%$ | 327 |
| $\Omega^-$ | $0(\frac{3}{2}^+)$[f] | 1672.5±.5 | $1.3^{+0.4}_{-0.3}\times10^{-10}$ | $\Xi^0\pi^-$ | Total of | 294 |
| | | $m^2$= 2.797 | $c\tau$ = 3.9 | $\Xi^-\pi^0$ | 28 events | 290 |
| | | | | $\Lambda K^-$ | seen | 211 |

*S = Scale factor = $\sqrt{\chi^2/(N-1)}$, where N ≈ number of experiments. S should be ≈ 1. If S > 1, we have enlarged the error of the mean, $\delta x$, i.e., $\delta x \to S\delta x$. This convention is still inadequate, since if S >> 1, the experiments are probably inconsistent, and therefore the real uncertainty is probably even greater than $S\delta x$. See text and ideogram in Stable Particle Data Card Listings.

a. Quoted upper limits correspond to a 90% confidence level.

b. In decays with more than two bodies, $P_{max}$ is the maximum momentum that any particle can have.

c. See Stable Particle Data Card Listings for energy limits used in this measurement.

d. Theoretical value; see also Stable Particle Data Card Listings.

e. See note in Stable Particle Data Card Listings.

f. P for $\Xi$ and $J^P$ for $\Omega^-$ not yet measured. Values reported are SU(3) predictions.

g. Assumes rate for $\Xi^-\to\Sigma^0 e^-\nu$ small compared with $\Xi^-\to\Lambda e^-\nu$.

## ADDENDUM

| | Magnetic moment | |
|---|---|---|
| e | 1.001 159 6577 ± .000 000 0035 | $\frac{e\hbar}{2m_e c}$ |
| μ | 1.001 166 16 ±.000 000 31 | $\frac{e\hbar}{2m_\mu c}$ |

**μ Decay parameters** [a]

$\rho = 0.752 \pm 0.003$ $\eta = -0.12 \pm 0.21$
$\xi = 0.972 \pm 0.013$ $\delta = 0.755 \pm 0.009$ $h = 1.00 \pm 0.13$
$|g_A/g_V| = 0.86^{+0.33}_{-0.11}$ $\phi = 180° \pm 15°$

**$K^\pm$**

| Mode | Partial rate (sec$^{-1}$) | |
|---|---|---|
| $\mu\nu$ | $(51.55 \pm 0.25)10^6$ | S=1.2* |
| $\pi\pi^0$ | $(16.91 \pm 0.24)10^6$ | S=1.2* |
| $\pi\pi^+\pi^-$ | $(4.51 \pm 0.02)10^6$ | S=1.1* |
| $\pi\pi^0\pi^0$ | $(1.36 \pm 0.04)10^6$ | |
| $\mu\pi^0\nu$ | $(2.59 \pm 0.09)10^6$ | S=1.8* |
| $e\pi^0\nu$ | $(3.92 \pm 0.06)10^6$ | S=1.2* |

**$\Delta I = \frac{1}{2}$ rule for $K^\pm \to 3\pi$**

$\pi^+\pi^+\pi^-$ $c_g = -.206 \pm .007$
$\pi^-\pi^+\pi^-$ $c_g = -.194 \pm .007$
$\pi^+\pi^0\pi^0$ $c_g = .527 \pm .017$
See also Stable Particle Data Card Listings and Appendix I

**Form factors for leptonic decays**

See Stable Particle Data Card Listings for λ and ξ

**$K^0_S$**

| Mode | Partial rate (sec$^{-1}$) | |
|---|---|---|
| $\pi^+\pi^-$ | $(0.799 \pm .006)10^{10}$ | S=1.2* |
| $\pi^0\pi^0$ | $(0.361 \pm .004)10^{10}$ | S=1.2* |

**$K^0_L$**

| Mode | Partial rate (sec$^{-1}$) | |
|---|---|---|
| $\pi^0\pi^0\pi^0$ | $(4.13 \pm 0.13)10^6$ | S=1.1* |
| $\pi^+\pi^-\pi^0$ | $(2.43 \pm 0.06)10^6$ | |
| $\pi\mu\nu$ | $(5.18 \pm 0.12)10^6$ | |
| $\pi e\nu$ | $(7.54 \pm 0.13)10^6$ | |
| $\pi^+\pi^-$ | $(3.03 \pm 0.10)10^4$ | |
| $\pi^0\pi^0$ | $(1.81 \pm 0.36)10^4$ | S=1.5* |

**CP violation parameters**

$|\eta_{+-}| = (1.96 \pm 0.03)10^{-3}$, $\phi_{+-} = (43 \pm 3)°$
$|\eta_{00}| = (2.09 \pm 0.12)10^{-3}$, $\phi_{00} = (43 \pm 19)°$
S=1.2*

Charge asymmetry:

$$\delta = \frac{\Gamma(K^0_L \to \ell^+) - \Gamma(K^0_L \to \ell^-)}{\Gamma(K^0_L \to \ell^+) + \Gamma(K^0_L \to \ell^-)} = .32 \pm .03 \times 10^{-2}$$

**$I = \frac{1}{2}$ rule for $K^0_L \to 3\pi$**

$\pi^+\pi^-\pi^0$ $c_g = .60 \pm .03$ S=3.1*
See Data Cards & App. I

**$\Delta S = -\Delta Q$**

Re x = −.003± .026 S=1.5*
Im x = −.007± .039 S=1.2

**Form factors for leptonic decays**

See Stable Particle Data Card Listings for λ, ξ

**η**

| Mode | Asymmetry parameter |
|---|---|
| $\pi^+\pi^-\pi^0$ | (1.2±0.5)% S=1.3* |
| $\pi^+\pi^-\gamma$ | (1.1±1.3)% |

| | Magnetic moment ($e\hbar/2m_p c$) | Mode | Decay parameters [b]: Measured α | Measured φ(degree) | Derived γ | Derived Δ (degree) | $g_A/g_V$ [b] | $g_V/g_A$ [b] |
|---|---|---|---|---|---|---|---|---|
| p | 2.792782 ±.000017 | | | | | | | |
| n | −1.913148 ±.000066 | $pe^-\nu$ | | | | | −1.242±0.008 S=1.2* [δ=(178.6±0.9)°] | |
| Λ | −0.67 ±.06 | $p\pi^-$ | 0.645±0.016 | (−6.3±3.5)° | 0.76 | $(7.4^{+4.0}_{-4.1})°$ | | |
| | | $n\pi^0$ | 0.649±0.046 | | | | | |
| | | $pe\nu$ | | | | | −0.66±0.06 S=1.2* | |
| $\Sigma^+$ | 2.59 ±.46 | $p\pi^0$ | −0.991±0.019 | (22±90)° | 0.12 | $(183^{+11}_{-12})°$ | | |
| | | $n\pi^+$ | +0.066±0.016 | (167±20)° S=1.1* | −0.97 | $(-73^{+136}_{-10})°$ | | |
| | | $p\gamma$ | $-1.03^{+.52}_{-.42}$ | | | | | |
| $\Sigma^-$ | | $n\pi^-$ | −0.069±0.008 | (10±15)° | 0.98 | $(249^{+12}_{-115})°$ | | |
| | | $ne^-\nu$ | | | | | See Data Cds. | |
| | | $\Lambda e^-\nu$ | | | | | | 0.35±0.18 |
| $\Xi^0$ | | $\Lambda\pi^0$ | −0.35±0.08 | (25±21)° S=1.3* | 0.85 | $(228^{+16}_{-37})°$ | | |
| $\Xi^-$ | | $\Lambda\pi^-$ | −0.40±0.03 | (−4±8)° S=1.1* | 0.91 | $(170^{+18}_{-17})°$ | | |

*S = scale factor. Quoted error includes scale factor; see footnote to main Stable Particle Table for definition.

a. $|g_A/g_V|$ defined by $g_A^2 = |C_A|^2 + |C'_A|^2$, $g_V^2 = |C_V|^2 + |C'_V|^2$, and $\Sigma\langle \bar{e}|\Gamma_i|\mu\rangle\langle\bar{\nu}|\Gamma_i(C_i + C'_i\gamma_5)|\nu\rangle$;
φ defined by $\cos\phi = -R_e(C^*_A C'_V + C'_A C^*_V)/g_A g_V$ [for more details, see text Section IV E]

b. The definition of these quantities is as follows [for more details on sign convention, see text Section IV H]:

$\alpha = \frac{2|s||p|\cos\Delta}{|s|^2+|p|^2}$; $\beta = \sqrt{1-\alpha^2}\sin\phi$; $g_A/g_V$ defined by $\langle B_f|\gamma_\lambda(g_V - g_A\gamma_5)|B_i\rangle$;

$\beta = \frac{-2|s||p|\sin\Delta}{|s|^2+|p|^2}$. $\gamma = \sqrt{1-\alpha^2}\cos\phi$. δ defined by $g_A/g_V = |g_A/g_V|e^{i\delta}$.

c. The definition of the slope parameter of the Dalitz plot is as follows: $|M|^2 = 1 + g\left(\frac{s_3 - s_0}{m^2_{\pi^+}}\right)$.

# Meson Table

April 1972

*Quantities in italics have changed by more than one (old) standard deviation since April 1971*

| Name (I: 0, 1; G −: φ, π; G +: η, ρ) | $I^G(J^P)C_n$ ⊢⊣ estab. | Mass M (MeV) | Full Width Γ (MeV) | $M^2$ $\pm\Gamma M$[a] $(GeV)^2$ | Partial decay mode: Mode | Fraction % | | p or $p_{max}$[b] (MeV/c) |
|---|---|---|---|---|---|---|---|---|
| $\pi^{\pm}(140)$ | $1^-(0^-)+$ | 139.58 | 0.0 | 0.019483 | See Stable Particle Table | | | |
| $\pi^0(135)$ | | 134.97 | 7.8 eV ±.9 eV | 0.018217 | | | | |
| $\eta(549)$ | $0^+(0^-)+$ | 548.8 ±0.6 | 2.63 keV ±.58 keV | 0.301 ±.000 | All neutral | 71 | See Stable Particle Table | |
| | | | | | $\pi^+\pi^-\pi^0$ + $\pi^+\pi^-\gamma$ | 29 | | |
| ε | $0^+(0^+)+$ | $\delta_0^0$ is near 80°-90° in mass region 800-1000 MeV, with probably only slow variation below and cusp at $K\bar{K}$ threshold. Inelasticity ≈0 below 2 $m_K$. | | | | | | |
| See note on ππ S wave¶. | | | | | | | | |
| $\rho(765)$ | $1^+(1^-)-$ | 765 ±10(c) | 135 ±20(c) | 0.585 ±.103 | ππ | ≈100 | | 356 |
| | | | | | $e^+e^-$ | *0.0042±.0004* (d) | | 382 |
| | | | | | $\mu^+\mu^-$ | 0.0067±.0012 (d) | | 368 |
| | | | | | For upper limits, see footnote (e) | | | |
| $\omega(784)$ | $0^-(1^-)-$ | 783.9 ±0.3 S=1.3* | *10.0* *±0.6* | 0.614 ±.008 | $\pi^+\pi^-\pi^0$ | 89.7±4.0 | S=1.1* | 328 |
| | | | | | $\pi^+\pi^-$ | *1.2±0.3* | *S=1.4** | 366 |
| | | | | | $\pi^0\gamma$ | 9.0±1.0 | | 380 |
| | | | | | $e^+e^-$ | 0.0075±.0016 | S=1.8* | 392 |
| | | | | | For upper limits, see footnote (f) | | | |
| $\eta'(958)$ or $X^0$ | $0^+(\ ^-)+$; $J^P = 0^-$ or $2^-$ | 957.1 ±0.6 | < 4 | 0.916 <.004 | $\eta\pi\pi$ | 68.1±2.2 | | 230 |
| | | | | | $\pi^+\pi^-\gamma$ (mainly $\rho^0\gamma$) | 30.1±2;3 | S=1.1* | 458 |
| | | | | | γγ | *1.8±0.3* | | 479 |
| | | | | | For upper limits, see footnote (h) | | | |
| $\pi_N(975)$ | $1^-(0^+)+$ | ~ 975 | ~ 60 | 0.950 | ηπ | (g) | | 304 |
| Possibly related to the I=1 $K\bar{K}$ threshold enhancement | | | | | | | | |
| $S^*$ | $0^+(0^+)+$ | ~ 1000 | | 1.000 | Seen as I=0 $K\bar{K}$ threshold enhancement; appears coupled to the ππ channel. | | | |
| See notes on ππ and $K\bar{K}$ S wave¶. | | | | | | | | |
| $\Phi(1019)$ | $0^-(1^-)-$ | 1019.1 ±0.5 S=1.8* | *4.4* *±.3* | 1.039 ±.004 | $K^+K^-$ | 49.1±2.0 | | 126 |
| | | | | | $K_LK_S$ | *30.7±2.4* | *S=1.4** | 109 |
| | | | | | $\pi^+\pi^-\pi^0$ (incl. ρπ) | 17.5±2.5 | S=1.1* | 461 |
| | | | | | ηγ | *2.6±1.2* | *S=1.9** | 362 |
| | | | | | $e^+e^-$ | .032±.003 | S=1.9* | 510 |
| | | | | | $\mu^+\mu^-$ | .025±.003 | | 498 |
| | | | | | For upper limits, see footnote (i) | | | |
| $A_1(1070)$ | $1^-(1^+)+$ | ~ 1070 | | 1.14 | ρπ | ~ 100 | | 232 |
| Broad enhancement in the $J^P=1^+$ ρπ partial wave; not clear if resonant¶. | | | | | | | | |
| B(1235) | $1^+(1^+)-$ | 1233 ±10§ | 100 ±20§ | 1.52 ±.12 | ωπ | ≈ 100 | | 348 |
| | | | | | ππ | < 30 | Absence suggests $J^P$ = Abnormal | 600 |
| | | | | | $K\bar{K}$ | < 2 | | 369 |
| | | | | | For other upper limits, see footnote (j) | | | |
| f(1260) | $0^+(2^+)+$ | 1269 ±10§ | 156 ±25§ | 1.60 ±.20 | ππ | ≈ 80 | | 619 |
| | | | | | $2\pi^+2\pi^-$ | 6±2 | | 556 |
| | | | | | $K\bar{K}$ | ≈ 6 | | 393 |

| Name | $I^G(J^P)C_n$ estab. | Mass M (MeV) | Full Width Γ (MeV) | $M^2$ $\pm\Gamma M$[a] $(GeV)^2$ | Partial decay mode: Mode | Fraction % | p or $p_{max}$[b] (MeV/c) |
|---|---|---|---|---|---|---|---|
| D(1285) | $0^+$(A )+ | 1286 ±4 S=1.3* | *21 ±10 S=1.3** | 1.65 ±.03 | $\eta\pi\pi$ | Probably seen | 484 |
| | | | | | $K\bar{K}\pi$ | Seen | 305 |
| | | | | | †[$\pi_N(975)\pi$ | Seen] | 246 |
| | $J^P = 0^-, 1^+, 2^-$, with $1^+$ favoured | | | | $2\pi^+2\pi^-$ (prob. $\rho^0\pi^+\pi^-$) | Seen | 565 |
| $A_2$(1310) | $1^-(2^+)+$ | 1310 ±10 § | 100 ±20 § | 1.72 ±.13 | $\rho\pi$ | 76.8±1.8 | 416 |
| | | | | | $\eta\pi$ | 16.3±1.5 | 529 |
| | | | | | $K\bar{K}$ | 5.8±0.8 | 428 |
| Controversy whether unsplit or split¶. | | | | | $\eta'(958)\pi$ | 1.1±1.1 S=1.3* | 280 |
| E(1422) | $0^+$( )+ | 1422 ±4 | 69 ±8 | 2.02 ±.10 | $K^*\bar{K} + \bar{K}^*K$ | 50±10 | 154 |
| | | | | | $\pi_N(975)\pi$ | 50±10 | 357 |
| → | $J^P = 0^-$ or $1^+$ | | | | $\pi\pi\eta$ | < 60 | 568 |
| f′(1514) | $0^+(2^+)+$ | 1514 ±5 | 73 ±23 S=1.8* | 2.29 ±.11 | $K\bar{K}$ | 72±12 (k) | 570 |
| | | | | | $K^*\bar{K} + \bar{K}^*K$ | 10±10 (k) | 294 |
| | | | | | $\pi\pi$ | < 14 (k) | 744 |
| | | | | | $\eta\pi\pi$ | 18±10 (k) | 624 |
| | | | | | $\eta\eta$ | < 40 (k) | 521 |
| $F_1$(1540) | 1 (A ) | 1540 ±5 | 40 ±15 | 2.37 ±.06 | $K^*\bar{K} + \bar{K}^*K$ | Only mode seen | 321 |
| Evidence based on only one experiment | | | | | | | |
| $A_3$(1640) | $1^-(2^-)+$ | ~ 1640 | | 2.69 | $f\pi$ | Dominant | 306 |
| | | | | | $\omega\pi\pi$ (m) | | 597 |
| Broad enhancement in the $J^P = 2^-$ $f\pi$ partial wave; not clear if resonant¶. | | | | | | | |
| Φ(1675) | $0^-$(N )− | 1664 ±13 S=1.2* | 141 ±17 | 2.77 ±.23 | $\rho\pi$ | Dominant | 647 |
| | | | | | $3\pi$ | Possibly observed | 804 |
| | | | | | $5\pi$ | 10±10 | |
| g(1680) | $1^+(3^-)-$ | 1680 ±20 § | 160 ±30 § | 2.82 ±.27 | $2\pi$ | ~ 40 (ℓ) | 828 |
| | | | | | $4\pi$ (incl. $\pi\pi\rho, \rho\rho, A_2\pi, \omega\pi$) | ~ 50 (ℓ) | 781 |
| | | | | | $K\bar{K}$ | ~ 3 (ℓ) | 677 |
| $J^P$, M and Γ from the $2\pi$ mode (ℓ). | | | | | $K\bar{K}\pi$ (incl. $K^*\bar{K}$) | ~ 3 (ℓ) | 617 |
| → S(1930) | $1^+$( )− | ~ 1930 | 30 ±20 § | 3.72 ±.06 | $\bar{p}p$ | Possibly seen | 226 |
| | | | | | $\pi\pi$ | Possibly seen | 955 |
| Seen in $\pi^-p \to (MM)^-p$ ; may be related to the structure seen in 460 MeV/c $\bar{p}p$ backwards scattering¶. | | | | | | | |

→ See note (p) for other possible heavy states.

| Name | $I(J^P)$ | Mass M (MeV) | Full Width Γ (MeV) | $M^2$ $\pm\Gamma M$ $(GeV)^2$ | Mode | Fraction % | p or $p_{max}$ (MeV/c) |
|---|---|---|---|---|---|---|---|
| $K^+$(494) | $1/2(0^-)$ | 493.84 | | 0.244 | See Stable Particle Table | | |
| $K^0$(498) | | 497.79 | | 0.248 | | | |
| $K^*$(892) | $1/2(1^-)$ | *891.7 ±0.5* | 50.1 ±1.1 | 0.795 ±.045 | $K\pi$ | ≈ 100 | 288 |
| | | | | | $K\pi\pi$ | < 0.2 | 216 |
| → | | (Charged mode; $m^0 - m^\pm = 6.1\pm1.5$ MeV) | | | | | |
| κ | $1/2(0^+)$ | $\delta_0^{1/2}$ is near 90°, with slow variation, in mass region 1200-1400 MeV. In addition, $\delta_0^{1/2}$ may be resonant at M ~ 890 MeV, Γ ~ 30 MeV. | | | | | |
| See note on $K\pi$ S wave¶. | | | | | | | |

| Name (G \ I: 0, 1; −: φ, π; +: η, ρ) | $I^G(J^P)C_n$ ⊢⊣ estab. | Mass M (MeV) | Full Width Γ (MeV) | $M^2$ $\pm\Gamma M^{(a)}$ $(\mathrm{GeV})^2$ | Partial decay mode: Mode | | Fraction % | p or $p_{max}^{(b)}$ (MeV/c) |
|---|---|---|---|---|---|---|---|---|
| Q { $K_A(1240)$ or C | 1/2(1⁺) | 1242 ±10 | 127 ±25 | 1.54 ±.16 | Kππ | | Only mode seen | |
| | | seen in $\bar{p}p$ at rest | | | †[K*π | | Large] | |
| $K_A$(1280 to 1400) | 1/2(1⁺) | 1280 to 1400 | | | †[Kρ | | Seen ] | |
| →→ Resonance interpretation unclear(n). | | | | | | | | |
| $K_N(1420)$ | 1/2(2⁺) | *1421* ±*5* § | 100 ±10 § | 2.02 ±.14 | Kπ | | < 56.3±3.0 | 616 |
| | | | | | K*π | Inequalities | > 27.8±2.7 | 415 |
| | | See note (o). | | | Kρ | explained in | > 9.5±2.5 | 324 |
| →→ | | | | | Kω | note (o). | > 4.5±1.7 | 304 |
| | | | | | Kη | | > 2.0±2.0 | 482 |
| L(1770) | 1/2(A) | 1763 ±10 § | $100^{+100}_{-50}$ § | 3.11 ±.18 | Kππ | | Dominant | 787 |
| | | | | | Kπππ | | Seen | 756 |
| →‡ | $J^P=2^-$ favoured, $1^+$ and $3^+$ not excluded. | | | | †[$K_N(1420)\pi$ and other subreactions¶] | | | |

→ Data on the following candidates, excluded above, are listed among the data cards¶ :

M(953), H(990), $\eta_N(1080)$, $A_{1.5}(1170)$, $X_{I=0}(1430)$, $X_{I=1}(1440)$, $X^-(1795)$, η/ρ(1830), φ/π(1830), ρ(2100), T(2200), ρ(2275), $N\bar{N}(2350)$, U(2375), $N\bar{N}(2375)$, $X^-(2500)$, $X^-(2620)$, X(2800), $X^-(2880)$, X(3030), $X^-(3075)$, $X^-(3145)$, $X^-(3475)$, $X^-(3535)$; $K_A^{I=3/2}(1175)$, $K_A^{I=3/2}(1265)$, $K_N(1370)$, $K_N(1660)$, $K_N(1760)$, $K_N(1850)$, $K^*(2200)$, $K^*(2800)$.

¶ See Meson Data Card Listings.

* Quoted error includes scale factor $S = \sqrt{\chi^2/(N-1)}$. See footnote to Stable Particle Table.

† Square brackets indicate a subreaction of the previous (unbracketed) decay mode(s).

§ This is only an educated guess; the error given is larger than the error of the average of the published values. (See Meson Data Card Listings for the latter.)

(a) ΓM is approximately the half-width of the resonance when plotted against $M^2$.

(b) For decay modes into ≥ 3 particles, $p_{max}$ is the maximum momentum that any of the particles in the final state can have. The momenta have been calculated by using the averaged central mass values, without taking into account the widths of the resonances.

(c) The values given for M(ρ) and Γ(ρ) and their errors are not average values from various experiments, but rather are intended to give the range where we believe the actual values are most likely to fall. Contrast the results tabulated in this note (references in the Meson Data Card Listings).

| | M(MeV) | Γ(MeV) | |
|---|---|---|---|
| $\rho^0$ | 775±7 | 149±23 | } From $e^+e^- \to \pi^+\pi^-$, fitted to Gounaris-Sakurai formula. |
| $\rho^0$ | 768±10 | 140±14 | |
| $\rho^-$ | 764±2 | 147±4 | } From physical region fits to $\pi N \to \pi\pi N$, using energy-dep. width. |
| $\rho^0$ | 775±3 | 145±9 | |
| $\rho^0$ | 768±2 | 132±13 | |
| $\rho^0_0$ | 759±7 | 119±20 | } From pole extrapol. in $\pi N \to \pi\pi N$ |
| $\rho^-$ | 760 | 131 | |

(d) The $e^+e^-$ branching ratio is from $e^+e^- \to \pi^+\pi^-$ experiments only. The ωρ interference is then due to ωρ mixing only, and is expected to be small. See note in Meson Data Card Listings. The $\mu^+\mu^-$ branching ratio is compiled from 3 experiments; each possibly with substantial ωρ interference. The error reflects this uncertainty; see notes in Meson Data Card Listings. If eμ universality holds, $\Gamma(\rho^0 \to \mu^+\mu^-) = \Gamma(\rho^0 \to e^+e^-) \times$ phase space correction.

(e) Empirical limits on fractions for other decay modes of ρ(765) are $\pi^\mp\gamma < 0.5\%$, $\pi^\mp\eta < 0.8\%$, $\pi^+\pi^+\pi^-\pi^- < 0.15\%$, $\pi^\pm\pi^+\pi^-\pi^0 < 0.2\%$.

(f) Empirical limits on fractions for other decay modes of ω(784) are $\pi^+\pi^-\gamma < 5\%$, $\pi^0\pi^0\gamma < 1\%$, η + neutral(s) < 1.5%, $\mu^+\mu^- < 0.02\%$, $\pi^0\mu^+\mu^- < 0.2\%$.

(g) See Meson Data Card Listings for a typed note and an entry "π(950-1020)", which contains the data referred to as δ(962), $\pi_N(975)$, and $\pi_N(1016)$ in our April 1971 edition.

(h) Empirical limits on fractions for other decay modes of η′(958): $\pi^+\pi^- < 2\%$, $\pi^+\pi^-\pi^0 < 5\%$, $\pi^+\pi^+\pi^-\pi^- < 1\%$, $\pi^+\pi^+\pi^-\pi^-\pi^0 < 1\%$, $6\pi < 1\%$, $\pi^+\pi^-e^+e^- < 0.6\%$, $\pi^0e^+e^- < 1.3\%$, $\eta e^+e^- < 1.1\%$, $\pi^0\rho^0 < 4\%$, $\pi^0\omega < 8\%$.

(i) Empirical limits on fractions for other decay modes of $\phi(1019)$ are $\pi^+\pi^- < 0.03\%$, $\pi^+\pi^-\gamma < 4\%$, $\omega\gamma < 5\%$, $\rho\gamma < 2\%$, $\pi^0\gamma < 0.35\%$.

(j) Empirical limits on fractions for decay modes of B(1235): $\pi\pi < 30\%$, $K\bar{K} < 2\%$, $4\pi < 50\%$, $\phi\pi < 1.5\%$, $\eta\pi < 25\%$, $(K\bar{K})^{\pm}\pi^0 < 8\%$, $K_SK_S\,\pi^{\pm} < 2\%$, $K_SK_L\,\pi^{\pm} < 6\%$.

(k) There is only a weak indication for a $K^*\bar{K} + \bar{K}^*K$ mode of the f′(1514). If this mode does not exist, the $K\bar{K}$ branching fraction will have to be reported as 80 ± 13% (rather than 72 ± 12% as given in the table), and $\eta\pi\pi$ as 20 ± 13%.

(ℓ) We assume as a working hypothesis that peaks with $I^G = 1^+$ observed around 1.7 GeV all come from g(1680). For indications to the contrary see Meson Data Card Listings.

(m) A possible $\omega\pi\pi$ decay mode of the A3 has mass 1690 MeV and width 80 MeV.

(n) See Q-region note in Meson Data Card Listings. Some investigators see a broad enhancement in mass $(K\pi\pi)$ from 1250-1400 MeV (the Q region), and others see structure. Only the $K_A(1240)$ or C seems well established, whereas possible structures from 1280 to 1400 MeV cannot be disentangled. For the whole Q region the decay rate into $K^*(892)\pi$ is large, and a $K\rho$ decay is seen. The $K\eta$, $K\omega$, and $K\pi$ are less than a few percent.

(o) $K_N(1420)$ properties are uncertain because both principal modes have energy-dependent backgrounds:-$K\pi$ mode: Firestone et al. (LBL 516, subm. Phys. Rev. 1972) find a large S-wave phase shift with $\sin^2\delta_0$ peaking at ≈1350 MeV, which probably caused older experiments to overestimate both Γ and the $K\pi$ branching fraction. Instead of our average of 56%, $K\pi$ fraction could be 40-50%, with other fractions raised accordingly. $K\pi\pi$ mode is contaminated with diffractively produced $Q^{\pm}$.
The tabulated mass of 1421 MeV comes only from charged $K_N(1420) \to K\pi$ measurements (to avoid $Q^{\pm}$ contamination); the average of the neutral $K_N(1420)$ mass is also 1420 MeV (i.e., $m^0 - m^{\pm} \approx 0$) but see typed note "K*(892) Mass" in Meson Data Card Listings.

(p) We tabulate here Y = 0 bumps with M ≳ 1700 MeV, for which no satisfactory grouping into particles is yet possible. See Meson Data Card Listings.

| Name | $I^G$ | $J^P$ | M (MeV) | Γ (MeV) | Decay modes observed | Tentative grouping |
|---|---|---|---|---|---|---|
| $K\bar{K}(1740)$ | 1 | | 1740 | ≈ 120 | $K^0K^{\pm}$ | R(1750) |
| R3(1750) | 1,2 | | 1748 ± 15 | ≤ 38 | $(MM)^-$ | |
| $\pi\pi(1764)$ | | | 1764 ± 15 | $87^{+14}_{-20}$ | $\pi^+\pi^-$ | |
| $K\bar{K}\pi(1820)$ | 0,1,2 | | 1820 ± 12 | 50 ± 23 | $K_SK^0\pi^0$ | 1830 region |
| R4(1830) | 1,2 | | 1830 ± 15 | ≤ 30 | $(MM)^-$ | |
| $\eta/\rho(1830)$ | + | | 1832 ± 6 | 42 ± 11 | $\pi^+\pi^-\pi^+\pi^-$ | |
| $\phi/\pi(1830)$ | - | | 1848 ± 11 | 67 ± 27 | $\omega\pi^+\pi^-$, possibly $\omega\rho^0$ | |
| $X^-(2086)$ | 1,2 | | 2086 ± 38 | ≈ 150 | $(MM)^-$ backward | ρ(2100) |
| ρ (2120) | $1^+$ | $3^-$(?) | 2120 | < 249 | $\pi^+\pi^-$, $\bar{p}p$ | |
| $\pi\pi(2157)$ | $1^+$ | (odd)- | 2157 ± 10 | 68 ± 22 | $\pi^+\pi^0$ | T region Seems to require >1 resonance |
| $K\bar{K}\omega(2176)$ | $0^-,1^+$ | | 2176 ± 5 | $20^{+16}_{-2}$ | $K_SK_S\omega$ | |
| $N\bar{N}(2190)$ | $1^-$ | | 2190 | 20-80 | $\rho^0\rho^0\pi^0$, $\bar{p}p$ | |
| | 1 | | 2190 ± 10 | ≈ 85 | Structure in $N\bar{N}$ total σ | |
| T (2195) | 1,2 | | 2195 ± 15 | ≤ 13 | $(MM)^-$ | |
| $3\pi(2207)$ | ≤ $3^-$ | | 2207 ± 13 | 62 ± 52 | $\pi^+\pi^-\pi^0$ | |
| $4\pi(2207)$ | $1^+,2^+,3^+$ | | 2207 ± 22 | ≈ 130 | $\rho^-\pi^+\pi^-$ | |
| $X^-(2260)$ | 1,2 | | 2260 ± 18 | ≤ 25 | $(MM)^-$ backward | ρ(2275) |
| ρ (2290) | $1^+$ | $5^-$(?) | 2290 | < 165 | $\pi^+\pi^-$, $\bar{p}p$ | |
| $N\bar{N}(2350)$ | 1 | | 2350 ± 10 | ≈ 140 | Structure in $N\bar{N}$ total σ | U region |
| $X^-(2370)$ | 1,2 | | 2370 ± 17 | ≈ 57 | $(MM)^-$ backward | |
| $N\bar{N}(2375)$ | 0 | | 2375 ± 10 | ≈ 190 | Structure in $N\bar{N}$ total σ | |
| U (2380) | 1,2 | | 2382 ± 24 | < 30 | $(MM)^-$ | |
| $X^-(2500)$ | 1,2 | | 2500 ± 32 | ≈ 87 | $(MM)^-$ backward | |
| $X^-(2620)$ | 1,2 | | 2620 ± 20 | 85 ± 30 | $(MM)^-$ | 2650 region |
| $4\pi(2676)$ | (1,2,3)+ | | 2676 ± 27 | ≈ 150 | $\rho^-\pi^+\pi^-$ | |
| $X^-(2800)$ | 1,2 | | 2800 ± 20 | 46 ± 10 | $(MM)^-$ | 2850 region |
| $X^+(2820)$ | 1,2 | | 2820 ± 10 | 50 ± 10 | $(K\bar{K}\pi\pi)^+$ | |
| $X^-(2880)$ | 1,2 | | 2880 ± 20 | ≤ 15 | $(MM)^-$ | |
| $X^+(3013)$ | $1^-$ | | 3013 ± 5 | < 40 | $7\pi$ | 3020 region |
| $X^-(3025)$ | 1,2 | | 3025 ± 20 | ≈ 25 | $(MM)^-$ | |
| NN(3035) | + | | 3035 ± 25 | 200 ± 60 | $4\pi$, $6\pi$ | |
| $X^-(3075)$ | 1,2 | | 3075 ± 20 | ≈ 25 | $(MM)^-$ | |
| $X^-(3145)$ | 1,2 | | 3145 ± 20 | ≤ 10 | $(MM)^-$ | |
| $X^-(3475)$ | 1,2 | | 3475 ± 20 | ≈ 30 | $(MM)^-$ | |
| $X^-(3535)$ | 1,2 | | 3535 ± 20 | ≈ 30 | $(MM)^-$ | |

# Baryon Table

April 1972

[ See notes on N's and $\Delta$'s, on possible $Z^{*}$'s, and on $Y^{*}$'s at the beginning of those sections in the Baryon Data Card Listings; also see notes on individual resonances in the Baryon Data Card Listings.]

| Particle[a] | I $(J^P)$ ⊢⊣ estab. | $\pi$ or K Beam<br>T(GeV)<br>p(GeV/c)<br>$\sigma = 4\pi\lambda\!\!\!^{-2}$ (mb) | Mass<br>$M^b$<br>(MeV) | Full Width<br>$\Gamma^b$<br>(MeV) | $M^2$<br>$\pm \Gamma M^c$<br>$(GeV^2)$ | Partial decay mode: Mode | Fraction % | p or $p_{max}$ [d] (MeV/c) |
|---|---|---|---|---|---|---|---|---|
| p<br>n | $1/2(1/2^+)$ | | 938.3<br>939.6 | | 0.880<br>0.883 | See Stable Particle Table | | |
| N' (1470) | $1/2(1/2^+)$ $P'_{11}$ | T=0.53πp<br>p=0.66<br>σ=27.8 | 1435 to<br>1505 | 165 to<br>400 | 2.16<br>±0.34 | $N\pi$<br>$N\pi\pi$ | 60<br>40 | 420<br>368 |
| N'(1520) | $1/2(3/2^-)$ $D'_{13}$ | T=0.61<br>p=0.74<br>σ=23.5 | 1510 to<br>1540 | 105 to<br>150 | 2.31<br>±0.18 | $N\pi$<br>$N\pi\pi$<br>$[\Delta(1236)\pi]^e$<br>$N\eta$ | 50<br>50<br>[dominant][e]<br>~0.6 | 456<br>410<br>224 |
| N'(1535) | $1/2(1/2^-)$ $S'_{11}$ | T=0.64<br>p=0.76<br>σ=22.5 | 1500 to<br>1600 | 50 to<br>160 | 2.36<br>±0.18 | $N\pi$<br>$N\eta^n$<br>$N\pi\pi^n$ | 35<br>55<br>~10 | 467<br>182<br>422 |
| N'(1670)[i] | $1/2(5/2^-)$ $D'_{15}$ | T=0.87<br>p=1.00<br>σ=15.6 | 1655 to<br>1680 | 105 to<br>175 | 2.79<br>±0.24 | $N\pi$<br>$N\pi\pi$<br>$[\Delta(1236)\pi]^e$<br>$\Lambda K$<br>$N\eta$ | 40<br>60<br>[44][e]<br>< .3<br>< 1[j] | 560<br>525<br>357<br>200<br>368 |
| N'(1688)[i] | $1/2(5/2^+)$ $F'_{15}$ | T=0.90<br>p=1.03<br>σ=14.9 | 1680 to<br>1692 | 105 to<br>180 | 2.85<br>±0.21 | $N\pi$<br>$N\pi\pi$<br>$[\Delta(1236)+\pi]^e$<br>$\Lambda K$<br>$N\eta$ | 60<br>40<br>[26][e]<br>< .2<br>< .5[j] | 572<br>538<br>371<br>231<br>388 |
| N''(1700)[i] | $1/2(1/2^-)$ $S''_{11}$ | T=0.92<br>p=1.05<br>σ=14.3 | 1665 to<br>1765 | 100 to<br>400 | 2.89<br>±0.42 | $N\pi$<br>$\Lambda K$<br>$N\eta$ | -65<br>5 | 580<br>250<br>340 |
| N''(1780)[i] | $1/2(1/2^+)$ $P''_{11}$ | T=1.07<br>p=1.20<br>σ=12.2 | 1650 to<br>1860 | 50 to<br>450 | 3.17<br>±0.51 | $N\pi$<br>$\Lambda K$<br>$N\eta$ | 30<br>~ 7<br>~10[j] | 633<br>353<br>476 |
| N(1860) | $1/2(3/2^+)$ $P_{13}$ | T=1.22<br>p=1.36<br>σ=10.4 | 1770 to<br>1900 | 180 to<br>330 | 3.46<br>±0.57 | $N\pi$<br>$N\pi\pi$<br>$\Lambda K$<br>$N\eta$ | 25<br><br>~ 5<br>~ 4[j] | 685<br>657<br>437<br>545 |
| N(2190) | $1/2(7/2^-)$ $G_{17}$ | T=1.94<br>p=2.07<br>σ=6.21 | 2000 to<br>2260 | 270 to<br>325 | 4.80<br>±0.67 | $N\pi$<br>$N\pi\pi$ | 25 | 888<br>868 |
| N(2220) | $1/2(9/2^+)$ $H_{19}$ | T=2.00<br>p=2.14<br>σ=5.97 | 2200 to<br>2245 | 260 to<br>330 | 4.93<br>±0.65 | $N\pi$<br>$N\pi\pi$ | 15 | 905<br>887 |
| N(2650) | $1/2(\ ?^-)$ | T=3.12<br>p=3.26<br>σ=3.67 | 2650 | 360 | 7.02<br>±0.95 | $N\pi$<br>$N\pi\pi$ | (J+1/2)x<br>=0.45[f] | 1154<br>1140 |
| N(3030) | 1/2( ? ) | T=4.27<br>p=4.41<br>σ=2.62 | 3030 | 400 | 9.18<br>±1.21 | $N\pi$<br>$N\pi\pi$ | (J+1/2)x<br>=0.05[f] | 1366<br>1354 |
| $\Delta'(1236)$[m] | $3/2(3/2^+)$ $P'_{33}$ | T=0.195 (++)<br>p=0.304<br>σ=91.8 | 1230 to<br>1236 | 110 to<br>122 | 1.53<br>±0.14 | $N\pi$<br>$N\pi^+\pi^-$<br>$N\gamma$ | 99.4<br>0<br>~0.6 | 231<br>90<br>262 |
| | | Pole Position[m]: | 1211 | ± i50 | | | | |
| $\Delta(1650)$ | $3/2(1/2^-)$ $S_{31}$ | T=0.83<br>p=0.96<br>σ=16.4 | 1615 to<br>1695 | 130 to<br>200 | 2.72<br>±0.28 | $N\pi$<br>$N\pi\pi$ | 28<br>72 | 547<br>511 |

| | Particle[a] | I ($J^P$) estab. | $\pi$ or K Beam T(GeV) p(GeV/c) $\sigma = 4\pi\lambda\!\!\!^{-2}$ (mb) | Mass $M^b$ (MeV) | Full Width $\Gamma^b$ (MeV) | $M^2 \pm \Gamma M^c$ ($GeV^2$) | Partial decay mode: Mode | Fraction % | p or $p_{max}$ [d] (MeV/c) |
|---|---|---|---|---|---|---|---|---|---|
| Δ | Δ(1670) | 3/2(3/2⁻) $D_{33}$ | T=0.87<br>p=1.00<br>σ=15.6 | 1650 to 1720 | 175 to 300 | 2.79<br>±0.40 | Nπ<br>Nππ | 15 | 560<br>525 |
| | Δ(1890) | 3/2(5/2⁺) $F_{35}$ | T=1.28<br>p=1.42<br>σ=9.88 | 1840 to 1920 | 135 to 350 | 3.57<br>±0.49 | Nπ<br>Nππ | 17 | 704<br>677 |
| | Δ(1910) | 3/2(1/2⁺) $P_{31}$ | T=1.33<br>p=1.46<br>σ=9.54 | 1780 to 1935 | 230 to 420 | 3.65<br>±0.62 | Nπ<br>Nππ | 25 | 716<br>691 |
| | Δ(1950) | 3/2(7/2⁺) $F_{37}$ | T=1.41<br>p=1.54<br>σ=8.90 | 1930 to 1980 | 140 to 220 | 3.80<br>±0.39 | Nπ<br>Δ(1236)π<br>ΣK<br>Σ(1385)K | 45<br>≈50<br>~ 2<br>1.4 | 741<br>571<br>460<br>232 |
| | Δ(2420) | 3/2(11/2⁺) | T=2.50<br>p=2.64<br>σ=4.68 | 2320 to 2450 | 270 to 350 | 5.86<br>±0.75 | Nπ<br>Nππ | 11<br>>20 | 1023<br>1006 |
| | Δ(2850) | 3/2( ?⁺ ) | T=3.71<br>p=3.85<br>σ=3.05 | 2850 | 400 | 8.12<br>±1.14 | Nπ<br>Nππ | (J+1/2)x =0.25[f] | 1266<br>1254 |
| | Δ(3230) | 3/2( ? ) | T=4.94<br>p=5.08<br>σ=2.25 | 3230 | 440 | 10.4<br>±1.4 | Nπ<br>Nππ | (J+1/2)x =0.05[f] | 1475<br>1464 |
| Z* | Z* | Evidence for states with hypercharge 2 is controversial. See the Baryon Data Card Listings for discussion and display of data. | | | | | | | |
| Λ | Λ | 0(1/2⁺) | | 1115.6 | | 1.24 | See Stable Particle Table | | |
| | Λ'(1405) | 0(1/2⁻) $S'_{01}$ | p<0 K⁻p | 1405 ±5[g] | 40 ±10[g] | 1.97<br>±0.06 | Σπ | 100 | 142 |
| | Λ'(1520) | 0(3/2⁻) $D'_{03}$ | p=0.389<br>σ=84.5 | 1518 ±2[g] | 16 ±2[g] | 2.30<br>±0.02 | N$\bar{K}$<br>Σπ<br>Λππ<br>Σππ | 45±1<br>42±1<br>9.6±.7<br>1.0±.1 | 234<br>258<br>250<br>140 |
| | Λ''(1670) | 0(1/2⁻) $S''_{01}$ | p=0.74<br>σ=28.5 | 1670 | 15 to 38 | 2.79<br>±0.04 | N$\bar{K}$<br>Λη<br>Σπ | ~20<br>~35<br>~45 | 410<br>64<br>393 |
| | Λ''(1690) | 0(3/2⁻) $D''_{03}$ | p=0.78<br>σ=26.1 | 1690 | 27 to 85 | 2.86<br>0.09 | N$\bar{K}$<br>Σπ<br>Λππ<br>Σππ | ~20[h]<br>~60<br>~ 2<br>≤18 | 429<br>409<br>415<br>352 |
| | Λ'(1815) | 0(5/2⁺) $F'_{05}$ | p=1.05<br>σ=16.7 | 1820 ±5[g] | 64 to 104 | 3.30<br>±0.15 | N$\bar{K}$<br>Σπ<br>Σ(1385)π | 62<br>11<br>4 | 542<br>508<br>362 |
| | Λ'(1830) | 0(5/2⁻) $D'_{05}$ | p=1.09<br>σ=15.8 | 1835 | 74 to 150 | 3.37<br>±0.20 | N$\bar{K}$<br>Σπ<br>Λππ | ~10<br>~30<br>~11 | 554<br>519<br>536 |
| | Λ(2100) | 0(7/2⁻) $G_{07}$ | p=1.68<br>σ=8.68 | 2100 | 60 to 140 | 4.41<br>±0.22 | N$\bar{K}$<br>Σπ<br>Λη<br>ΞK<br>Λω | 25<br>~ 5<br>< 3<br><br><10 | 748<br>699<br>617<br>483<br>443 |
| | Λ(2350) | 0( ? ) | p=2.29<br>σ=5.85 | 2350 | 140 to 324 | 5.52<br>±0.55 | N$\bar{K}$ | (J+1/2)x =0.7[f] | 913 |

| Particle[a] | I ($J^P$) ⊢⊣ estab. | π or K Beam: T(GeV), p(GeV/c), $\sigma = 4\pi\lambda\!\!\!^{-2}$ (mb) | Mass $M^b$ (MeV) | Full Width $\Gamma^b$ (MeV) | $M^2$ $\pm\Gamma M^c$ ($GeV^2$) | Partial decay mode: Mode | Fraction % | p or $p_{max}$[d] (MeV/c) |
|---|---|---|---|---|---|---|---|---|
| Σ | 1(1/2⁺) | | (+)1189.4<br>(0)1192.5<br>(−)1197.3 | | 1.41<br>1.42<br>1.43 | See Stable Particle Table | | |
| Σ′(1385) | 1(3/2⁺)$P'_{13}$ | p<0 $K^-p$ | (+)1383±1<br>S=1.3*<br>(−)1386±2<br>S=2.2* | (+)36±3<br>S=1.9*<br>(−)36±6<br>S=3.5*! | 1.92<br>±0.05 | Λπ<br>Σπ | 89±5<br>11±5<br>S=1.9* | 208<br>117 |
| Σ′(1670)[k] | 1(3/2⁻) $D'_{13}$ | p=0.74<br>σ=28.5 | 1670 | 50 | 2.79<br>±0.08 | $N\bar{K}$<br>Σπ<br>Λπ<br>Σππ<br>[Λ(1405)π][e]<br>Λππ | ~8 | 410<br>387<br>447<br>326<br>207<br>397 |
| | Mass, width and elasticity are the values obtained in partial wave analyses for a $D_{13}$ resonance. For more results see the Baryon Data Card Listings and footnote k. | | | | | | | |
| Σ″(1750) | 1(1/2⁻) $S''_{11}$ | p=0.91<br>σ=20.7 | 1750 | 50 to<br>80 | 3.06<br>±0.11 | $N\bar{K}$<br>Λπ<br>Ση | ~15<br>seen<br>seen | 483<br>507<br>54 |
| Σ(1765) | 1(5/2⁻) $D_{15}$ | p=0.94<br>σ=19.6 | 1765<br>±5[g] | ~120 | 3.12<br>±0.21 | $N\bar{K}$<br>Λπ<br>Λ(1520)π<br>Σ(1385)π<br>Σπ | ~42<br>~15<br>~14<br>~ 4<br>~1 | 496<br>518<br>187<br>315<br>461 |
| Σ′(1915)[i] | 1(5/2⁺) $F'_{15}$ | p=1.25<br>σ=13.0 | 1910 | 70 | 3.65<br>±0.13 | $N\bar{K}$<br>Λπ<br>Σπ | ~11 | 612<br>619<br>568 |
| | Formation and production experiments do not agree on Σπ/Λπ ratio. | | | | | | | |
| Σ(2030) | 1(7/2⁺) $F_{17}$ | p=1.52<br>σ=9.93 | 2030 | 100 to<br>170 | 4.12<br>±0.27 | $N\bar{K}$<br>Λπ<br>Σπ<br>ΞK | ~ 20<br>~ 20<br>~ 3<br>< 2 | 700<br>700<br>652<br>412 |
| Σ(2250) | 1( ? ) | p=2.04<br>σ=6.76 | 2250 | 100 to<br>230 | 5.06<br>±0.37 | $N\bar{K}$<br>Σπ<br>Λπ | (J+1/2)x<br>=0.3[f] | 849<br>842<br>799 |
| Σ(2455) | 1( ? ) | p=2.57<br>σ=5.09 | 2455 | ~120 | 6.03<br>±0.29 | $N\bar{K}$ | (J+1/2)x<br>=0.2[f] | 979 |
| Σ(2620) | 1( ? ) | p=2.95<br>σ=4.30 | 2620 | ~175 | 6.86<br>±0.46 | $N\bar{K}$ | (J+1/2)x<br>=0.3[f] | 1064 |
| Ξ[l] | 1/2(1/2⁺) | | (0)1314.7<br>(−)1321.3 | | 1.73<br>1.75 | See Stable Particle Table | | |
| Ξ(1530)[l] | 1/2(3/2⁺)<br>p-wave | | (0) 1531.3±0.5<br>S=1.4*<br>(−) 1535.8±1.0 | (0) 9.2±0.8<br>(−) 16.2±4.6 | 2.34<br>±0.01 | Ξπ | 100 | 144 |
| Ξ(1820)[l] | 1/2( ? ) | | 1795 to<br>1870 | 12 to<br>99 | 3.31<br>±0.10 | $\Lambda\bar{K}$<br>Ξπ<br>Ξ(1530)π<br>Σ$\bar{K}$ | | 396<br>413<br>234<br>306 |
| | All four decay modes have been seen. Branching ratios not quoted because there may be more than one state here. | | | | | | | |
| Ξ(1940)[l] | 1/2( ? ) | | 1894 to<br>1961 | 42 to<br>140 | 3.72<br>±0.18 | Ξπ<br>Ξ(1530)π | | 499<br>336 |
| | Seen in both final states; not clear if one, or more, states present. | | | | | | | |
| Ω⁻ | 0(3/2⁺) | | 1672.5 | | 2.80 | See Stable Particle Table | | |

* Quoted error includes an S(scale)factor. See footnote to Stable Particle Table.

→ An arrow at the left of the Table indicates a candidate that has been omitted because the evidence for the existence of the effect and (or) for its interpretation as a resonance is open to considerable question. See the Baryon Data Card Listings for information on the following: N(1700) $D''_{13}$, N(1990) $F_{17}$, N(2040) $D'''_{13}$, N(2100) $S'''_{11}$, N(2100) $D''_{15}$, N(2175) $F''_{15}$, N(3245), N(3690), N(3755), $\Delta$(1690) $P''_{33}$, $\Delta$(1960) $D_{35}$, $\Delta$(2160) $P'''_{33}$, $Z_0$(1780), $Z_0$(1865), $Z_1$(1900), $\Lambda$(1330), $\Lambda$(1750) $P_{01}$, $\Lambda$(1860) $P_{03}$, $\Lambda$(1870) $S'''_{01}$, $\Lambda$(2010) $D'''_{03}$, $\Lambda$(2020) $F_{07}$, $\Lambda$(2110) $F''_{05}$ or $D''_{05}$, $\Lambda$(2585), $\Sigma$(1440), $\Sigma$(1480), $\Sigma$(1620) $S'_{11}$, $\Sigma$(1620) $P'_{11}$ , $\Sigma(1670)^k$, $\Sigma$(1690), $\Sigma$(1880) $P''_{11}$, $\Sigma$(1940) $D''_{13}$, $\Sigma$(2070) $F''_{15}$, $\Sigma$(2080) $P''_{13}$, $\Sigma$(2100) $G_{17}$, $\Sigma$(3000), $\Xi$(1630), $\Xi$(2030), $\Xi$(2250), $\Xi$(2500).

a. For the baryon states, the name [such as N'(1470)] contains the mass, which may be different for each new analysis. The value chosen is the rounded average from Table I of the note on N's and $\Delta$'s in the Baryon Data Card Listings. For $Y^*$'s and $\Xi^*$'s, the mass is an educated guess obtained by looking at the reported values. The convention for using primes in the names is as follows: when there is more than one resonance on a given Argand diagram, the first has been designated with a prime, the second with a double prime, etc. The name (col. 1) is the same as can be found in large print in the Baryon Data Card Listings.

b. See note on N's and $\Delta$'s in the Baryon Data Card Listings. For M and $\Gamma$ of most baryons we report here an interval instead of an average. Averages are appropriate if each result is based on independent measurements, but inappropriate here where the spread in parameters arises because different models or procedures have been applied to a common set of data. Where only one value is given it is either because only one experiment reports that state or because the various experiments agree. An error is quoted only when the various experiments averaged have taken into account the systematic errors.

c. For this column M is the rounded average which also appears in the name column. For the N's and $\Delta$'s, $\Gamma$ is the average quoted on Table II of the N's and $\Delta$'s note in the Baryon Data Card Listings; for the $Y^*$'s and $\Xi^*$'s, $\Gamma$ is taken as the center of the interval given in the column labeled "$\Gamma$".

d. For decay modes into ≥3 particles $p_{max}$ is the maximum momentum that any of the particles in the final state can have. The momenta have been calculated using the averaged central mass values, without taking into account the widths of the resonances.

e. Square brackets indicate a sub-reaction of the previous unbracketed decay mode.

f. This state has been seen only in total cross sections. J is not known; x is $\Gamma_{el}/\Gamma$.

g. This is only an educated guess; the error given is larger than the error of the average of the published values (see the Baryon Data Card Listings for the latter).

h. In previous editions we quoted a larger elasticity. It was required by unitarity because of the large value of $x \cdot x_e$ for the $\Lambda\pi\pi$ decay. A partial wave analysis of new data of the $\Lambda\pi\pi$ channel yields smaller values of $x \cdot x_e$ allowing a smaller elasticity, which is more consistent with partial-wave analysis in the elastic channel.

i. Only information coming from partial-wave analyses has been used here. For the production experiments results see the Baryon Data Card Listings.

j. Value obtained in an energy-dependent partial-wave analysis which uses a t-channel-poles-plus-resonance parametrization. The values of the couplings obtained for the resonances may be affected by double counting.

k. In this energy region the situation is still confused. Formation experiments suggest two states: $P_{11}$(1620) decaying mainly into $\Sigma\pi$, and $D_{13}$(1670) with branching fractions $\Sigma\pi$(40%), $\Lambda\pi$(10%), $\Sigma\pi\pi$ (< 14%). Production experiments report four states: $\Sigma$(1620) seen only in the $\Lambda\pi$ mode, $\Sigma_1$(1660) with appreciable $\Lambda\pi$ and $\Sigma\pi$ modes. $\Sigma_2$(1660) with main decay mode $\Lambda(1405) + \pi$ (that is, $\Sigma\pi\pi$), and $\Sigma$(1690) seen in the $\Lambda\pi$ mode. Of these four, $\Sigma_1$ and $\Sigma_2$ seem to be on firmer ground than the other two and both seem to have $J^P = 3/2^-$ like the $D_{13}$(1670) seen in formation experiments. Two resonances of the same spin and parity have been hypothesized as the origin of much of the complexity observed in production experiments. With the addition of the $P_{11}$(1620), there are three candidates that eventually might be required to clarify the situation.

ℓ. Only $\Xi$(1530) is firmly established; information on the other states comes from experiments that have poor statistics due to the fact that the cross sections for S = -2 states are very low. For $\Xi$ states, because of the meager statistics, we lower our standards and tabulate resonant effects if they have at least a four-standard-deviation statistical significance and if they are seen by more than one group. So $\Xi$(2030), with main decay mode $\Sigma\bar{K}$, reported as a 3.5-standard-deviation effect, is not tabulated. See the Baryon Data Card Listings for the other states.

m. See note on $\Delta$(1236) in the Baryon Data Card Listings. Values of mass and width are dependent upon resonance shape used to fit the data. The pole position appears to be much less dependent upon the parametrization used.

n. The preliminary results of DIEM 70 quoted in the Baryon Data Card Listings have been revised so that they are now in agreement with the values quoted in the present table (G. Smadja, private communication).

# PHYSICAL AND NUMERICAL CONSTANTS*

## PHYSICAL CONSTANTS

| | |
|---|---|
| $N$ | $= 6.022169(40)\times10^{23}\ \text{mole}^{-1}$ (based on $A_{C^{12}} = 12$) |
| $c$ | $= 2.9979250(10)\times10^{10}\ \text{cm sec}^{-1}$ |
| $e$ | $= 4.803250(21)\times10^{-10}\ \text{esu} = 1.6021917(70)\times10^{-19}\ \text{coulomb}$ |
| 1 MeV | $= 1.6021917(70)\times10^{-6}\ \text{erg}$ |
| $\hbar$ | $= 6.582183(22)\times10^{-22}\ \text{MeV sec}$ |
| | $= 1.0545919(80)\times10^{-27}\ \text{erg sec}$ |
| $\hbar c$ | $= 1.9732891(66)\times10^{-11}\ \text{MeV cm} = 197.32891(66)\ \text{MeV fermi}$ |
| | $= 0.6240088(21)\ \text{GeV mb}^{1/2}$ |
| $\alpha$ | $= e^2/\hbar c = 1/137.03602(21)$ |
| $k_{\text{Boltzmann}}$ | $= 1.380622(59)\times10^{-16}\ \text{erg K}^{-1}$ |
| | $= 8.61708(37)\times10^{-11}\ \text{MeV K}^{-1} = 1\ \text{eV}/11604.85(49)\text{K}$ |
| $m_e$ | $= 0.5110041(16)\ \text{MeV} = 9.109558(54)\times10^{-31}\ \text{kg}$ |
| $m_p$ | $= 938.2592(52)\ \text{MeV} = 1836.109(11)\,m_e = 6.72211(63)\,m_{\pi^\pm}$ |
| | $= 1.00727661(8)\,m_1$ (where $m_1 = 1\ \text{amu} = \frac{1}{12} m_{C^{12}} = 931.4812(52)\ \text{MeV}$) |
| $r_e$ | $= e^2/m_e c^2 = 2.817939(13)$ fermi (1 fermi $= 10^{-13}$ cm) |
| $\lambda\!\!\!^{-}_e$ | $= \hbar/m_e c = r_e \alpha^{-1} = 3.861592(12)\times10^{-11}\ \text{cm}$ |
| $a_{\infty\ \text{Bohr}}$ | $= \hbar^2/m_e e^2 = r_e \alpha^{-2} = 0.52917715(81)\ \text{Å}$ ($1\ \text{Å} = 10^{-8}$ cm) |
| $\sigma_{\text{Thomson}}$ | $= \frac{8}{3}\pi r_e^2 = 0.6652453(61)\times10^{-24}\ \text{cm}^2 = 0.6652453(61)$ barns |
| $\mu_{\text{Bohr}}$ | $= e\hbar/2m_e c = 0.5788381(18)\times10^{-14}\ \text{MeV gauss}^{-1}$ |
| $\mu_{\text{nucleon}}$ | $= e\hbar/2m_p c = 3.152526(21)\times10^{-18}\ \text{MeV gauss}^{-1}$ |
| $\frac{1}{2}\omega^e_{\text{cyclotron}}$ | $= e/2m_e c = 8.794014(27)\times10^{6}\ \text{rad sec}^{-1}\ \text{gauss}^{-1}$ |
| $\frac{1}{2}\omega^p_{\text{cyclotron}}$ | $= e/2m_p c = 4.789484(27)\times10^{3}\ \text{rad sec}^{-1}\ \text{gauss}^{-1}$ |

Hydrogen-like atom (nonrelativistic, $\mu$ = reduced mass):

$$\left.\frac{v}{c}\right)_{\text{rms}} = \frac{ze^2}{n\hbar c};\quad E_n = \frac{\mu}{2}v^2 = \frac{\mu z^2 e^4}{2(n\hbar)^2};\quad a_n = \frac{n^2\hbar^2}{\mu z e^2}$$

$R_\infty = m_e e^4/2\hbar^2 = m_e c^2\alpha^2/2 = 13.605826(45)$ eV (Rydberg)

$pc = 0.3\ H\rho$(MeV, kilogauss, cm); 0.3 (which is $10^{-11}c$) enters because there are ≈ 300 "volts"/esu volt.

| | |
|---|---|
| 1 year (sidereal) | $= 365.256$ days $= 3.1558\times10^7$ sec ($\approx \pi\times10^7$ sec) |
| density of dry air | $= 1.205\ \text{mg cm}^{-3}$ (at 20°C, 760 mm) |
| acceleration by gravity | $= 980.62\ \text{cm sec}^{-2}$ (sea level, 45°) |
| gravitational constant | $= 6.6732(31)\times10^{-8}\ \text{cm}^3\,\text{g}^{-1}\,\text{sec}^{-2}$ |
| 1 calorie (thermochemical) | $= 4.184$ joules |
| 1 atmosphere | $= 1033.2275\ \text{g cm}^{-2}$ |
| 1 eV per particle | $= 11604.85(49)$ °K (from $E = kT$) |

## NUMERICAL CONSTANTS

| | | | | | |
|---|---|---|---|---|---|
| $\pi$ | $= 3.1415927$ | 1 rad | $= 57.2957795$ deg | $\sqrt{\pi}$ | $= 1.7724539$ |
| $e$ | $= 2.7182818$ | $1/e$ | $= 0.3678794$ | $\sqrt{2}$ | $= 1.4142136$ |
| $\ln 2$ | $= 0.6931472$ | $\ln 10$ | $= 2.3025851$ | $\sqrt{3}$ | $= 1.7320508$ |
| $\log_{10} 2$ | $= 0.3010300$ | $\log_{10} e$ | $= 0.4342945$ | $\sqrt{10}$ | $= 3.1622777$ |

*Compiled by Stanley J. Brodsky, based mainly on the adjustment of the fundamental physical constants by B. N. Taylor, W. H. Parker, and D. N. Langenberg, Rev. Mod. Phys. 41, 375 (1969). The figures in parentheses correspond to the 1 standard deviation uncertainty in the last digits of the main number.

# CLEBSCH-GORDAN COEFFICIENTS AND SPHERICAL HARMONICS

Note: A $\sqrt{\ }$ is to be understood over <u>every</u> coefficient; e.g., for -8/15 read $-\sqrt{8/15}$.

Notation:

| $m_1$ | $m_2$ | J | J | ... |
|---|---|---|---|---|
| | | M | M | ... |
| $m_1$ | $m_2$ | Coefficients | | |
| . | . | | | |
| . | . | | | |

$$Y_1^0 = \sqrt{\frac{3}{4\pi}}\cos\theta$$

$$Y_1^1 = -\sqrt{\frac{3}{8\pi}}\sin\theta\, e^{i\phi}$$

$$Y_2^0 = \sqrt{\frac{5}{4\pi}}\left(\frac{3}{2}\cos^2\theta - \frac{1}{2}\right)$$

$$Y_2^1 = -\sqrt{\frac{15}{8\pi}}\sin\theta\cos\theta\, e^{i\phi}$$

$$Y_2^2 = \frac{1}{4}\sqrt{\frac{15}{2\pi}}\sin^2\theta\, e^{2i\phi}$$

$$Y_\ell^{-m} = (-1)^m Y_\ell^{m*}$$

## 1/2 × 1/2

| $m_1$ | $m_2$ | J = 1 |
|---|---|---|
| | | M = +1 |
| +1/2 | +1/2 | 1 |

| $m_1$ | $m_2$ | 1 | 0 |
|---|---|---|---|
| | | 0 | 0 |
| +1/2 | -1/2 | 1/2 | 1/2 |
| -1/2 | +1/2 | 1/2 | -1/2 |

| $m_1$ | $m_2$ | 1 |
|---|---|---|
| | | -1 |
| -1/2 | -1/2 | 1 |

## 1 × 1/2

| $m_1$ | $m_2$ | 3/2 |
|---|---|---|
| | | +3/2 |
| +1 | +1/2 | 1 |

| $m_1$ | $m_2$ | 3/2 | 1/2 |
|---|---|---|---|
| | | +1/2 | +1/2 |
| +1 | -1/2 | 1/3 | 2/3 |
| 0 | +1/2 | 2/3 | -1/3 |

| $m_1$ | $m_2$ | 3/2 | 1/2 |
|---|---|---|---|
| | | -1/2 | -1/2 |
| 0 | -1/2 | 2/3 | 1/3 |
| -1 | +1/2 | 1/3 | -2/3 |

| $m_1$ | $m_2$ | 3/2 |
|---|---|---|
| | | -3/2 |
| -1 | -1/2 | 1 |

## 2 × 1

| $m_1$ | $m_2$ | 3 |
|---|---|---|
| | | +3 |
| +2 | +1 | 1 |

| $m_1$ | $m_2$ | 3 | 2 |
|---|---|---|---|
| | | +2 | +2 |
| +2 | 0 | 1/3 | 2/3 |
| +1 | +1 | 2/3 | -1/3 |

| $m_1$ | $m_2$ | 3 | 2 | 1 |
|---|---|---|---|---|
| | | +1 | +1 | +1 |
| +2 | -1 | 1/15 | 1/3 | 3/5 |
| +1 | 0 | 8/15 | 1/6 | -3/10 |
| 0 | +1 | 6/15 | -1/2 | 1/10 |

| $m_1$ | $m_2$ | 3 | 2 | 1 |
|---|---|---|---|---|
| | | 0 | 0 | 0 |
| +1 | -1 | 1/5 | 1/2 | 3/10 |
| 0 | 0 | 3/5 | 0 | -2/5 |
| -1 | +1 | 1/5 | -1/2 | 3/10 |

| $m_1$ | $m_2$ | 3 | 2 | 1 |
|---|---|---|---|---|
| | | -1 | -1 | -1 |
| 0 | -1 | 6/15 | 1/2 | 1/10 |
| -1 | 0 | 8/15 | -1/6 | -3/10 |
| -2 | +1 | 1/15 | -1/3 | 3/5 |

| $m_1$ | $m_2$ | 3 | 2 |
|---|---|---|---|
| | | -2 | -2 |
| -1 | -1 | 2/3 | 1/3 |
| -2 | 0 | 1/3 | -2/3 |

| $m_1$ | $m_2$ | 3 |
|---|---|---|
| | | -3 |
| -2 | -1 | 1 |

## 1 × 1

| $m_1$ | $m_2$ | 2 |
|---|---|---|
| | | +2 |
| +1 | +1 | 1 |

| $m_1$ | $m_2$ | 2 | 1 |
|---|---|---|---|
| | | +1 | +1 |
| +1 | 0 | 1/2 | 1/2 |
| 0 | +1 | 1/2 | -1/2 |

| $m_1$ | $m_2$ | 2 | 1 | 0 |
|---|---|---|---|---|
| | | 0 | 0 | 0 |
| +1 | -1 | 1/6 | 1/2 | 1/3 |
| 0 | 0 | 2/3 | 0 | -1/3 |
| -1 | +1 | 1/6 | -1/2 | 1/3 |

| $m_1$ | $m_2$ | 2 | 1 |
|---|---|---|---|
| | | -1 | -1 |
| 0 | -1 | 1/2 | 1/2 |
| -1 | 0 | 1/2 | -1/2 |

| $m_1$ | $m_2$ | 2 |
|---|---|---|
| | | -2 |
| -1 | -1 | 1 |

## 3/2 × 1

| $m_1$ | $m_2$ | 5/2 |
|---|---|---|
| | | +5/2 |
| +3/2 | +1 | 1 |

| $m_1$ | $m_2$ | 5/2 | 3/2 |
|---|---|---|---|
| | | +3/2 | +3/2 |
| +3/2 | 0 | 2/5 | 3/5 |
| +1/2 | +1 | 3/5 | -2/5 |

| $m_1$ | $m_2$ | 5/2 | 3/2 | 1/2 |
|---|---|---|---|---|
| | | +1/2 | +1/2 | +1/2 |
| +3/2 | -1 | 1/10 | 2/5 | 1/2 |
| +1/2 | 0 | 3/5 | 1/15 | -1/3 |
| -1/2 | +1 | 3/10 | -8/15 | 1/6 |

| $m_1$ | $m_2$ | 5/2 | 3/2 | 1/2 |
|---|---|---|---|---|
| | | -1/2 | -1/2 | -1/2 |
| +1/2 | -1 | 3/10 | 8/15 | 1/6 |
| -1/2 | 0 | 3/5 | -1/15 | -1/3 |
| -3/2 | +1 | 1/10 | -2/5 | 1/2 |

| $m_1$ | $m_2$ | 5/2 | 3/2 |
|---|---|---|---|
| | | -3/2 | -3/2 |
| -1/2 | -1 | 3/5 | 2/5 |
| -3/2 | 0 | 2/5 | -3/5 |

| $m_1$ | $m_2$ | 5/2 |
|---|---|---|
| | | -5/2 |
| -3/2 | -1 | 1 |

## 2 × 1/2

| $m_1$ | $m_2$ | 5/2 |
|---|---|---|
| | | +5/2 |
| +2 | 1/2 | 1 |

| $m_1$ | $m_2$ | 5/2 | 3/2 |
|---|---|---|---|
| | | +3/2 | +3/2 |
| +2 | -1/2 | 1/5 | 4/5 |
| +1 | +1/2 | 4/5 | -1/5 |

| $m_1$ | $m_2$ | 5/2 | 3/2 |
|---|---|---|---|
| | | +1/2 | +1/2 |
| +1 | -1/2 | 2/5 | 3/5 |
| 0 | +1/2 | 3/5 | -2/5 |

| $m_1$ | $m_2$ | 5/2 | 3/2 |
|---|---|---|---|
| | | -1/2 | -1/2 |
| 0 | -1/2 | 3/5 | 2/5 |
| -1 | +1/2 | 2/5 | -3/5 |

| $m_1$ | $m_2$ | 5/2 | 3/2 |
|---|---|---|---|
| | | -3/2 | -3/2 |
| -1 | -1/2 | 4/5 | 1/5 |
| -2 | +1/2 | 1/5 | -4/5 |

| $m_1$ | $m_2$ | 5/2 |
|---|---|---|
| | | -5/2 |
| -2 | -1/2 | 1 |

## 3/2 × 1/2

| $m_1$ | $m_2$ | 2 |
|---|---|---|
| | | +2 |
| +3/2 | +1/2 | 1 |

| $m_1$ | $m_2$ | 2 | 1 |
|---|---|---|---|
| | | +1 | +1 |
| +3/2 | -1/2 | 1/4 | 3/4 |
| +1/2 | +1/2 | 3/4 | -1/4 |

| $m_1$ | $m_2$ | 2 | 1 |
|---|---|---|---|
| | | 0 | 0 |
| +1/2 | -1/2 | 1/2 | 1/2 |
| -1/2 | +1/2 | 1/2 | -1/2 |

| $m_1$ | $m_2$ | 2 | 1 |
|---|---|---|---|
| | | -1 | -1 |
| -1/2 | -1/2 | 3/4 | 1/4 |
| -3/2 | +1/2 | 1/4 | -3/4 |

| $m_1$ | $m_2$ | 2 |
|---|---|---|
| | | -2 |
| -3/2 | -1/2 | 1 |

$$\langle j_1 j_2 m_1 m_2 | j_1 j_2 J M\rangle = (-1)^{J - j_1 - j_2}\langle j_2 j_1 m_2 m_1 | j_2 j_1 J M\rangle$$

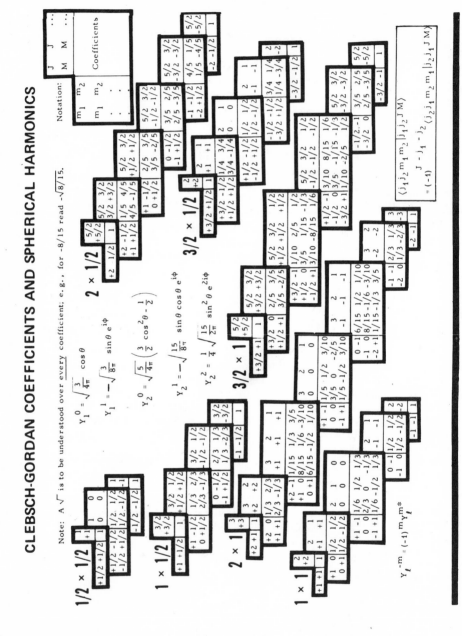

# Subject Index